D0710761

Journey into the Heart

Journey into the Heart

A Tale of Pioneering Doctors and Their Race to Transform Cardiovascular Medicine

DAVID MONAGAN

With David O. Williams, M.D., Medical Editor

GOTHAM BOOKS

GOTHAM BOOKS
Published by Penguin Group (USA) Inc.
375 Hudson Street, New York, New York 10014, U.S.A.
Penguin Group (Canada), 90 Eglinton East, Toronto, Ontario M4P 2Y3, Canada (a division of Pearson Penguin Canada Inc.); Penguin Books Ltd, 80 Strand, London WC2R 0RL, England; Penguin Ireland, 25 St Stephen's Green, Dublin 2, Ireland (a division of Penguin Books Ltd); Penguin Group (Australia), 250 Camberwell Road, Camberwell, Victoria 3124, Australia (a division of Pearson Australia Group Pty Ltd); Penguin Books India Pvt Ltd, 11 Community Centre, Panchsheel Park, New Delhi – 110 017, India; Penguin Group (NZ), cnr Airborne and Rosedale Roads, Albany, Auckland 1310, New Zealand (a division of Pearson New Zealand Ltd); Penguin Books (South Africa) (Pty) Ltd, 24 Sturdee Avenue, Rosebank, Johannesburg 2196, South Africa

Penguin Books Ltd, Registered Offices: 80 Strand, London WC2R 0RL, England

Published by Gotham Books, a member of Penguin Group (USA) Inc.

First printing, February 2007
1 3 5 7 9 10 8 6 4 2

LIBRARY OF CONGRESS CATALOGING-IN-PUBLICATION DATA

Monagan, David.
Journey into the heart : a tale of pioneering doctors and their race to transform cardiovascular medicine / by David Monagan ; with David O. Williams, medical editor.
p. ; cm.
Includes bibliographical references and index.
ISBN-13: 978-1-592-40265-6 (hardcover)
ISBN-10: 1-592-40265-8 (hardcover)
1. Cardiology—History. 2. Heart—Endoscopic surgery—History. I. Williams, David O. (David Owen), 1943– II. Title.
[DNLM: 1. Cardiology—Biography. 2. Cardiology—history. 3. History, 20th Century. 4. Thoracic Surgery—history. WZ 112.5.C2 M734j 2007]
RC666.5.M66 2007
616.1'2009—dc22 2006026948

Printed in the United States of America
Set in Minion
Designed by Elke Sigal

To Nancy Ziomek,
my sister with an immense heart.

Contents

Journey into the Heart

Introduction

The human heart has always been an object of mysticism and awe. The ancient Greeks believed it to be the seat of the soul, a chamber that processed inspired air or "pneuma," which supposedly cooled the brain. The Christians in turn taught that the heart was sacred; the sacrament of communion, after all, signifies nothing less than partaking of the body and blood of the suffering heart of Jesus. African cannibals and Aztec priests exalted above all in the power to be gained by devouring their victims' hearts.

Inviolate behind a cage of ribs, and almost never glimpsed until after death, the heart's machinery was shrouded in darkness and taboo before 3,000 years of fumbling gave way to the astonishing precision of modern medicine. Reflect for a moment: Less than a century ago, no doctor would dare even to touch a human heart. Only five decades ago, the basic therapy for a heart attack was to lie down and bear it.

All this, of course, has changed. Now, heart attacks can often be stopped in their midst, and the need for traumatic surgery—once celebrated as akin to miracle working—is dwindling year by year. The arteries, valves, congenital malformations, and pacing of the heart can all be repaired with unprecedented intricacy. The journey of but a few inches into the chest entails an epic story, involving a cast of thousands and complexities that long boggled the most brilliant minds on earth. Ultimately, the final triumphs in this saga can be distilled into the lives of a handful of still-little-known figures whose quests encompassed all the promise and anguish of the last century.

The first was a German named Werner Forssmann, who in the 1920s performed death-defying experiments on himself, the most famous being to twist a crude catheter into his own heart when everyone said this would be the kiss of death. Next came a blustery American, Mason Sones, who stormed through the Cleveland Clinic in the 1950s with a passion to illuminate the invisible recesses of the coronary arteries as no one had before. Maniacally obsessed, he created road maps into the unknown that led to the emergence of coronary bypass surgery—among much else. One night he put it this way: "Doing research is like kicking down a door. Every time you kick down one door, you find seven more doors waiting to be kicked down." Sones got only halfway home.

During the 1960s, a still wilder eccentric named Charles Dotter jerry-rigged medical instruments out of gadgets like Volkswagen steering cables and guitar strings and feathered them through every artery and vessel that surgeons told him to leave alone. "Crazy Charlie," as the Oregon mountaineer was called, was a man who respected no limits. Even worldwide ridicule could not diminish his zeal to find an alternative to the knife. Dotter called his novel treatments "ream jobs" and likened himself to a "body plumber." The restless inventor fairly begged to be lampooned. Yet he could never be shaken from his conviction that surgery, first practiced in Europe by barbers who draped their white poles with bloody rags, was indeed "barbaric" and unfit for modern times. Dotter's experiments sometimes resembled a flailing in the dark. Yet he and Sones lived to savor the last triumphal laugh, since their daring and dedication inspired another seeker who turned their visions into a kind of medical wand.

Thanks to an East German refugee named Andreas Roland Gruentzig, the dreams of Forssmann, Dotter, and Sones have been transformed into a stunning therapeutic reality that benefits millions of patients today. Thanks to Gruentzig, any person with cardiovascular disease—which means about every fourth adult in Europe and North America—now stands a good chance of undergoing life-enhancing treatments that could not even have been imagined a generation ago.

Gruentzig, once derided as another charlatan, changed the course of medicine. At first lonely and futile, his work inspired an arc of discovery that has never stopped rising. From the brain to the heart, and the kidneys to the ligaments in the knees, critical human organs and vessels are now routinely cleared and repaired with miniature probes—nearly all of which took their inspiration from Gruentzig's improbable experiments. The man's life reads like an Icarus tale. A flight to fame; then the fall.

Having fled the Iron Curtain to study in West Germany, Gruentzig found his way in 1969 to a junior doctor's position at the University of Zürich, where he became possessed by a vision. The most disease-ridden arteries, he decided, could be reconfigured by inflating miniature balloons—not Volkswagen steering cables—at the ends of slender catheters that would be slipped into the vessels depths. Right. To most colleagues, the notion sounded delusional, perhaps half mad. But nothing would stop Andreas Gruentzig. Recruiting a small cadre of believers, he spent years in a dark Zürich apartment fussing with plastic tubing, Krazy Glue, and thread. Step by step, he began to assemble ever more refined balloon probes, which he first wove into dogs' hearts and then into human leg arteries. Ultimately, Gruentzig defied the skeptics' dire warnings and worked his instruments into patients coronary arteries and then quickly pushed aside angina-inducing blockages of plaque. His methodical but risky experimenting christened the new age of knifeless surgery that his predecessors had blindly groped after.

Gruentzig's first success with the artery-clearing technique drew baffled attention. Nearly painless and promising virtually overnight recovery, his procedure promised to trump coronary artery bypass surgery with its six months of hard recuperation. But few caught the gospel initially. Heart surgeons, at the time celebrated as being on par with the astronauts probing outer space, derided the radical simplicity of his alternative. Most laughed at the idea that a cardiologist could ever work inside the heart with half the precision with which a plumber's snake weaves through a blocked drain. Yet other doctors, including Dotter and Sones, could sense the dawning of a new medical era.

Soon, Gruentzig was invited to Emory University in Atlanta, newly flush with a $105 million gift from Coca-Cola. There, he turned heads. One nurse described him as being so stunningly handsome that he resembled "Omar Sharif, Clark Gable, and Errol Flynn rolled into one." Venture capitalists did their own double takes before the meteoric rise of a technique that would eventually spawn vast corporations out of garage workshops. A few of those chased the beckoning profits so hard that they ended up in trouble with the law. The medical device industry, now generating nearly $300 billion in annual revenues, burst into prominence, often throwing off Silicon Valley–like earnings and pursuing sketch-pad start-ups with gold-rush fever. Today, vast corporations and hundreds of smaller companies compete to follow the insights that Gruentzig first laid forth. But their profits do not capture the essence of the saga of discovery, triumph, and tragedy embodied by the life of Andreas Gruentzig, one of the most charismatic doctors to grace the twentieth century.

After struggling in the wilderness, this once impoverished refugee began savoring the fruits of the classic American success story—a gorgeous young wife, boundless wealth and fame, a new mansion, and a hot personal plane, which he flew with abandon. Some say that, like John F. Kennedy, he created a kind of Camelot. Others claim that Andreas became intoxicated by his own success. He had every reason, for the world of medicine was changing before his eyes—and at his hand.

Despite this swashbuckling medical pioneer's shocking end, his legacy survives to this day, affecting countless patients around the world, emblematic of mankind's quest through the last decades. The man always said he stood on the shoulders of giants, that his quest was theirs. Each of these towering figures had a common point of inspiration—they regarded the human heart with awe.

Chapter 1

The long journey into the human heart is an epic tale of heroism and risk, but also marked with vengeance and pettiness, and peppered with more than a little black comedy.

Picture poor Charles II of England suddenly falling off his barber's chair one February morning in 1685. Whether the cause was a massive heart attack or a cerebral hemorrhage will never be known, but the sixty-five-year-old king had been through a lot. During the English Civil War, Charles was forced into youthful exile on the continent, a time he rather seemed to enjoy. Consorting with at least fifteen mistresses—one named Moll was termed "the most impertinent slut in the world"—Charles developed a lifelong affection for gambling, swilling, and fathering illegitimate children. Fighting Cromwell and losing in 1653, he was eventually restored to the throne in 1660. Five years later, the sovereign writhed on the floor of Whitehall Palace, thrashing his legs and moaning in agony. For all his power, the king, with fourteen physicians and wizards rushing to his side, might as well have been attended by a pack of court jesters.

Secret potions, sorcery, and superstition were what the wisdom of the ages had to offer the great king, whose affliction was attributed to the "bad humors of the blood." England's most esteemed physicians set to slicing open Charles's right wrist to drain away sixteen ounces of "imbalanced" blood. Three cuts were then made into his shoulder, to get closer to the suspected contamination of his heart. Out spewed another eight ounces into flaming hot vials. Next, he was treated to a "voluminous emetic" of poison to induce vomiting. As the Irish sometimes say, this alone would be "enough to kill a hardened sinner."

But the textbooks of the ancients, especially Rome's revered Galen, ordained that to restore "the good humors of the blood," you must not only rid the vascular system of diabolical elements, but introduce life-renewing restoratives. And what better way could there possibly be to treat a heart in peril than to ram healing potions up a dying man's rectum? Hollow tubes with squeezable animal bladders shot the potions—containing aloe, antimony, beet root, violets, camomile flowers, fennel seed, saffron, cinnamon, cardamom seed, linseed, rock salt, and mallow leaves—upward, presumably toward the heart.

The healers rushed about the palace, cradling vials of every antidote to cardiovascular maladies known at that time. In a further effort to recharge the "good blood," the wizards now hit home with curatives of barley water, licorice, thistle, and absinthe. Despite all this firepower, Charles II languished.

If ever there was a Monty Pythonesque attempt to prevent death, this was it. The experts consulted ancient texts. Pigeon dung was applied to Charles II's feet, blistering camphor and mustard plasters containing Spanish fly were smeared upon his shaved head. Poison was blown up the monarch's nose. And, lo, Charles actually seemed to revive. "The blessing of God being approved by the application of proper and seasonable remedies," proclaimed the king's chief physician, Sir Charles Scarburgh.

The next day, the royal patient sank back into unconsciousness. Melon seeds, peony, lavender, and pearl julep joined the therapeutic fray. Then came a more potent elixir, an extract of forty drops of brine from a human skull. Alas, all of the king's men could not put him together again.

Columbus was busy discovering America in 1492, when Pope Innocent VIII underwent an even more radical fix for an apparently failing heart. Two young Roman boys had hollow catheters stuck between their carotid arteries and the right and left jugular veins of the Pontiff, in order to flood his cardiovascular system with perfectly balanced young blood. Naturally, they died, as did the Pope.

In the seventeenth century, as medical research grew more inventive, curious souls experimented with alternative means to restore the ailing constitution. For example, the famous English architect and astronomer Christopher Wren intravenously injected opium and *crocus metallorum* (an extract from iron-rich chalk) into several dogs, which variously vomited to death or fell into a stupor. But Wren had at least validated the idea that medicaments could be delivered by injection into the blood.

In 1665, the London physiologist Richard Lower conducted some novel research of his own, drawing off half the blood in a dog's cardiovascular system and replacing it with an equal amount of beer mixed with wine, creating a literal "booze hound." Next, he devised the first purpose-built vascular catheter—consisting of two silver pipes with a quill valve in between—to draw blood from the carotid artery of a sheep and deliver it into the jugular vein of a young subject who, unlike Pope Innocent VIII, survived. In 1667, Jean-Baptiste Denis, court physician to the fabulous French "Sun King" Louis XIV, succeeded in transfusing blood between dogs and then from sheep to a series of perilously weak individuals—one of whom had been subjected to twenty bloodlettings, a supposed cure-all of that era. The next patient was a wanton adulterer. In this case, transfusions were engineered from a calf, the idea being that placid bovine blood would cool the man's runaway ardor. The technique flopped, and the randy Frenchman went insane and died.

Foolish stuff, we mutter, medieval in its ignorance. Yet consider: Even half a century ago, there was no meaningful intercession—*none*—for the scourge of coronary artery disease, for vascular maladies from the brain to the toes. Moreover, the standard treatment for heart attacks amounted to sedatives and prayers. President Dwight Eisenhower discovered this while on a Colorado vacation during the height of the Cold War.

During a round of golf on Friday, September 23, 1955, the leader of the Free World began to experience a burning discomfort, which he attributed to the onions on his lunchtime hamburger. In reality, an incipient calamity was in progress. A thrombus, or filmy clot of blood, was already congealing in his right coronary artery, thereby slowing the flow of oxygen-carrying blood to the President's heart. Ike's personal physician, Major General Howard McCrum Snyder, came by at dinnertime and, seeing that the Commander in Chief's upset seemed minor, mixed cocktails. Around 9:30 P.M., an increasingly uncomfortable Eisenhower was sent to bed with a bottle of Phillips' Milk of Magnesia. After all, Ike had had a long history of gastrointestinal problems.

But the bromide worked no magic. Instead, Eisenhower awoke with excruciating chest pain. The clot had bloomed into the agonizing tourniquet of *myocardial infarction*—the medical term for a heart attack—triggering the process whereby thousands of heart muscle cells die in every successive minute of inaction. Plenty of those idle minutes lay ahead. Mamie Eisenhower at last telephoned Snyder around 2:30 A.M., and the general practitioner who was also a general raced to the scene. What happened next was a scene of rank futility, only marginally updated from the waning days of Charles II. Seeing that the president was in agony, his chief physician administered morphine. Then, and for nine hours more, he did *nothing.* Meanwhile, fissures of permanent destruction spread through Eisenhower's left ventricle, a vital pumping chamber of the heart. Instead of administering anticoagulants to stave off the clot's inexorable thickening, the doctor merely sat by the presidential bedside, offering comfort as the drugged patient finally drifted off to sleep. It was in fact nearly noon the next day before Snyder called for an EKG machine that at last confirmed the destruction that had been wreaked. An hour later, Eisenhower was rushed to an oxygen tent at Fitzsimons Army Hospital.

The worried Denver physicians next summoned Paul Dudley White, an eminent Harvard professor and a founder of the American Heart Association, who struggled with therapeutic options that were not all that much more advanced than the purging and reaming of Charles II. Eisenhower was confined to seven weeks of in-hospital bed rest, with long-term administration of anticoagulants and constant monitoring. The world grew anxious and the stock market crashed. It was not until the symbolically chosen Veteran's Day, November 11, 1955, that a barely mobile Eisenhower was at last flown back to Washington, D.C., still not close to functioning normally.

Although Eisenhower ultimately recovered sufficiently to serve a second term, he would pay for the medical ignorance of his era, suffering a massive stroke and further heart attacks as the years wore on. Ultimately, his ruined cardiovascular system killed him in 1969, the same year Andreas Gruentzig set his sights on a better alternative.

Triumph over the heart required a vast quest that sometimes resembled a blind man stumbling through a maze. A few early Alexandrians were so eager to behold a heart in action that these "researchers" actually opened the chests of condemned criminals. The first-century Roman scholar, Celsus, described their adventures this way: "They open men whilst alive, criminals received out of prison from the kings, whilst these were still breathing, observed parts which beforehand nature concealed, their position, colour, shape, size, arrangement, hardness." Whatever they learned from such butchery is not clear. The ancient Greeks decided, in any case, that many ailments sprang from imbalances of the four controlling humors—the white phlegm (mucus) and the black bile, the ethereal *pneuma* and the laudable pus. What they couldn't figure out was how these vital fluids supposedly charged the blood with health.

In the second century A.D., Galen, physician to Roman Emperor Marcus Aurelius, effectively launched medical science with a festival of vivisection. Apes, horses, pigs, asses, cows, sheep, camels, deer, bears, wolves, weasels, mice, snakes, and various fish and birds all fell under the Greek's knife in an attempt to divine the mysterious mechanics of life. Scores of treatises fell from his pen and these dominated medical thinking for the next 1,400 years.

What Galen got right was remarkable—among countless things the connections of nerves to muscles. But what he misconceived spread confusion until the Renaissance and beyond. At last, a reborn fascination with anatomy, led by the pioneering dissections of Leonardo da Vinci, reawakened curiosity about the body's mechanics. The master of this macabre art proved to be a Belgian named Andreas Vesalius. Climbing gallows and robbing graveyards to retrieve cadavers, Vesalius performed dissections before hundreds of awestruck students at the University of Padua. In 1543, he published an incomparably beautiful set of engravings in a discourse called *The Fabric of the Human Body*. That same year Copernicus reconstructed man's view of the universe with the publication of his *De Revolutionibus.*

Vesalius's impact was just as profound. No longer was the body to be regarded with age-old complacency. Vesalius presented it as a construction containing miracles of engineering. As a commentator put it 400 years later, "He stood before life as if the conqueror of death itself, because it was out of the dead and decomposing body that he read the mysteries of living functioning perfectly."

Unfortunately, for all his inspiration, Vesalius failed to solve the puzzle of circulation. A perhaps apocryphal story holds that he eventually split the chest of a newly deceased noblewoman in Spain, only to discover in horror that her heart was still beating. Vesalius fled to the Holy Land and eventually died in a shipwreck. But a rage for vivisection was his legacy.

The Spanish anatomist Michael Servetus could next be heard postulating exquisite interconnections between the lungs and far-flung vascular tree, the branches of the circulation that feather to the fingers and toes. Alas, this pioneer had his own fatal flaw—his ego. With less than keen judgment, he sent John Calvin an impassioned treatise that also happened to denounce the Blessed Trinity as a monster with three heads. For this, Servetus was burned at a Swiss stake in 1553.

The seventeenth century arrived with an unquenchable thirst for free inquiry. Galileo, Newton, Kepler, Descartes, and Spinoza electrified the age's imagination. The mechanics of the heart at last became the subject of intensive investigation by another tower of the Enlightenment named William Harvey (1578–1657). Disparagingly dubbed "The Circulator" by slower-witted contemporaries, the Padua-trained English anatomist is considered the father of modern medicine.

Harvey single-handedly rid the heart of mysticism. With controlled experiments that remain a model of clarity to this day, he methodically demonstrated the interactions of the heart's valves and chambers, and their sublime synchrony with the lungs, and then he exquisitely detailed the regulation of blood flow from the head to the foot. Harvey's methods foreshadowed those of great medical explorers for centuries to come. In April 1616, a week before the death of Shakespeare, Harvey first sketched out to a London audience his notion that the heart was the most profound machine in the universe—and, ergo, potentially knowable to the smallest details. These he would ponder for the rest of his years.

An eccentric in his own right, "The Circulator" was so inspired by contemplation in darkness that he had an underground "thinking chamber" dug under his Surrey residence and had a fondness for wearing daggers on his belt. But there was no whimsy in Harvey's relentless detective work in exposing the circulation's intricate genius. Of this, he would write:

> I found the task so truly arduous, so full of difficulties, that I was almost tempted to think . . . that the motion of the heart was only to be comprehended by God. For I could neither rightly perceive at first when the systole [contraction] and when the diastole [relaxation] took place, nor when and where dilatation and contraction occurred, by reason of the rapidity of the motion, which in many animals is accomplished in the twinkling of an eye, coming and going like a flash of lightning; so that the systole presented itself to me now from this point, now from that; the diastole the same; and then everything was reversed, the motions occurring, as it seemed, variously and confused together. My mind was therefore greatly unsettled, nor did I know what I should myself conclude nor what to believe from others . . .

Today, it is difficult to grasp the thunderclap of transformation that Harvey's theories represented. Where lay the soul, and what was man, and what comprised the true nature of disease? These questions burned through Harvey's ideas. The correlative of his heart-as-machine theory was to abandon quasi-spiritual remedies for every human malady.

Seven years after Harvey's death in 1657, the English scientist-philosopher, Henry Power, summed up the spirit of a new age:

> These are the days that must lay a new foundation of a more magnificent Philosophy, never to be overthrown, that will empirically and sensibly canvass the phenomena of nature, deducing the causes of things from such originals in nature, as we observe are producible by Art and the infallible demonstration of mechanics. And certainly this is the way, and no other, to build a true and permanent Philosophy.

The Circulator's ideas sunk in slowly. His first important successor was the Reverend Stephen Hales. In 1733, this Englishman made a name for himself by inserting long glass and copper catheters, with baffles made of the flexible windpipes of geese, into the neck arteries of insensate horses. In this way, he conducted further groundbreaking explorations of the heart's pumping mechanics, which resounded with the quantifying power of what is essentially today's scientific method. A stickler for detail, Hales discovered reliable measures of blood pressure that were fundamental to the study of hypertension.

Late in the century, inquisitive minds set to unraveling still more delicate questions—if the heart cycled blood to the lungs with every pulse, what processes actually transpired to make this interchange so vital? What did exposure to fresh air really do to the blood?

The answers came from pioneering explorations into the invisible world. The first great chemists devised experiments to divine what the air itself was made of. Foremost among them was Joseph Priestly, who isolated oxygen in 1774. Next, the French chemist Antoine Lavoisier demonstrated that the *pas de deux* between the heart and lungs introduced renewing oxygen to the bloodstream, while sucking away not "bad humors" but noxious carbon dioxide. That insight would of course eventually transform the understanding of the function of every animal and plant on earth. Lavoisier paid a frightful price for his acumen, however, when an intellectual-hating mob of the French Revolution stormed his house and chopped off his head.

Ignorance was scarcely the only impediment—heart surgery itself remained unthinkable without the discovery of effective anesthesia. The twin "laughing gases" of nitrous oxide (another Priestly discovery) and ether were essentially toyed with and abused for seventy years before finally entering medical practice in Boston in 1846. Even then, the three Americans who ushered in this advance all went insane in their final years, perhaps from having indulged in too much self-experimentation.

Another hurdle waited. Nineteenth-century operating theaters—usually stuck in far-off hospital precincts to muffle the echoes of patients' screams—remained foul incubators for lethal infections until the principles of basic hygiene and antiseptic fields were comprehended and solved.

Afterward, surgeons of the 1860s and beyond began venturing across previously unimaginable thresholds of achievement—quickly excising tumors or removing gallstones and impacted teeth without unbearable pain or risk of mortal infection. Inevitably, certain pioneers itched to get their hands on wounded hearts, usually at first to do nothing more than to remove a bullet lodged in the outer protective sac called the *pericardium*. Yet nine out of ten heart patients died on the spot, leading experts at the century's end to write off the whole field. In 1896, the British surgeon Stephen Paget proclaimed: "Surgery of the heart has probably reached the limits set by nature to all surgery; no new method and no new discovery can overcome the natural discoveries that attend a wound to the heart."

"Any surgeon who wishes to preserve the respect of his colleagues would never attempt to suture a human heart," concurred Theodor Billroth, the German pioneer of abdominal surgery.

More depressing was the experience of a respected Danzig, Germany, physician named H.M. Block, who had proposed a bold new era of heart surgery based on his research into arterial suturing in rabbits. In 1892, this pioneer chose to split open the chest cavity of a young relative believed to be dying from tuberculosis—only to have her succumb as he worked. Block killed himself the next day.

So the heart still dangled beyond medicine's reach, a thing of eternal beauty, fascination, and fear. It would take another seventy years before Andreas Gruentzig approached the ancient riddle with an idea of breathtaking simplicity and a personality that shined like a torch.

Chapter 2

Curious developments were in progress by 1904, the year Werner Forssmann was born in Berlin. The new century had wafted in like a fantasy out of Jules Verne. The Wright Brothers had just sputtered aloft in the first successful heavier-than-air machine. The same year, the ice cream cone had been invented in St. Louis. One of the first motion pictures—*Voyage to the Moon*—had debuted around the world, soon to be followed by the thrilling spread of inexpensive neighborhood cinemas, called "nickelodeons" in the U.S. Even as young Forssmann, destined to become either the most heroic or foolish heart pioneer of all time, learned to kick a ball in the park, self-propelled cars and trucks were replacing the horse and dog carts that delivered coal to his door. Dirigibles floated like dreams through the clouds.

There seemed to be no limits to what mankind could achieve now. In a world medical center like Berlin, the evening talk would turn to further scientific discoveries. How could anybody guess what might come next? The ever-widening implications of William Roentgen's discovery of the X-ray in 1895 remained a hot topic for years, promising untold revelations. The first eerie image exposing every skeletal filament of Roentgen's wife's hand inspired awe around the world. Enhancements to Roentgen's all-penetrating rays kept capturing fresh headlines. Surgeons began removing tumors with astonishing speed, once they were equipped with foreknowledge of the previously invisible world beneath the skin. Soon, the deepest reaches of the body became ripe for illumination by injections of special heavy-metal-containing "contrast" fluids that glowed when subjected to X-ray beams.

Medicine was center stage again with the award of the 1912 Nobel Prize to Alexis Carrel of the Rockefeller Institute in New York. His trick was to find a way to stitch together pulsing blood vessels as deftly as his French seamstress mother completed a hem. In one experiment, Carrel successfully removed a dog's neck, or carotid, artery and grafted it into the same vessel of a second animal. Next, the visionary Frenchman transplanted kidneys, thyroid glands, ovaries, and even the combined lungs and heart from one animal to another. In 1910, he sewed a graft vessel into the left coronary artery of a dog, which constituted nothing less than the first coronary artery bypass.

Forssmann's uncle Walter Hindenberg, a country doctor sixty miles north of Berlin, told the young boy about such discoveries, and even brought his nephew on house calls in a yellow horse-drawn carriage. But other wonders were abroad—namely, Emperor Kaiser Wilhelm leading ever more warlike processions of plumed cavalry down Berlin's avenues.

Werner was still too young to grasp the menace in these scenes. Indeed, no adult in the summer of 1914 could have predicted the apocalyptic destruction set in motion by the assassination of Archduke Ferdinand in Sarajevo, which led to the first clashes of the Great War.

But Forssmann's idyll was doomed. On August 2, 1914, his father Julius, a reserve captain in the Imperial infantry, waved goodbye from a trainload of singing recruits embarking for the gathering Russian front. Two years later, Julius Forssmann was killed in Galicia, the western part of modern Hungary—one of ten million combatants and civilians to succumb to the carnage and epidemics of World War I. His widowed wife and son were left to subsist on rutabagas, potatoes, and the odd slice of fermenting dark bread, with special Sunday dinners featuring baked crows.

Forssmann sought refuge wherever he could, collecting small animals and fish for his aquarium, and exploring the invisible world with a Leitz microscope his uncle gave him. By the war's end, Germany was a sorry place for inspiring visions in a fourteen-year-old. The Kaiser abdicated, Bolshevik sympathizers took to the streets, and such chaos reigned that Forssmann, like other adolescents, was recruited to assist in maintaining the railroads. By the time the French invaded the Ruhr, bourgeois Prussian families like the Forssmanns yearned for a return to the rigid social order of the Imperial era. Forssmann, meanwhile, plodded along with his classical studies.

With his graduation examinations approaching, the only child felt the weight of expectation sliding onto his shoulders. A concerned teacher asked what he intended to go into next. "Business!" replied Forssmann.

The droll response: "Forssmann, you go into business and you'll be the only one who won't make money. You're much too thick to be a businessman. You ought to study. Study medicine, that's where your talent lies."

In October of 1922, Forssmann matriculated at Berlin's prestigious Friedrich-Wilhelm University. So much for an exalted rite of passage: The eighteen-year-old nearly vomited on his first day of class. A gnarled anatomy professor, standing before a row of embalmed cadavers, enlisted the new student to help peel back flaps of skin from the first instructional corpse. By the stink of formaldehyde and putrefying flesh, he was baptized. Soon enough, Werner would be earning pfennigs by retrieving fresh corpses from their zinc-lined basement containers, helping to pay his tuition and put food on the table for his mother and grandmother. Meanwhile, he acquired a working knowledge of the newest attempts to manipulate the heart and great vessels.

There was much to learn, for a fresh cast of explorers had been weighing in since the turn of the century. In Norway in 1901, and England the next year, surgeons succeeded in massaging a pulse back into two patients' exposed hearts, proving that at least the exterior of the taboo-ridden organ could be handled without mortal consequences. In 1908, Friedrich Trendelenburg of Leipzig capped years of animal experimentation with the first successful attempt to expose the human chest cavity and remove death-inducing clots from the great pulmonary artery that connects the heart to the lungs. In less than forty-five seconds, Trendelenburg incised the imperiled vessel, pulled out the clot with a forceps, and stitched the artery whole. Unfortunately, the next three hundred attempts to replicate this stunning breakthrough had a death rate that exceeded 98 percent.

Others pressed on with daring attempts to open blocked valves in children, a consequence of the rheumatic fever that was still endemic in Europe. A historic moment occurred in 1912 when a French physician, Théodore Tuffier, opened the chest of a young patient with every intent to transform the future of cardiac surgery. But Tuffier couldn't pull the trigger; in fact, he dropped his scalpel. Fearful of killing the patient with an incision, the surgeon elected to try something less perilous. Pressing his index finger against the outer wall of the aorta, Tuffier felt his way to the narrowed ring of the great trunk vessel's valve: the sweat dripped off his brows as the heart writhed to his touch. Was he about to bestow or end life? Tuffier took the risk. Never physically penetrating the interior of the patient's heart, he nonetheless exerted a crushing dilating force to the center of the valve. Voila! The knotted blockage, or "stenosis," opened neatly and a profound resurgence of blood flow followed. However crudely, the inner depths of the heart had at last been touched and healed by the hand of man.

Forssmann was constantly handling human hearts himself—albeit dead ones. The organ fascinated him, thanks to the inspiration of an anatomy professor named Rudolf Frik, whose father had been a leading nineteenth-century investigator of cardiac physiology. Frik regarded the heart with wonder, explaining to his students that this little bundle of hollowed-out muscle, weighing less than a pound and about the size of a clenched fist, was a thing of marvels. In a symphony of precision, a healthy heart beats forty-eight million times a year, 2.6 billion times a lifetime, and with the might to move seventy million gallons of blood before lapsing into silence. The heart of even a ninety-eight-pound weakling has sufficient pumping power to lift ten tons—the weight of an armored tank—ten feet every month, or ten miles into the sky in a lifetime; or, for that matter, to flood the Roman Colosseum.

In his captivating autobiography, *Experiments on Myself,* Forssmann attributes his life's turning point to his professor's lectures about nineteenth-century physiologists, especially a group of Frenchmen who sought to calibrate the workings of the heart by probing its pumping chambers. In this experiment, Jean-Baptiste Auguste Chaveau had inserted a catheter through a horse's jugular vein and then advanced it to the depths of the right ventricle. The instrument had a balloon on its tip whose trembling before every pulse was fed back to create intricate graphs of tiny changes in blood pressure. The image mesmerized Forssmann.

All the while, alas, darker visions worked through his university, where several professors proved to be vicious protégés of Adolf Hitler's latest Munich demagoguery. "Just remember one thing. People are evil and stink, corpses just stink," an odious anatomist named Friedel advised Forssmann. The proto-Nazi constantly disparaged Jews, Catholics, and especially Slavs, while drawing poor and hungry students like Forssmann to his free luncheon club. The future Nobel Prize winner would long rue this and graver misjudgments to come.

The young Forssmann was wildly impetuous, absolutely the last student any teacher would predict to transform the treatment of the human heart. No intellectual, he bounced through life according to whatever cards were dealt his way—sometimes with blundering stupidity. His worst early sins seem to have involved a penchant for downing beer with his friends.

In 1924, a group of them hatched a plan to bicycle from Bavaria to Venice on summer break. The merry wanderers trained to Nürnberg and pedaled and drank onward to Munich, where their finances promptly fell apart. Never one to give up, Forssmann made it by his lonesome to Innsbruck in Austria, where he slept rough and guzzled cheap wine with a tramp. Soon he was befriending peasants in the high Alps, sleeping in their huts, and climbing a peak without much forethought—where he nearly killed himself. This was his signature style.

Back in medical school in the autumn, Forssmann spent the next years rounding out the academic side of his education. By now, he would have learned about the latest surgical progress on the heart, much of it pushed forward in America.

In Boston, surgeons had attempted to perfect a far more scientific instrument than Tuffier's finger for splitting open calcified heart valves. Called a cardioscope, this ambitious gadget had an outer viewing lens that telescoped into a remotely controlled cutting apparatus at the far or distal end. The idea was prescient, except it didn't work. An imperative nonetheless beckoned, and this was to find ever bolder ways to penetrate the heart's depths.

In 1925, a London surgeon, Henry Souttar, embarked on a startlingly close approximation of a modern cardiac valve operation. Cutting his way through the pericardium and into the atrial chamber of a languishing fifteen-year-old girl, he clamped aside the obscuring tissue flaps, then stuck his naked finger through the rheumatic, petrified mitral valve and into the left ventricle. Here were surgical heroics on the grand scale. But just when he was about to open the encrusted valve, the heart's infernal force confounded medical progress once again. The sutures split, geysers of blood shot forth, and Souttar was left flailing to patch up the mess. The girl survived, but another bout of rheumatic fever further crippled her valves' function within a year. The medical establishment had seen enough. For a quarter century, nobody dared replicate Souttar's radical methods—which are commonplace today.

Broad-shouldered, heavy-cheeked, and thick-necked—but with curiously intense dark-brown eyes—Forssmann scarcely cut the figure of a future Nobelist. He rotated through various hard-knock residency tours, work that included helping to deliver babies and cleaning up botched abortions in the seedy depths of a Berlin that was immortalized by Bertolt Brecht. Next, Forssmann directly transfused 900 milliliters of blood from his arm to a burly transportation worker in danger of bleeding to death while awaiting surgery for a perforated gastric ulcer. Ages after Jean-Baptiste Denis's eighteenth-century experiments, things were still being done that way.

In February of 1929, Forssmann received his license to practice medicine. He then enrolled as an intern in the understaffed Red Cross Hospital in Eberswalde, a provincial town northeast of Berlin. Nonacademic, with crises popping up constantly, the atmosphere fit Forssmann's personality to a T. He could cut here and patch there with scarcely anyone to stop or guide him—all this while still only twenty-four. In no time, Forssmann was back in the beer hall boasting about a truly head-turning scheme. Why, he was going to advance a catheter, by then commonly used to inject X-ray dye into the urinary tract, into a brachial vein in the arm and then direct it all the way into a human heart. Forssmann's autobiography suggests the dialogue:

"Hadn't the French firmly demonstrated how easy it is to do this with a horse? What was a man but a mammal built much the same? Okay, pushing through the human jugular could introduce dangerous air bubbles that might travel to the brain. Fair enough. I'll just drive forward to the heart through the arm instead."

This appears to be as far as the future medical legend's thinking advanced. When he summoned the nerve—rarely a problem for Forssmann—to announce his plan to the hospital's kindly chief and family friend, Peter Schneider, Forssmann was greeted by one of the more incredulous looks in the history of science.

"I cannot possibly allow you to carry out such an experiment on a patient," Schneider guffawed.

"Well, there's another way to prove it's not dangerous. I'll experiment on myself," Forssmann replied, brushing aside the need for prior animal research thanks to a few experiments by his long-dead French heroes.

"My no is final and absolute," Schneider repeated. He had no stomach for having to explain to a widowed mother and personal friend how he allowed her only child to kill himself.

Forssmann pretended to acquiesce. But he immediately began scheming to circumvent the genial boss. The daredevil's own words are eloquent:

> I let a few days go by and then started to prowl around Nurse Gerda Ditzen like a sweet-toothed cat around the cream jug. I knew I'd be able to carry out my black deed only during the afternoon siesta while everyone in the hospital was dozing, so I made a point of dawdling in the canteen after lunch, hoping to meet Nurse Gerda as she left the nurses' dining room. We'd often lent each other books, so it was easy to find something to gossip about; and she'd invite me back to her little office . . . Gradually, carefully, I steered the conversation round to my hobbyhorse, and found she was interested . . . So, little by little, I won over my essential disciple. When, about a fort-night after my conversation with Schneider, she said with a sigh, 'What a pity we can't do this together!' I decided the time had come.

Forssmann had already showed the affable nurse the engravings of the magical equine procedure of seventy years before, assuring her the job would be a piece of cake in humans, too. Swiftly he convinced her to participate, on the sly, in his harrowing experiment.

It was the hospital's quaint siesta time; nobody would notice a thing. She followed him to a procedure room and unlocked all the requisite equipment under her domain—a venesection kit for cutting into the arm, a local anesthetic, and an urethral catheter.

There, the brave and perhaps smitten nurse offered herself as the target for the first catheter ever to be threaded into a human heart. Meanwhile, bold Werner poured forth bouquets of devious charm. Feigning agreement to the nurse's every objection, he coaxed Gerda to lie back suppliant on the operating table:

> With the speed of light I strapped her down so tightly that she couldn't reach the buckle; I then tied down her hands. Amazingly enough she accepted my explanation that I had to take all these precautions against her falling off the table since I had no one to assist me. I'd pushed the instrument tray behind her head so she couldn't see what I was doing. In the twinkling of an eye I had anesthesized my left elbow. Now I went back to her and began to iodize her elbow ceremoniously, and then to lay a sterile cloth over it, all very slowly and deliberately in order to kill time. When my anesthetic began to take effect I quickly made an incision into my skin, inserted a Deschamps aneurism needle under the vein, opened it and pushed the catheter about a foot inside. I packed it with gauze and laid a sterile split over it. Then I released Nurse Gerda's right hand and loosened the straps around her knees.

Gerda gesticulated wildly. Here stood her esteemed young doctor dawdling beside her with a catheter stuck through the length of his arm and reaching to some unimaginable crossroads of mortal danger. With a glazed look in his eyes, Forssmann kept twisting the thing ever deeper. He commanded that

Gerda now help him to the basement X-ray room. It was imperative, Forssmann insisted, that they document for all time what he intended to do next.

To the amazement of passing colleagues, the bizarre catheter-trailing duo crept through the corridors, weaved down the steps, and pushed open the door to the cellar chamber. A stunned nurse named Eva dropped her jaw. But this was a Germany where subordinates did what they were told. Eva quickly obeyed and readied the equipment.

By this point, the hospital was abuzz. The radiology lab's doors flew open again, and in burst Forssmann's drinking partner Peter Romeis, his hair tousled from siesta sleep. "You idiot, what the hell are you doing?" the young doctor demanded.

Romeis had heard his pal's late-night boasts all right, but what of it? You heard all kinds of things when you socialized with this self-pronounced friend of the tramps. Yet here was madness incarnate—Forssmann grinning before him with a catheter crammed into his arm and beckoning toward a rendezvous with death. Romeis shouted, then made a grab to pull the instrument free.

"I had to give him a few kicks on the shin to calm him down," Forssmann recalled blithely.

Romeis, Gerda, and the X-ray technician all stood in a ring of awe. They well knew that they were witnessing a leap into forbidden ground, an act so absurdly melodramatic it could have been cooked up by Richard Wagner for the Bayreuth stage. Yet Forssmann kept issuing his methodical Prussian commands to set everything in order to confirm the final push toward the center of his heart. The critical thing was that he be able to view the telltale X-rays himself.

> I had a mirror placed so that by looking over the top of the screen I could see in it my thorax and upper arm. As I'd expected the catheter had reached the head of the humerus. Romeis wanted me to stop at this point and remove it. But I wouldn't hear of it. I pushed the catheter in further, almost to the two foot mark. Now the mirror showed the catheter inside the heart, with its tip at the right ventricle, just as I'd envisioned it.

Easy as pie. You stick a feather in your cap and call it macaroni. You put one foot in, one foot out, and drive a catheter into the dead center of the organ that had defied understanding since the beginning of time. Word of what had happened shot through the Eberswalde Clinic; crazy Forssmann, meanwhile, was all but doing the jig. The phone rang: it was the chief's secretary, conveying the normally implacable hospital head's outrage and demanding an immediate appearance in his office. Direct disobedience of the chief is countenanced in no medical institution, and especially not in the Prussian-dominated Germany of 1929. Schneider told Forssmann—barely twenty-five—that he had unforgivably undermined his personal authority and risked making a laughingstock of his clinic.

In walked the boy-man clutching his historic X-rays. To a radiologist working with current technology, they would be a blurry joke, showing little more than a dark line arrowing behind the ribs toward a vague after image of the heart. The revelation in fact lay in where the instrument pointed, not in what it displayed. Yet the implications were breathtaking. Struggling to make sense of the weirdest afternoon in his sleepy hospital's history, Schneider grasped enough to congratulate Forssmann. A bit later, the chief even allowed him to perform a second heart catheterization on a young woman dying from a botched abortion, this latter merely a test on a patient who had no hope of survival. Heady stuff, but unconscionably dangerous. Schneider advised him to seek a job elsewhere.

Forssmann's next stop, at Berlin's renowned Charité Hospital, was a thing of comedy. Here lay the fiefdom of one of the most famous heart surgeons in the world, perhaps the most esteemed doctor in Germany: the all-powerful Ferdinand Sauerbruch—whose name means "sour brook." A tyrant upon whom Hitler bestowed a gold medal in Nürnberg a decade later, Sauerbruch had a mean streak as long as Forssmann's catheter. His reputation was based upon unrivaled surgical bravery and dexterity, and upon a contraption that epitomized the blundering progress of heart surgery in the first decades of the twentieth century.

Sauerbruch piloted a kind of medical U-boat. He had invented a glass-walled box that could be so prodigiously drained of air that it created a powerful low-pressure chamber—like a training capsule for astronauts that simulated conditions in outer space. The point was to keep the lungs from collapsing as he removed tumors without having to race against the clock. The stern physician sat sweating in the thing as he worked with a kind of periscope shooting fresh air towards his mouth. The patient's neck stuck out of the far end, a pressure seal provided by leather flaps encircling the chin. So cramming a catheter into a heart was dunce-simple to the master surgeon's eyes.

In Germany at that time, a *Herr Geheimrat Professor* of such stature routinely surrounded himself with deep moats of bureaucracy and subservience. The imperious Sauerbruch therefore ignored the lowly and ever itchier Forssmann for weeks. A craven understudy meanwhile sneered, "So you're the gentleman from the provinces who's going to teach us all about science, are you? Well, we'll see about that."

In mid-October, the rising arc of youthful ambition inevitably smacked into the wall of entrenched authority. Overnight coverage of Forssmann's breakthrough was splayed across a lowbrow Berlin tabloid newspaper, and this was the kiss of scientific death by the lights of Herr Geheimrat Sauerbruch, a stain upon the worldwide eminence of his citadel of exacting experimentation.

Forssmann, a man with a null sense of politics and a spectacular penchant for sticking his foot in his mouth, was summoned to at last face the almighty "Sour Piss"—as the understudies called him. Never mind that his breakthrough would shortly be splashed across newspapers from Vienna to New York. The verdict on his worth was finally to be uttered by the Über Surgeon. In strutted Forssmann, attempting pleasantries. Herr Geheimrat drummed his fingers on his vast desk, frowned and sneered. Then he let the upstart have it, demanding that Forssmann explain on the spot what exactly he intended to do with the rest of his twenty-five-year-old life.

To this, the young doctor hesitated, stammering that he was still figuring out the answer and that, er, ah, maybe he should be a university lecturer.

The Herr Geheimrat uncoiled with a vengeance. "You might lecture in a circus about your little tricks, but never in a respectable German university! What do you really want to be, an internist or a surgeon?"

Backed against the ropes, Forssmann protested that he was still unsure.

"There we have it!" Sauerbruch exploded. "The real Forssmann who can't make up his mind about anything! *Every inch an internist!* A true surgeon thinks of only one thing: *Operate! Operate! Operate!*' "

The impetuous Forssmann readied a counterpunch, opening a mouth that would have been better left shut in this and countless conflicts to come. Bang! Forssmann issued one of the most intemperate ripostes ever uttered by an apprentice to a master with the power to damn the rest of his working life. "Herr Geheimrat Sauerbruch, there are hunters and shooters." The double entendre implied wily potency to himself, and perhaps the opposite to his chief.

Sauerbruch's beady eyes bulged behind his thick spectacles as he screamed, "Get out! Leave my department immediately!"

The next day Forssmann, tail between his legs, arranged to retrieve his old job at Eberswalde. The door to greatness seemed to have closed forever.

In the comforting backwater of Forssmann's twice-tried community hospital, his sympathetic chief at least encouraged further investigations. By now, Black Tuesday, October 29, 1929, had sunk the New York Stock Exchange. The spreading ripples of panic shortly snuffed the "roar" out of the twenties. An era of economic expansion and good hope was plummeting toward the Great Depression.

But straight-ahead Werner never dwelled too deeply on such distractions as economics or politics. He still had his singular mission. So he began racing to work on a newly purchased BMW motorcycle, often clenching a potato sack stuffed with a drugged dog. His new passion was to demonstrate that the heart could be injected with X-ray dye in order to illuminate the actual inner pumping chambers. These experiments were in fact critical on two fronts—first, to prove that his self-experiments were more than a stunt, and second, to allay universal fears that the toxic solutions would prove lethal in the heart. Although increasingly routine in the kidneys and urinary tract, X-ray dye visualizations often provoked excruciating reactions, although the locally delivered, intravenous injections in those regions rarely induced seizures or death. Nobody—except, of course, Werner Forssmann—dared dump the still evolving brew of heavy metals into the heart's chambers, where they might pool in thick concentrations and perhaps induce cardiac arrest.

Young Werner didn't see it that way. He would refine his concept through animal experiments and never mind that he had done nothing similar before. For the sake of medical progress, and his own thin wallet, Forssmann made a deal with a local dog breeder to purchase mutts for twenty marks, then fetch fifteen marks back if they could be returned in good health.

The young researcher proceeded swiftly, shooting the fickle X-ray dye into the hearts of his canine subjects, while praying that his reputation and wallet would remain intact. Once again, luck shined. Forssmann produced a series of historic images, the details much more clinically promising than the crude initial one of his own heart.

Forssmann next sought to prove that the same degree of visualization could be realized, without grave side effects, in a human. So it was time for a volunteer. Naturally, a certain candidate popped up—Werner Forssmann himself, not yet twenty-six. The inveterate risk-taker well knew that a dog's heart and a man's differed profoundly, and that this time, there were no ghosts of French physiologists to nod him forward.

For starters, he held an open X-ray dye-filled test tube within his mouth for hours in order to check for violent reactions. Experiencing none, he made the next leap—catheterizing his own heart with the so-called "contrast solutions" a total of eight times in the next year or so, without suffering any major side effects.

Unfortunately, the primitive radiological equipment at Eberswalde still yielded fairly dim images. Instead of providing an indelible blueprint of the heart's chambers, Forssmann's triumph was essentially a symbolic one of encouraging curious thinkers elsewhere to advance his explorations under far more scientific conditions. Forssmann, who rarely read professional journals, did not realize how soon that call would be answered—in a few German medical centers, in distant Cuba, and in a variety of university hospitals in the United States.

But a principle that launched the field of modern cardiology had been established. Two physicians at Columbia University in New York, André Cournand and Dickinson W. Richards, spent the late 1930s mapping impairments of the atria and ventricles—respectively, the heart's reservoirs and pumping chambers—with exquisite precision. This work reopened the door to cardiac valve surgery and led to their sharing the Nobel Prize with Forssmann in 1956.

Personal misjudgment and the ugliness of history ensured that Forssmann himself would flounder in the wilderness for most of the intervening years. In another one of his classically impulsive moves, the explorer leapt at the opportunity when Herr Geheimrat Sauerbruch suddenly invited him to rejoin the Charité Hospital, this time at the lowest of levels. Toiling away at minor surgery, he fell under the sway of a professor who extolled the Nazi dream of crushing the era's rampant social disorder and returning Germany to world power. In 1931, Forssmann signed up as a Party member. His autobiography spends a lot of ink apologizing for this decision, and relating countless future acts of personal compassion. His war years were indeed a harrowing saga of sacrifice and courage in treating maimed human beings. But he had made the sinister choice.

At the outbreak of World War II, the young surgeon was called to war. Ridden with lice, covered with blood and sometimes operating for days on end thanks to the occasional help of amphetamines, he spent the next years in hell, a couple of them in tents on the Russian front where the dead stacked up in rigor mortis at the end of his operating table. The encircled "Spandau E" Battalion and SS Deathwatch Division "Eicke" had run out of dynamite to blast holes for graves.

By the end of June of 1945, the Nazis' game was up. Forssmann fled from the advancing Red Army, desperate to seek imprisonment by the reportedly more benevolent Americans. He dragged himself to the great dividing line of the River Elbe, said his prayers, and dove into the churning waters. As a final thank you, die-hard SS men strafed the future Nobelist with a machine gun as he swam in terror across the river. He was captured as he crawled onto the far shore. The GIs who frisked this exhausted specimen of the Wehrmacht found only waterlogged photographs of Forssmann's beloved wife and children, and a drenched copy of Goethe's *Faust*.

A great pioneer of the heart had been reduced to squirming like a reptile in the mud, but with a book of visions clutched to his chest.

Chapter 3

By the end of World War II, the balance of not only military but also medical power had shifted firmly to America, especially in pioneering work on the heart. A new generation of seekers had pushed aside their predecessors' trepidations and set out to conquer coronary disease with a variety of radical innovations. Out of this quest arose one of the most fantastic inventions in the history of medicine: the heart-lung machine.

Back in February of 1931, a young Harvard Medical School trainee named John Gibbon sat in a dusk-till-dawn vigil beside the bed of a middle-aged woman suffering the death throes of a pulmonary embolism choking off the flow of blood to her lungs. Her chest pains were excruciating, her pulse racing, her nausea turning her insides out. No American surgeon had ever successfully excised such a ticking bomb, but if ever the attempt seemed advised, this was it. Yet the odds were miserable. Once the chest was cracked open and the woman's pulmonary artery was clamped off so that the cutting and stitching could begin, her circulation would be in mortal suspension. If the procedure took more than six and a half minutes, she would perish. All through that night as he sat beside the languishing woman's bed, Gibbon dreamed of a changed future. He dreamed and he prayed for a day when there would be ample time for the unfolding techniques of "thoracic" (chest-cavity) surgery to be employed with some semblance of calm. In that long vigil, he formed a vision of a machine that might one day temporarily seize control of the circulation so that freshly oxygenated blood could keep surging through the arteries even when the heart's pumping was stopped. As it happened, the patient's dawn procedure took a mere seven minutes: but that was thirty seconds too long. Gibbon watched over the woman's demise and wept.

He then poured his life's passion into realizing his vision. He gradually constructed the first prototype out of baffles, hand-carved cork valves, jerry-rigged tubes, and a cranky electric motor. The earliest version coughed to life like a Rube Goldberg contraption, with a trail of dead animals left as laboratory roadkill whenever it was employed. Failures of every description haunted the project, but still the contemplative doctor, steeped in poetry and philosophy, persisted.

Finally, in 1952, Gibbon was confident, after twenty years of experimentation and anguishing failures, that he had it right. He stood by as a fifteen-month-old baby girl's heart was stopped for surgery to correct her severe congestive heart failure. To his horror, she died on the table, although the machine seemed to work all right. A year later, his rickety creation, with six attendants busy at its controls, was fired up again for an operation on an eighteen-year-old girl with an atrial-septal defect—a congenital hole in a wall of her heart—who had suffered three calamitous heart failures in the preceding six months. For the entire twenty-seven minutes of the perilous surgery, the Gibbon jalopy rattled along on all cylinders. The defect was closed, the girl survived, and in fact flourished. But two more deaths followed and Gibbon could never again bring his crushed soul into an operating room.

In the end, Gibbon's visionary technology prevailed spectacularly, because others had been swept into the saga. One of these was Charles Lindbergh. After traversing the Atlantic in 1927 in his impossibly frail *Spirit of St. Louis*, and after flying to South America and winging the frozen Antarctic, you might think that the man had conquered every horizon he needed. Not quite.

Distressed by the apparently needless death of his sister-in-law at the hands of cardiac surgeons, he soon found his way to a long collaboration with Alexis Carrel, the 1912 Nobel Prize winner, in an attempt to perfect a rival heart-lung machine, which unfortunately never quite worked out.

Another recruit joined the quest. He was a can-do Midwestern surgeon named C. Walton Lillehei. Despite nearly succumbing to cancer a few years earlier, Lillehei was the Answer Man. Among many achievements, he designed the heart's first cardiac pacemaker with the help of a local television repairman, who then founded the now multibillion-dollar Minnesota corporation called Medtronic. In short order, Lillehei created a vastly more effective heart-lung device than Gibbon's prototype, and this advance redefined the possibilities for surgery within the heart for all time.

Given time to divert the circulation, cardiac surgeons were itching to perform procedures that no previous generation would have dared. But an all-important hurdle blocked the path: The heart still remained half shrouded in darkness. Doctors could still not really see or understand the workings of the arteries that ruled its functioning, and they didn't have a clue how to fix what was sheer murk to them. The catheter techniques that Werner Forssmann started had more or less crested, and the coronary tree with all its intricate branches was considered a no-go area for X-ray visualization. So for millions of patients suffering from the emerging post-war pandemic of coronary artery disease, little could be done. The man who began to change all that, and who would pave the way for Andreas Gruentzig's breakthrough, was named Frank Mason Sones. He was one unlikely hero.

Born in 1919, Sones suffered through his formative years as a fat, ugly kid with thick glasses, which made him a target for ridicule whenever he ventured out to play. But he was driven from the start, and took to beating on drums and joining marching bands to make his presence felt. After a childhood in Noxapater, Mississippi, his parents moved to Baltimore, where he studied medicine at the University of Maryland. There a professor informed him that cardiology was "a nothing specialty," and added, "There'll never be any great discoveries in cardiology." To a free spirit like Mason Sones—or for that matter, Carrel, Lindbergh, or Werner Forssmann—such nay-saying served as a starter's flag for a lifetime quest.

Sones did a two-year stint in the U.S. Army Air Force at the end of

World War II, then found his way to a medical residency in Detroit. From there he joined the Cleveland Clinic in 1950, which was one of the most respected medical centers in the United States. His early career was devoted to diagnosing the cardiovascular troubles of children at death's door. His Cleveland Clinic colleagues soon realized they were dealing with a maverick, and one who was willing to work fourteen to eighteen hours a day, seven days a week. The young doctor was often seen spending the night at the bedsides of his tiny patients. And his talk—exuberant, endless, exhausting—was inimitable.

Before long, Sones set to storming through the hallowed Cleveland Clinic's corridors in a sweat-stained white T-shirt that made him resemble the lowliest janitor. Ruddy-faced, he bellowed, cursed, and cajoled, and left trails of assistants and nurses recoiling and guffawing by turns. When desperate to dictate case reports, he was known to kick in the ladies' room door and roar that it was time for some secretary to get off the pot. "Type, type, type!" was his only half-self-mocking staff greeting in the morning. But Sones was also brilliant, relentless, and rock honest, with each of those qualities delivered unto him in spigot force; the combination made him one singular engine of discovery.

One of the earliest specialists in pediatric cardiology, Sones sought to move mountains for the sake of young children afflicted by congenital malformations of the cardiac valves and embryonic ducts connecting the great vessels between the lungs to the heart. He bellowed because he knew these kids were destined for crippling infections or early death unless they received swift diagnoses and treatment. Eisenhower's heart attack in 1955 may have played out in utter futility, but the future was already brightening for children who needed repairs to their malformed heart valves. New techniques were being rapidly perfected, and Sones believed they could be advanced dramatically, if surgeons could only be equipped with the right foreknowledge through systematic X-ray studies. He set to unlocking the puzzles of congenital heart disease, ignoring the grave trepidation others advised in such explorations. At the age of thirty-five, he stood before a national medical meeting and told the assembly of doctors that they were hesitating like cowards. He would erupt stage center many times in the years that followed, telling one audience of esteemed doctors that their plans for a study randomizing patients to proven heart therapy or none at all for comparison purposes was the reincarnation of the deliberate inoculation of black prisoners in Tuskegee, Mississippi, with syphilis bacteria, just to track the progression of the disease:

> I told them, 'I've been listening to you all day and watching you shrug your shoulders over this infamous proposal we've been invited to consider. All of you think this is a charade, all of you think you're anonymous. But this entire session has been documented, and I want you to know that if you affirm this diabolical plot to treat people like guinea pigs, every one of you will be known to the American public; I will see to that, I'll take the story of this insane plan everywhere in the country.'

Early in his career, Sones began steering diagnostic catheters straight into the aortic trunk arteries of children in their first year of life, even as top cardiologists called his work pointless, and perhaps criminal. If stricken infants didn't survive on their own power until they became toddlers, they were marked for death anyway—or so went the thinking. Sones, some said, was doing the devil's work, spreading needless trauma, and destined to break desperate parents' hearts.

But Mason proved his adversaries wrong—something he would make a habit of all his life. He produced such intricately detailed diagnoses that the Cleveland Clinic's chief cardiac surgeon, Donald Effler, began operating on young patients no one else would touch. Sones and Effler had a unique safety net in the person of Willem Kolff, another pioneer bent on perfecting the workings of the heart-lung machine. A seeker, a doer, and a mechanical wizard set to working in unison, playing a futuristic concerto of collaboration when at their best. At their worst, they screamed and hissed like demented tomcats.

Sones's retired colleagues still shake their heads about the antics of those days, when the esteemed doctors stopped speaking to one another. By the mid-1950s, the Cleveland Clinic was nonetheless regarded as a temple of medicine. Legions of sick kids stood an unprecedented chance of salvation there, and desperate parents flooded the cardiology department with babes in arms and pale, moaning toddlers.

Elaine Clayton, Sones's secretary of seventeen years, managed the miniature conga line assembled outside her office, while doctors-in-training tried to slap stethoscopes onto squirming little chests. "Somebody would shout, 'Here, hold this kid so I can hear his heart!' " recalled the still fine-featured secretary, with a startling power of memory. "I'd be rocking a buggy and making phone appointments and trying to take dictation."

Mason Sones burrowed into a back office where he studied admittance sheets and operative results and dictated four-page, every-detail-under-the-sun reports. He was an oddball for the ages. His furious smoking created mountains of smoldering butts in twin oversize ashtrays, and his ash-flicking sometimes ignited surrounding papers. When a report caught fire, he'd shove it onto the floor and stamp out the flames. He occasionally erupted, Wizard of Oz–like, through the blue haze of tobacco. Coughing and hacking, Sones would look around wildly, plow past pretty nurses and goggle-eyed parents, and steam off to the X-ray laboratories to take pictures of some newly imperiled, walnut-size heart.

This wheezing Vesuvius could have been a poster boy for R. J. Reynolds as he smoked in corridors and used surgical forceps to hold his cigarettes butts while he explored the coronary anatomy. Minutes later, he might race off to the operating room to observe and complain about whatever was in progress. "Stop! Give him 2 cc's!" he shouted in the middle of one procedure where he was most definitely supposed to shut up.

"Two cc's of what?" gasped the bewildered technician.

Strangely, cardiologists to this day almost never bother to watch the cutting and stitching to which they consign patients. But Sones did constantly, wanting to grasp every intricacy. He quickly made a name for himself as a colossal irritant—especially to Don Effler, the suave, gifted, and equally egotistical chief cardiac surgeon. Sones flooded him with torrents of second guesses.

A three-ring circus played out every day. Arguments in the hallways grew so vitriolic that Elaine Clayton remembers her glass bookcases shuddering. The masters of the heart altogether refused to speak with each other: What else could giants do in their moment of triumph? The former department chairman, William Proudfit, still rangy and droll when interviewed at the age of eighty-nine, chuckles about that era. The clinic's top authorities demanded that Proudfit preside over the menagerie in a daily staff meeting.

"It was an impossible situation. It was terrible. They would have a daily explosion," recalled the self-effacing Proudfit. "The board of governors finally appointed me as chairman of a committee, not because of my ability, but because I was on speaking terms with the three participants. So we had a meeting every morning at eight o'clock, which was not Mason's favorite time of day."

There were various reasons the early hour would challenge Sones. One was that he worked until midnight or two A.M. as the need arose. Another was that he had a prodigious affinity for Scotch whiskey. After putting aside his catheters and charts, Sones would head off to the downtown Theatrical Grill on Short Vincent Street, or the bar in the nearby Bolton Square Hotel. There he would bellow and laugh into the wee hours. At home, his growing brood of four kids saw little of dear old dad.

Proudfit shuddered to remember the mornings after. He clearly relished the opportunity to reveal his insights into one singular human story, never fully told. "Sones would say, 'Tell that son of a bitch Effler such and such,' when I was sitting right between the two of them. Then he would say, 'And tell that son of a bitch Kolff that he screwed up on the heart-lung machine.' Then Effler or Kolff would shout the same stuff back, without a one of them ever talking to each other. It was totally childish."

"I swear, if they would have ever met each other in some dark alley, one of them wouldn't have survived the encounter," remembered another colleague, Earl Shirey.

Now over eighty and still boyishly crew-cut, Shirey had no idea what he was getting into as a doctor-in-training. A quiet figure, he would soon be recruited as the intrepid Robin to Sones's cantankerous Batman. That transforming day in 1956 started out as ordinarily as any other. "I was making rounds with another intern and here comes Sones down the hall—he was referred to as 'Little Napoleon.' He had this bounce to his walk and he surged ahead with speed; he never went along leisurely. His white coat flew out, and he came right up to me and barked, 'Shirey, I want to talk to you!' . . . There was a little bathroom nearby with room for about one-and-a half-people. He says, 'Get in there!' I walked in, feeling practically pushed in the door, and I thought, 'What a way to go.' But he backed me up and said, 'Shirey, I want you to join me!' Of course, I almost went into shock. He said, 'I understand the gastroenterologists are interested in you. You don't want to listen to all that gas and smell all that shit. Get into something real like cardiology!' "

In that ludicrous moment, a historic duo was formed. Shirey didn't smoke and scarcely drank, but he managed to keep pace with his insanely demanding mentor. He proved to be the perfect foil to the Irascible One. Proudfit, meanwhile, nurtured Sones and his fellow quarreling roosters back onto speaking terms; indeed, he massaged painstaking thoroughness into much of their written work for years to come.

Sones embarked upon a new obsession, working with physicists and engineers at corporations like Philips and Eastman Kodak. He goaded their technicians to vastly increase the power of X-ray amplification and double the size of the viewing screens that guided fluoroscopic study of the heart. His big dream was to find a way to display the hidden secrets of the coronary arteries. The problem was, nobody knew how to get inside the vessels without provoking a heart attack or exposing patients to other dangers. Sones started cautiously, experimenting with various means of forcing backwashes of illuminating dye into the coronary tree, including injecting big doses of the heavy metals into a potentially perilous region, called the sinus of Valsalva, near the mouth of all the coronary arteries. But the results all proved useless.

Sones kept throwing his weight around nonetheless. He cajoled the Cleveland Clinic's authorities to create a vastly expanded catheterization laboratory in the basement. He also demanded that a special chamber be dug four feet beneath the lab's floor, a pit to house his humongous new image amplifier. In this pit lay the rub. It had to do with an underground tributary of the Cuyahoga River, which gurgled toward Lake Erie about two miles downstream. This drainage system was reviled as one of the most contaminated on the planet. *Time* magazine said the Cuyahoga "oozes rather than flows," and the infernally toxic waterway sometimes spontaneously bursts into flame. Its underground channels threatened to bubble up into Sones's chamber, greeting his patients with percolating slime.

But somehow the ooze was blocked off, and Sones had his pit of revelation. Standing down there, he would raise a kind of periscope to view the progress of each catheter snaking into a patient's heart as he or she lay on a table across the room. Sones cut quite a figure as he barked out his commands. With a cigarette clenched in his forceps, his gaze flashed from his periscope to the images fluttering onto his specially ordered viewing screen. Helped on by Earl Shirey, he kept probing, measuring, and calibrating like a man obsessed.

Mason Sones was no Forssmann. He was brusque and impetuous, but his mind was a trap that could calculate risks in an instant. So when the next great moment of the journey into the heart occurred unexpectedly on the morning of October 30, 1958, he reacted not with bravado but wild dismay. Down in his pit, gazing into the depths of his periscope, Sones watched as a diagnostic catheter was easily inserted into the ascending aorta of Otis Dickey, a twenty-seven-year-old patient from West Virginia. A young Welsh cardiology trainee or "fellow" named Royston Lewis awaited the command to unleash a high-pressure 30-cc shot of dye to illuminate the patient's aorta. For the Sones team, this was as routine as flossing teeth.

"Okay, fire!" Sones barked, then shifted his periscope and glanced at the viewing screen. Instantly, he saw a nightmare unfolding.

Just as the interior image of Otis Dickey's heart coalesced, Sones watched in horror as the little catheter suddenly whipped around like an out-of-control garden hose and spat a flood of toxic contrast medium deep into the patient's right coronary artery. This was a recipe for instant death. The fear was that flooding such stuff into a vital artery would either choke off all normal flow and cause a heart attack, or trigger a potentially fatal misfiring in the central pacemaker of the heart. Tortured spasms in the pumping chambers, quivering chaos in the heart muscle, then the pulsing jelly called "ventricular fibrillation"—those were the expected results.

Sones was beside himself. He jumped out of his pit hollering for his assistant to wrench the wretched catheter back. "Pull it out!" he yelled. A man was about to die, and the fault was his alone. The electrocardiograph had gone flat, indicating not fibrillation but cardiac arrest.

"Cough, Goddamn it!" the frenzied Mason Sones shouted. He prayed that an explosive shudder from the lungs would produce enough pressure in the aorta to heave the burden of toxic fluid from the imperiled artery. The hospital squawk box screamed. A phalanx of the clinic's physicians raced down the halls, with Earl Shirey in the thick.

A desperate, cursing Mason Sones called for a scalpel to cut open the man's chest, hoping against hope he could shock the stopped heart back into normal rhythm. Cardiac defibrillators in the 1950s were impotent without direct access to an exposed heart. Considering how long it took to cut through the chest, a patient in jeopardy might as well kiss his life goodbye. So Sones bellowed for more coughs. Sweat dripping from his brow, he looked aside to the oscilloscope for the latest EKG reading. There he saw something remarkable. At first mincing back to life, now racing, and finally steadying, Otis Dickey's heart rhythms gradually took on the smoothness of a gentle harbor swell.

You might call this discovery serendipity. But few have the genius to recognize serendipity's promise. Once it was apparent that the crisis had run its course, Sones caught his breath and looked around at the expectant faces of the staff. Did they realize what had just happened? Did they know how important this near-death experience was? And he set to thinking. Here, the true historic moment unfolded. Having stared down chaos, Sones was already reaching for higher meaning. Within minutes, he left Otis Dickey ticking away happily, and strode back toward his office with Royston Lewis, with another cardiology resident named Hector Garcia nipping behind. Elaine Clayton, the imperturbable secretary, had heard about the commotion and asked what was up.

"We just revolutionized cardiology!" proclaimed Mason Sones, the former collegiate drum major, as he slapped his desk.

The secretary blithely responded, "Again?"

"He would tell me he had revolutionized cardiology about three times a week," she explained many years later. "But Dr. Lewis sat down in one of the office chairs, and said, 'No, Elaine. We really did.' I said, 'Did you really?' " Lewis nodded, and Sones demanded to know why she believed the understudy rather than himself. "I said, 'Dr. Lewis has never claimed to have revolutionized cardiology before. But you say it all the time.' "

She then asked Lewis whether the patient from Appalachia was okay.

"He's fine, but we're better."

Of her boss, Clayton recalled, "I would say he was already planning. His mind was going and he was planning—'What do I do next? What is the next step here? We have done it and we didn't kill him.' He and Dr. Lewis went over the details again and again, and asked Dr. Garcia what he thought. He was like a little Mexican jumping bean, moving all over the place. But Dr. Sones was fixed in thought, he wasn't hyper at all. He always moved on, he always took what had been done and moved on."

An hour after the historic accident, Sones pronounced that he intended to begin scheduling patients for a systematic program to study the depths of the coronary arteries.

Clayton had witnessed the cat's pounce of her boss's mind many times before. "He would go from point A to point G, and he would always land right on point G. He was brilliant and he would make this quantum leap, and he was always right."

William Proudfit concurs. "He recognized right away that this would revolutionize medicine. I thought it was amazing that instead of just being thankful that the patient hadn't died, he recognized immediately that this was a tremendous advance."

By the next morning, Sones was proposing an ambitious research protocol to Earl Shirey, his unflappable and ever-dedicated right-hand man. "If we could flood contrast down a patient's right coronary artery without hurting him, why can't we do this deliberately at lower doses?" he asked.

Sones quickly set to designing a purpose-built catheter to lower the risk of blocking vital blood flow and provoking heart attacks when probing the coronary arteries. The device, quickly fabricated by a small upstate New York enterprise called the United States Catheter Instrument Company (USCI), had an open, tapered tip and sieve-like side openings less than an inch back on its shaft, so that it could better deliver its X-ray dye. For this modest-sounding yet crucial advance, Sones received not a penny of royalties, not then, nor a million uses of the instrument later. The man was so consumed with penetrating the heart's secrets that he seems never to have given the matter a second thought.

Instead, Sones and Shirey worked out plans to perfect the new fluoroscopic analysis. Using much more diluted amounts of dye than in the serendipitous discovery, they tested the technique on another 1,000 patients, three of whom would die. Sones declared that not a single claim should be published until this systematic analysis was complete. Most doctors would have spiffed up a triumphant paper after a couple of dozen follow-up procedures. But not Mason Sones. Integrity was his byword, and he held a withering eye for complacent claims of all kinds. Of course, the dyslexic Sones hated to write. In fact, he despised the tedium of medical journals. He found reading such a chore that on certain mornings he summarily shoved huge stacks of correspondence into his trash, like Ignatius P. Reilly in the mad novel *A Confederacy of Dunces*, by John Kennedy Toole.

So the breakthrough that would flood light into the ancient darkness of the coronary arteries was quietly tested and refined for the next three years. The Batman and Robin of heart disease, one in a sweat-stained-T-shirt, the other starched and pressed, worked away, their patients scarcely realizing what was happening. It is hardly an exaggeration to compare their studies to a descent into a black cavern, a creeping forward into never-before-glimpsed passageways dripping with encrusted atherosclerotic disease. The eerie illumination of the X-rays showed the myriad branches and twists of every dye-filled artery in stunning mirror-image profile. These were as revelatory as the first skeletal outlines of Roentgen's wife's hand.

But each pocket of disease presented a profound puzzle, a curiosity of ghostly shadings where black was supposed to indicate the surging health of red blood. The idea sounds simple, but what meant white? And gray—what of that? Peering into their looking glass, the cardiologists struggled to make sense of a world where illusion and delusion toyed with reality at every turn.

The reverse illuminations of the tentacles and side whispers of an artery's course sometimes vanished into white nothingness. What did that mean? Had the X-ray machinery somehow failed, or the dye petered out? Was the tapestry of this particular patient's heart bizarrely anomalous? Or had they just pinpointed a massive atherosclerotic blockage that choked off crucial flow to the life-giving deltas beyond? How in the name of "The Circulator," William Harvey himself, could one decode these blurs? How did each aberration correlate with the agony of chest pain certain patients reported, while others with no history of angina showed a minefield of congealed coronary disease? At this point, no one had the vaguest idea of what normal coronary circulation looked like, and they knew still less about the strangulation of advanced disease.

Standing in the X-ray pit, Sones and Shirey might as well as have embarked on a journey into outer space. Or so at least Geraldine Sones thought as she retired night after night between lonely sheets, while her prodigal explorer husband passed out on some hospital bed, too exhausted to return home. Certain evenings, to be sure, Sones cut out early—often as not for a bit of carousing. His children's birthdays were forgotten, unless Elaine Clayton grabbed her boss's elbow and handed him the phone.

Sones worked his staff as though they were bent on a historic mission. Clayton and the nurses routinely slaved away until ten P.M. One night his loyal secretary completed the last patient reports at eight P.M. and announced that she was leaving.

Sones responded, "Well if you have to go home in the middle of the day, I suppose I can't stop it."

The congenial Proudfit interrupted, "Look, Mason, it's eight o'clock at night."

"That's what I said," Sones replied. "It's the middle of the day."

That was loveable Mason. The man's family, unfortunately, was not always assuaged by the idea that he was saving mankind. Within a few years, Sones was sued for divorce.

There were other trials by fire. At one point, three successive patients under study degenerated into ventricular fibrillation, the last spasms of hearts gone haywire. Here, many mortals would have given up in despair, just like John Gibbon in Philadelphia. But not Mason Sones. He enjoyed a risk-taking latitude that seems astounding today. Nowadays, medical research is regulated by ethical oversight committees called Institutional Review Boards, which are properly concerned with patient rights. Sones's juggernaut would be red meat for any modern IRB. But none existed in the early 1960s, and he pressed on as he saw fit.

One of his near catastrophes was vintage Sones. Standing in his procedural pit, he glanced at the EKG screen and noticed with alarm the sharpening spikes of imminent peril. In a fever to save the patient, Sones whipped aside the X-ray viewing screen on its adjustable armature, mistakenly smacking a young assistant named Bob Quint upside the head. The prematurely bald Quint crumpled to the floor. A frantic Sones looked down at his prone understudy, out cold, and shouted, "Quint! Get up off your lazy ass!" The story goes that the patient was in fact revived, and Quint's bruised pate looks none the worse to this day.

At medical meetings, Sones and Shirey freely aired their failures along with their triumphs, and audiences sat stunned as slide projectors revealed never-before-seen images of the coronary arteries. Word spread that a transformation in the understanding of the heart was blossoming in Cleveland. By 1960, cardiologists flocked like lemmings to behold the realization of a dream.

Martin Kaltenbach, a rising German kingpin in the diagnosis of heart disease and later to be a handmaiden to Gruentzig's breakthrough of balloon angioplasty, hurried over from Frankfurt. Arnoldo Fiedotin, an Argentinian who would become one of Gruentzig's earliest protégés, arrived as a research fellow. Struggling to keep up with the master's tireless pace, he eventually blurted, "I figure I am being paid twelve cents an hour for what I do here."

With that, the pugnacious Sones shoved his thick glasses into the understudy's face. "Would you like to be paid nothing?" he demanded.

Fiedotin, now retired and living in suburban Atlanta, still chuckles over that time of testing. "Bob Quint used to be upset at me for doing whatever Sones wanted. He would say, 'Why do you do that, Arnoldo?' I would say, 'Well, Bob, every morning when I come in here, I take off my shirt and my pants and hang up my balls. I am not going to carry my balls around all day for him to kick them. I put my balls back on when I leave.' "

Fiedotin in fact loved Sones, and grew tearful in an interview as he remembered him. When Sones began a long, adulterous affair with the wife of one of his patients—whom he would eventually marry—the Argentinian was one of the few to stick by him.

The buzz about the Cleveland Clinic revolution found its way to the popular press. The limitations were agonizing: The Sones method of fixing a route forward for his catheters required a nasty cut into the brachial artery above the wrist, with intricate clamps to control bleeding. This situation prevailed until a 300-pound Seventh-Day Adventist named Melvin Judkins perfected a kinder approach through little more than a needle puncture in the big femoral artery in the groin. A central fact remained: No one knew what to do about the choke points of coronary artery disease that Sones could now masterfully reveal.

The cardiac surgeons, backed up now by fail-safe heart-lung machines, new external defibrillators, and the vital studies of Mason Sones himself, were itching to extend their repertoires on many fronts. But coronary artery disease was still something they could not intercede against at all.

Early in the 1960s, a Montreal surgeon named Arthur Vineberg finally claimed that he had devised a solution. The first step, he said, was to excise the far or distal end of the internal mammary artery that fed from the heart to the chest wall. Next, he would tunnel the vessel back to the heart muscle to augment the impeded circulation in one or more coronary arteries. "Yeah, sure," Sones said, believing the technique to be a sham. Nonetheless, he agreed to perform coronary arteriography on two of the Canadian's patients, knowing his own post-procedure inspections were unlikely to cause any further harm and might just offer a new window on the workings of the cardiovascular system. In about twenty minutes, he was eating crow—there was a clear avenue of flow through the rerouted artery to the heart and evidence of tiny fresh channels spreading through the myocardium surrounding the implant's end. "Damn it, he's right," Sones muttered on his way out of the cath lab. The ever-twisting journey into the heart was nearing its fateful mark. Soon, the Cleveland Clinic was performing the Vineberg procedure with fervor.

The next turning point began with a kind of Charlie Chaplin story, unfolding right down the hall from Mason Sones in February of 1962. A man in a rumpled suit knocked on the door of the Cleveland Clinic's majordomo, George Crile, Jr., explaining that he had just arrived from Argentina and that his wife and suitcases were deposited at the nearby Bolton Square Hotel (its bar so well known to Sones). In thickly accented English, René Favaloro claimed to be an accomplished nuts and bolts general surgeon, and a refugee from the Perónist tyranny in his homeland. He had just flown 6,000 miles but was ready to jump into the operating theater the next day. And here is my letter of introduction, señor.

The supplicant was pointed toward the office of Don Effler, the suave foil to Mason Sones. Equipped with an equally oversize ego, Dapper Don tried to make sense of the Argentinian's spiel, explaining that one had to pass a rigorous examination to obtain a license to practice medicine in the United States.

Favaloro, having figured his Argentinian version would be good enough, shuffled off, crestfallen.

The next day, the supplicant returned to Effler's office and announced that he had unraveled the problem—he would immediately commence studying for the examination while helping out any way he could. After all, he had put his professional existence on the line just so that he could absorb all the new insights the Cleveland Clinic had to offer. I am at your service, and I will work for nothing but the privilege of learning, he said with a wave of his hand.

Who could say no?

Don Effler finally consented to let Favaloro assist him without pay for six months as a kind of glorified scrub nurse. The food in the cafeteria was cheap. The Bolton Hotel was at least close by, if you didn't mind the bedbugs. Sounds horrible? If you are a Werner Forssmann or René Favaloro, this was all the inroads you needed.

Favaloro soon became noticed as the first doctor in and the last one out. The cabinetmaker's son from La Plata did everything that was asked of him, and then doubled it, while studying for his American medical exams at night. Soon, he landed a staff assistant's job and proceeded to soak up every nuance of Effler's techniques. He also steadily absorbed Sones's magisterial knowledge of the circulation, his secret domain being the innermost threads of the human heart.

The Vineberg procedure had become a hot operation at the clinic, and Favaloro, increasingly adept, pushed the new coronary procedure to its limits, implanting the ends of not one but two internal mammary arteries into the cardiac wall. Follow-up X-rays showed that microchannels sprung to life around both implants, and patients reported marked relief to their angina. Still, no one was deluded about the operation's fundamental drawbacks—the procedure's success was obviously dependent upon an iffy and grossly indirect means of feeding vital flow to oxygen-starved sections of heart muscle. Certain prominent cardiologists were rightfully skeptical, since a subsequent clinical trial would show that the approach rarely had any lasting impact.

Favaloro, grasping all this, set out to solve the greater puzzle. He knew that vascular surgeons had been successfully excising lengths of saphenous vein from the leg to rebuild diseased groin, or "femoral," and kidney arteries. Why not try the same technique in the heart?

At the time, the Argentinian had no knowledge that three previous attempts had been made to do much the same thing. The first, in 1962, was by David Sabiston in Baltimore, whose patient promptly died of a cerebral hemorrhage after a clot washed into his brain. A couple of years later, a Houston group led by Edward Garret and the emerging celebrity surgeon Michael DeBakey had improvised a similar approach in an emergency, but subsequently abandoned the notion. A similar stab at rebuilding the coronary arteries had also been taken by George Green in New York, then dropped again.

So the heart remained ripe for exploration. Favaloro, in any case, had no idea what to expect on a May morning in 1967, when he boldly excised a section of a middle-aged woman's diseased right coronary artery and replaced it with an interposed length of saphenous vein retrieved from her leg. The clamping off and cutting away of a vital segment of living coronary artery was hair-raising in itself, fraught with the risk of triggering a full-blown emergency. The stitching together of wormlike foreign vessels into a seamless whole required exquisite precision to prevent the tiniest—and potentially fatal—leak fore and aft. Favaloro held his breath as he worked, but his sewing was magical and everything fell into place without complications. The first moment of truth came when the heart-lung machine was turned off. But the initial explosive rush of blood surged effortlessly through the rebuilt artery.

To Favaloro's relief, all held together. Now, the patient's chest cavity was sewn together, and the waiting began. Would she regain strength as the anesthesia wore off, or die? Fortunately, the recovery was no different than Favaloro had witnessed after hundreds of other operations on the heart. But it was still too early to proclaim success. Whether the conduit actually remained open and worked could not be known without a later examination by X-ray catheterization, once the patient had regained reasonable strength. Eight days later, Sones took a look, the intensity of his watchful gaze undoubtedly searing into Favaloro's consciousness.

Together, they walked back to Sones's office to view the all-revealing strips of film with several colleagues and trainees in tow. The collective mood quickly turned from anxiety to jubilation as the films revealed an astonishing transformation. The "before" was clearly a recipe for disaster: a vital coronary artery gone useless with disease and choking off flow. But the "after" was something none of these specialists had ever seen. The ruined length of artery had been cleanly excised—but they knew that already. Now, the film images showed that the graft vessel sewed in its place still surged with healthy flow, and most important, that this flow sluiced forward unimpeded, supplying the heart muscle beyond. To these doctors, the vision of a seamlessly rebuilt human artery was transfixing. The breakthrough that Alexis Carrel had proposed sixty years earlier beckoned.

Over the next few months, Favaloro, with Effler and a fellow surgeon named Laurence Groves, repeated the procedure whenever the right opportunity arose. On the fifteenth operation, the group decided to perform the first true coronary artery bypass graft, this time with a wrinkle that would earn that enduring name. Instead of excising the diseased section in the procedure, they simply detoured around it by stitching the saphenous vein segment fore and aft of the vessel's blockage, and let the heart pump to its joy via the newly liberated circulatory avenue. The technique worked beautifully, as Sones proceeded to demonstrate around the world with his revelatory pictures.

How far the long journey into the heart had come in a few decades! The image of Tuffier's fearful finger had already faded into oblivion. Thanks to Gibbon, Lillehei and the rest—William Harvey, Servetus, and Vesalius—miraculous pumping machines and myriad new techniques had transformed the horizons of heart surgery. Then Sones stepped forward with his wondrous torch of illumination. And now came Favaloro, revisiting the long dormant visions of Alexis Carrel. Within a couple of years, thousands of the Argentinian's coronary bypass procedures were being replicated wherever confident heart surgeons worked.

The 1960s were glory days for the Cleveland Clinic. Invitations poured in to speak about fathomless new possibilities in the diagnosis and treatment of the heart. Effler and Sones, with Favaloro sometimes in tow, accepted them on a global scale. To Brazil and Berlin, to New York and Paris they traveled. Singapore? Sure. Staff members likened their ever-ready road show to a famous vaudeville act from an earlier day called *The Pat and Mike Show*.

Eventually, the medals and testimonials became a distraction to the ever homespun Mason Sones. An eminent English doctor, Professor Sir Peter Tizard, recommended him for the esteemed Galen Medal of the ancient Worshipful Society of Apothecaries of London, whose members included the most luminary physicians in Europe. Tizard wrote Sones, the nonreader, two letters of lofty Etonian prose glittering with verbal tiaras. But he got no answer. Not a word. So the well-mannered English surgeon telephoned William Proudfit, asking him to resolve the matter.

"I asked Mason, 'You know who Galen was, don't you?' " Proudfit chuckled years later in referring to the father of medicine from Persia Minor, whose explorations wreath the first pages of about every medical history ever written.

Proudfit recalled, "I knew he didn't have the slightest idea who Galen was. He says, 'Oh yeah, Galen, I remember him. He was number sixty-two in my class in medical school, and I was number sixty-one.' " The award was presented in 1985 as Sones was dying.

A more memorable ceremony occurred in the early 1980s. The Stouffer Foundation, a cash-rich Cleveland charitable entity supported by the vast profits of a restaurant chain and pop-in-the-oven frozen-food products, decided to crown Mason Sones's achievements with a gala fete at a lavish hotel the corporation had built right in his hometown.

For this gritty city by Lake Erie, that celebration promised to rival the gaudiest New York society benefit. The jeweled ladies paraded in dress-to-kill gowns, their men festooned in crisp tuxedos. Cocktails flowed, silver trays of canapés were offered to dowagers and earnest Men of Medicine with arms folded across their learned chests. True to form, Mason Sones sought refuge in the bar.

The lights dimmed. The cream of Ohio's society took their seats in the ballroom, where Spain's Queen Isabella had received the previous Stouffer Medal: What could be more fitting, since her namesake of half a millennium earlier had underwritten the very discovery of America, with all its uncharted promise, such as was being celebrated again now?

Introductory speakers droned on about Mason Sones's achievements, his incredible devotion to patients beyond counting, his integrity, his selflessness—his illumination of the coronary arteries as no one ever had achieved before. Right here in Cleveland! Such a tribute to the human spirit! And with no further ado, I present you with none other than Mason Sones!

A slight problem: The Scotch-filled honoree was slumped in the front row, head on his first wife Geraldine's shoulder. She had taken him back out of sheer compassion, for sickness and loneliness were consuming his final years. The audience gaped in dismay. Geraldine delivered a nudge, then an emphatic growl into his ears. Finally, Sones arose like a disheveled bear.

The pillars of Ohio society watched as Mason wove toward the stage's podium. Once ensconced behind it, he plunked down his elbows and drew a breath. For a moment, his head seemed to list; then Mason pulled himself together, and the audience awaited a defining statement about bravery and determination, a future in which the killing force of coronary artery disease might be pushed back for all time—perhaps even a rousing plea for contributions from the beneficent souls gathered round. Never could there be a more uplifting Stouffer Foundation moment; an Ohio cardiologist remembers the hush of anticipation that befell the great room. The words that came forth were brief but quite remarkable.

"You know"—the doctor remembers the voice being hoarse, the pause long—"it's the environment."

"I thought, 'What the hell does he mean by that?' " Ralph Lach of Mount Carmel Hospital in Columbus recalled. "He paused for a few more minutes and said, 'It's the environment for sure. If somebody thinks you are an asshole, they call you an asshole.' "

No one knew how to react. Sones pushed the microphone away, and reeled off the dais. After a long pause, somebody in the audience starting clapping. Why not? Couldn't a world-famous explorer of the heart say whatever he pleased? Suddenly, there was a roar of applause, a surging Busby Berkeley standing ovation by the Ohio ladies, the real estate entrepreneurs, and learned professors. He spoke from the heart, he told it like it is! Medical pioneers could be quite some characters, people said back in the bar. Gaiety had returned.

Not for Mason Sones. He knew that his long battle with lung cancer would soon end. He had been pushed into the shadows by the governors of the Cleveland Clinic and would never re-emerge except for appearances like this night's oddball cameo. So Mason the fearless, Mason the proud, uttered his gnomic phrases to get done with the heartbreak of it all and staggered off.

Stage left, however, Mason had glimpsed a man advancing from the wings who was destined to realize his every dream. His name was Andreas Gruentzig.

Chapter 4

From the beginning of the historic quest, any would-be pioneer of the human heart needed to be equipped with passion and daring. Curiously, another prerequisite always seemed to be forbearance for personal heartbreak in pursuit of a fate that would leave one both blessed and cursed. The man who would extend Mason Sones's radical breakthroughs with his own stroke of genius matched all these requirements perfectly.

Andreas Roland Gruentzig's life started gently enough in Dresden, an ancient city on the River Elbe, fabled for its Gothic cathedrals, baroque palaces, and opera house, and long revered as the German Florence, a jewel of European high culture. At the time of his birth on June 25, 1939, a fleeting grace still surrounded the place, with summer flowers blooming outside the family's apartment. Even as his mother's piano concertos cast a genteel aura over the drawing room for the baby and his year-and-a-half-old brother, Johannes, black-on-red swastikas ominously hung over every municipal building outside.

Ten weeks later, Hitler's blitzkrieg into Poland broke Dresden's fragile peace. The invasions of the Low Countries, France, and then Russia altered every aspect of European civilization. Military-age fathers were swept en masse to the front from every German city and village. Before long, Andreas's father Wilmar, a secondary-school science teacher with a Ph.D. in chemistry, was conscripted into the meteorological service of the Luftwaffe, his job no longer to shape impressionable minds but to help forecast the prime conditions for aerial assault.

Like countless other mothers, Charlotta (Zeugner) Gruentzig was left alone to try to piece together the semblance of a happy childhood for her sons. The serenity of her beloved Dresden gave way to the rumble of troop trains that daily ferried 20,000 soldiers through the city and on to the carnage of the Eastern Front. Within a year, the Luftwaffe had rained terror upon London, and the Royal Air Force in turn set to bombing Berlin, only about a hundred miles north. It became obvious that every major German city would soon be a target. At the end of 1940, Charlotta took her two young sons to the house of a relative in the small town of Rochlitz in western Saxony. Nestled beside the meandering Zwickauer Mulde river, the town looked out on the low foothills of the Ore Mountains about forty miles from Dresden. For Andreas, this move was the beginning of a lifelong dislocation.

Within three years, the once almighty Reich was reeling. Allied bombers roared overhead as they chose. The young Andreas and Johannes Gruentzig rarely saw their father, who was by now a major and piloting his own reconnaissance plane, and they had plenty of reason to worry about his safety. The debacles of the Russian and North African campaigns were followed by the Italian and Normandy invasions. On February 13, 1945, the Allied operation dubbed "Plate Rack Force" struck the Gruentzigs' former home. Waves of British Lancaster bombers dropped 4,500 tons of incendiary bombs over the once exquisite city of Dresden, killing 30,000 civilians by the end of the follow-up American assault. The resultant inferno lit up the night sky clear to Rochlitz. As the bombing rained devastation from every point of the compass, the Russian and American armies stormed closer from the east and west. There would be no return to stability for six-year-old Andreas, not ever.

Things grew worse. The Rhine had been crossed, Prussia lost. Refugees from obliterated war zones filled the streets, a stream of misery without end. On April 14, three days after they liberated the infamous concentration camp at Buchenwald, General George Patton's 6th Armored Division appeared before the stone bridge to Rochlitz. SS brigades, joined by old men and youths on bicycles, fired from the hills in a last gasp defense of the Reich. The American regimental brass soon set up headquarters in the same house where the Gruentzigs had taken rooms. Records indicate that the Villa Corola, built circa 1895, was a stately three-story affair with sweeping lawns. By now, forced laborers from the East were busy looting shops, stealing cars, horses, and especially booze. Chaos reigned. Long into his adulthood, Andreas Gruentzig remembered the strange gifts handed to him by a few ornery GIs—live bullets.

Meanwhile, 1.5 million Germans were mowed down in the horrific final defense of Berlin. Andreas's father was nearly certainly one of them, although his body was never found. On May 8, Russian and American officers met in Rochlitz's Hotel Löwe to work out the region's fate. A few weeks later, the U.S. forces pulled back nearly a hundred miles to the west, making way for the Soviet occupation. On July 1, the hammer-and-sickle flag rose over Rochlitz's town square.

Charlotta, a dark-haired beauty, prayed day after day that her husband would limp out of some arriving railway carriage. The war ended, but there was no letter, no news. As the years passed, the stoic mother struggled to wear a brave face for her boys' sake. Every expectant trip to the train station brought heartbreak.

Being fatherless wasn't new for the family. Andreas's maternal grandfather perished in World War I; his great-grandfather was cut down in the Franco-Prussian War of 1870-71. It was almost as though a curse hung over the men of this family. Of course, all of Germany knew about cursedness. At the end of World War II, Werner Forssmann, medical daredevil and future Nobel laureate, struggled to sleep while standing upright in an American-run barbed-wire compound stuffed with 100,000 fellow Wehrmacht POWs. None wanted to lie down in the feces-ridden mud—which Forssmann calculated to be fouled with fifteen to twenty tons of fresh waste every day. So the captives leaned against each other in pathetic clusters and prayed for endurance, just as Charlotta Gruentzig was doing every day.

The Russian zone of occupation in the newly divided East Germany slowly grew into a prisoner of war camp of an unprecedented scale. Charlotta, desperate for work, fled Rochlitz to take up residence in a cramped Leipzig flat with her sons, her sister Alfreda Beier, and her aging mother. These were the days of stone soup, since this once exalted center of high culture had also been shattered by Allied bombing. The young Gruentzig boys scrounged eggs and potatoes at tattered farmers' markets patrolled by soldiers in red-starred caps. Once, Andreas splurged his few pfennigs to buy his mother a pair of nylon stockings, only to find that they were torn when his mother opened his gift. On weekends, he and Johannes followed their "Mutti" to nearby forests to pick mushrooms, the purpose being not only solace but sustenance.

At one point, Charlotta snuck off with her sons through a forested stretch of border to the Allied Sector to join another sister, Horta, who had begun to establish a marginally better life in Hanover. Her prospects proved scarcely better there. Despondent and homesick, Charlotta dragged the boys back to Leipzig. Occasionally, she tutored young piano students to augment her stints at low-paying clerical work. Mostly, the family's destitution was unrelieved.

Frau Gruentzig began to dream of a new escape. Her husband's brother had fled the Third Reich on the eve of World War II for Argentina, where a German subculture had been buoyed by the arrival of thousands of former Nazis. His letters spoke of a world rich in opportunity and free of ideological oppression, and he offered to help finance their passage. In 1950, Charlotta packed her boys on to a flight to Buenos Aires, the homeland of the heart surgeon René Favaloro. Andreas was then eleven.

So it was that the little family crossed the Andes, descended to a city of dreams, and met up with their waiting uncle and aunt and their growing brood of young children. At last, life could begin again.

The new immigrants, however, found themselves sharing tight quarters. Charlotta and her sister-in-law grated on each other's nerves. Taking on work as a seamstress, Charlotta scrimped and saved enough to send her sons to boarding school. But tensions festered, and in less than two years their homesick mother brought her boys back to their native Saxony. In Leipzig they took up residence in an apartment at 24 Paul Küstner Strasse, a quiet side street in the formerly stately Altlindenau district. Around the corner from their steeply gabled and terraced dwelling, the boys could enjoy kicking a ball in the city's central park. Leipzig was steeped in culture, its former residents including Mendelssohn, Schumann, Wagner, Goethe, and Nietzsche. Charlotta, with her own love of Germany's great composers, naturally enrolled her sons at Thomasschule Gymnasium, the secondary school beside the ancient St. Thomas Cathedral, which excelled in teaching both music and the natural sciences. Its famous Thomaner Choir, once led by Johann Sebastian Bach, soon had a new recruit named Andreas Gruentzig.

Charlotta wanted the world for her sons, whose intelligence by now was manifest. Andreas, already strikingly handsome and flashing with insights, immersed himself in his studies and music as he grew into adolescence. But his native imagination and wit were inevitably subjected to a daily bombardment of sloganeering, just like every other student in the *Deutsche Demokratische Republik* (DDR).

"There was no free exchange of ideas, but slogans of aspiration for the common good were plastered all over the newspapers," recalled Eberhard Zeitler, an East German–born doctor who would later become a close friend and inspiring figure in Gruentzig's quest to transform the treatment of the heart. "It was a driving force for many people from the time of their youth to live up to such slogans, to make better examinations than you thought you were capable of . . . and for every person to do more for the society around him."

Zeitler, now retired in Nürnberg, recalled one afternoon in his comfortable apartment in the formerly devastated citadel of Nazi power. "We used to have youth groups in the East called 'pioneers.' Members of the pioneers had to announce every morning in school, 'I intend to do better in my studies! I will strive to be a better person!' "

In any case, Gruentzig's teachers noticed the special talents of their rapidly self-bettering student. Andreas displayed a knack for Spanish, French, and English, as well as a keen interest in mathematics and science. He achieved outstanding results on the demanding "Abitur" or leaving examination in 1957. But the oppressive communist regime of the DDR was determined to root out every vestige of class privilege. So a question forever lurked: Were the achievements of a young striver a sufficient blessing to the common cause, or just a laurel unto himself? Never mind that three generations of the Gruentzig boys' ancestors had given their lives for the Fatherland; the authorities became convinced that the brothers' future must be reined in. It was common knowledge, after all, that their father Wilmar had been an intellectual, and that their pianist mother, if only a secretary now, was enamored of the elitist culture from discredited "bourgeois-fascist" days.

The social engineers smelled trouble. They decreed that Andreas's noble service to the DDR—seeing that he was so fond of precision—should consist of troweling wet cement onto block walls for the rest of his days. That dreary edict drove the young student to dream of unfettered opportunity across the border.

In 1956, the year the Soviets crushed the Hungarian uprising, his distraught mother advised the older Johannes to flee to her sister Horta's family in Hanover. A year later, Andreas, having completed his gymnasium studies, made his own escape. At that time, the menacing watch towers and machine gun nests of the Berlin Wall were but schemes for the future. The border remained porous, but the eyes of the secret police were everywhere when the eighteen-year-old Andreas made his way toward East Berlin. Hugging a satchel of rudimentary possessions, he cast his eyes about nervously as his streetcar creaked into the Western Sector. Charlotta was not up to facing the same wrenching dislocation as her sons, along with the risk of imprisonment for her sister and mother in reprisal after her flight. Instead, she decided to remain behind until she reached pension age, eventually considered sufficiently resource-draining to the commissars of the DDR that her own escape would be all but invited. Andreas's future now promised to be both motherless and fatherless.

Johannes enrolled as a medical student at the University of Heidelberg. Andreas followed him down the Rhine to spend the next year at the local Bunsen Gymnasium there. That step was also necessary in order to qualify with a West German "Abitur" for a medical course. For these two brothers, the days of stark deprivation were over. To refugees from the East like themselves, the regular-as-clockwork government stipends for students felt almost too good to be true. But the Ostis, as the East Germans were derisively termed by the rising middle class in West Germany, were regarded as equivalents to "Okies," the dust-bowl destitutes from Oklahoma who fled goggle-eyed to California during the Great Depression. Years later, Andreas Gruentzig would sometimes tell his colleagues that he was born in Bavaria, as if to skate away from the emotional baggage of hailing from such a brutally forlorn land.

Bernhard Meier, a Swiss cardiologist who assisted Gruentzig in his subsequent years in Switzerland and the United States, explained, "West Germans never looked at them as the enemy, but as a poor people, lower class, the cellar children who had never seen the sunlight. His dialect gave him away, everybody hearing him talk knew he was from the eastern part . . . This was a stigma."

Andreas Gruentzig began his medical studies in the autumn of 1958 and took to Heidelberg with zest. He embraced his free student digs as though he had just landed in the Ritz, and sang for extra change in beer halls. Meanwhile, he began mastering the rigorous demands of Heidelberg's highly pedigreed medical training program.

Eberhard Zeitler, the fellow Osti made good, noted that young East Germans at that time had no choice but to excel. "People born to even a little wealth and support from their parents do not have the same drive as refugees. People who came from the East into West Germany had a much stronger drive. They had lost everything; they had to start from nothing."

The East German brothers also fell into a heady social scene, with nightly revelry far different from that available in the dour DDR. In time, they scrounged enough cash to buy a tiny, two-seat red convertible. In the summer of 1961, the brothers set off to savor the now wide-open horizons at their doorstep. Roaring across France, they spun up the switchbacks to the Pyrenees and motored across the sprawling Spanish landscape to Andalusia. In exhilaration, they painted VIVA ESPAÑA on the car's door. Coming to the glistening Atlantic, Andreas and Johannes paused for a photograph that showed them both radiant with youth—one dark as a Spaniard himself, the other (Johannes) blond and classically Teutonic. Ostis no longer, they cavorted like citizens of a reborn Europe.

Back in Heidelberg, Andreas pushed on with his studies, completing his medical undergraduate program in 1964; he then rotated through a series of specialized internships in Mannheim, Hanover, Bad Harzburg, and Ludwigshafen. The last two eight-month residencies, in internal medicine and vascular surgery, intrigued him the most. At the end of 1966, Gruentzig returned to Heidelberg to take on a staff assistant job at the university's Institute for Social Medicine. There, he began investigating the underlying risk factors in patients suffering from either cardiovascular disease, chronic bronchitis, or liver degeneration.

Life was scarcely all work. This was the so-called Age of Aquarius after all. Hi-fi speakers blasted Jimi Hendrix, the Beatles, and the Rolling Stones—and other troubadours of the Sexual Revolution—from German student digs and bars every night. The dashing medical post-graduate found himself light years away from his repressive upbringing behind the Iron Curtain. A certain student in a neighboring apartment caught Andreas's eye, fell for his charisma, and in 1967 informed him that she was pregnant. With no interest in marriage, Gruentzig was still forced to acknowledge his paternity before the civil authorities to an out-of-wedlock daughter named Katrin Hoffman, whose embittered mother summarily cut off all contact with him. Around the same time, he met a willowy brown-haired student from Bavaria, then finishing her language studies with a goal toward becoming a translator. Michaela Seebrunner was tall, dark-haired, and keenly intelligent. She had a voice that lilted with the music of the high Alps and the two quickly became soul mates. It was not lost on Andreas that Mikki, as he called his new love, was ever solicitous toward his mother Charlotta, who settled in Heidelberg in 1968 once she was deemed an expendable drain on the East German state.

During these affairs of the heart, Gruentzig became involved in another unfolding revolution involving coronary mechanics. By the early 1960s, cardiac surgeons were tackling the once-forbidden organ with a gung-ho enthusiasm. The heart-lung machine and Mason Sones's pinpoint mappings of the coronary anatomy had opened spectacular new possibilities. The top guns of surgery were itching to make medical history—some quietly, and some not. None outdid a pair of Texans with a taste for dueling by scalpel, television camera, and raw power brokering: Michael DeBakey and Denton Cooley.

These were doctors ready for prime time. DeBakey, an only son of Lebanese immigrants who enhanced his short stature with platform shoes, was a protean physician. An acerbic, driven advocate of a fearless new surgical age, he busted past whoever and whatever got in his way, removing tumors, transplanting kidneys, and rebuilding shot heart valves with an almost assembly-line-like prowess. He plowed through a dozen operations every day, a great number of which would have been regarded as impossible a decade earlier. Meanwhile, DeBakey let everybody within earshot know exactly what he had achieved. He made a gaudy "bring 'em on" show of his supposedly invincible skills. Once, after operating nonstop from seven A.M. to 1:30 A.M. the next day, he flung open the door to the deserted hospital corridor and cried out through his surgical mask, "Anybody else out there want an operation?"

DeBakey reigned like a lord incarnate over Houston's Methodist Hospital, where he convinced the local oil barons to shell out $20 million to provide him with a palace of surgery such as existed nowhere else. He invariably got what he wanted. Indeed, as early as 1962, DeBakey landed a $2.5 million federal grant in order to collaborate with an Argentinian engineer named Domingo Liotta on the development of one of the most hubristic contraptions of the decade—a metal and plastic replacement heart.

When it came time for the first worldwide television broadcast by the "Early Bird" satellite, none other than DeBakey was invited to step to center stage. Two hundred million viewers watched as the program opened with a mariachi band strumming and bobbing in Mexico City, then cut to a lederhosen-strutting oompah performance on the cobbled streets of Essen, Germany, before the cameras closed in on the awesome sight of one hawk-nosed and obsessive surgeon preparing to open a human being's chest in Houston. Three cameras had been organized to render his every cut and stitch. Such was the glory that suddenly attended heart surgery in the late 1960s.

DeBakey should have been watching his back, because rival seekers for acclaim were by now leaping out of the woodwork. The most formidable of these proved to be a twelve-years younger and homegrown Texan doctor named Denton Cooley. The man had everything DeBakey lacked: striking good looks, an instinctive physical grace, society connections, and a seductive patrician aura that asserted itself with the ease of a cat waking from a nap. A graduate of Johns Hopkins medical school, this former all-Texas basketball player had the head-turning impact of a virtuoso, mastering the most difficult challenges with the seeming effortlessness of one performing sleights of hand. A nurse described him as "Doctor Wonderful."

DeBakey got wind of the man's reputation and invited Cooley to join him in creating a heart surgery program that would be second to none. The fox thus entered the henhouse, where at first he did as he was told. But Cooley soon set to reinventing procedures *his* way, and demonstrated such exquisite delicacy that he enjoyed likening surgery to "circumcising gnats."

DeBakey grew annoyed. One staff physician put it succinctly: "Denton was always the most popular kid in the crowd, the leader, the one with charisma. And the handsomest son of a bitch ever to pick up a scalpel. How'd you like to shave Mike DeBakey's face every morning and then have to look across the table at Denton Cooley?"

The tension boiled. In 1962, the Houston homeboy found his way to a rival palace of surgery that was christened the "Texas Heart Institute," where he quickly set to raising $4.5 million to outfit the place to his every desire. Now the Texan gunslinging began in earnest. A question snaked through dinner conversations wherever doctors gathered: Who was the fastest draw to the heart?

Some said DeBakey, for all his faults, was a pillar of honesty, if pumped sky high with adrenaline. Clearly, Cooley enjoyed a heady sense of his own worth: "Cool Acres" was the name he bequeathed upon the sprawling house he had erected outside town at no charge, thanks to the gratitude of a rich patient. His old chief, the famous Baltimore surgeon Alfred Blalock, once asked him, "Tell me, Denton, can you still knot a suture inside a matchbox?"

"With either hand," he replied, like some John Wayne.

A new kind of medical figure had arrived, and this was the heart surgeon self-fashioned as a keeper of mortality. Queried about the enormously risky procedures he seemed to improvise as he worked, Cooley dryly responded, "I'm often obliged to experiment inside the heart of a patient whose problem hasn't yet been worked out in a dog."

A local cardiologist begged him to take a look at an X-ray of a man's aorta with an ominous shadow, suggesting the possibility of an aneurysm about to explode. Cooley insisted it was no such thing, but the cardiologist pressed his case. Doctor Wonderful suggested a wager, finally blurting, "If that thing's an aneurysm, I'll eat it!"

The next morning Cooley received a phone call from the cardiologist's secretary advising him to get his knife and fork ready, because Methodist Hospital's surgeons had completed their work and most certainly excised an aneurysm.

Hesitation being unthinkable, Cooley told the secretary to preserve the aneurysm in alcohol rather than the usual formalin. A few minutes later, he sent an orderly to fetch a red-checkered tablecloth from a nearby Chinese restaurant and another to collect the excised vessel from Methodist. Then he called his hospital's resident video photographer. Once the scene was set for filming, Cooley bid a striking nurse in a hitched-up skirt to deliver the vessel of contention his way on a plate covered by a surgical basin. On cue, the nurse lifted the lid and proffered salt and pepper shakers; in high camp, Cooley blithely stared at the gruesome vessel, dissected off a bit, and ate it.

"Did you get that?" he asked the photographer.

"In sixteen-millimeter color," was the response.

"Okay. Develop it and deliver it to my cardiologist friend at Methodist. Tell him Denton Cooley is a man of his word."

Such was the chutzpah of heart surgeons in the 1960s.

On December 3, 1967, Christiaan Barnard performed the so-called Miracle of Cape Town—the first heart transplant—at an unsung hospital called Groote Schuur, which is Afrikaans for "great barn." The event shocked the world, for it pronounced for all time that the heart, so long wreathed in sanctity and taboo, could now be replaced like a spare automobile part.

The son of a veldt preacher, Barnard had gathered little prior fame, certainly nothing on the DeBakey or Cooley scale. That was just as well, because the most original thing he had done previously had been to graft an extra head on a hapless experimental dog. Barnard was driven. In the early 1960s he had taken a fellowship to the University of Minnesota, where he studied under the great C. Walton Lillehei, the perfecter of the heart-lung machine and pioneer of cardiac pacemakers. Lillehei even sent the South African home with his own free heart-lung machine, which promptly transformed the surgical horizons in South Africa. A couple of years later, Barnard managed another short stint in Virginia with a surgeon named James Lower, who was conducting rigorous animal studies of heart transplantation with Norman Shumway of Stanford University in California.

Back at Groote Schuur, Barnard began dabbling with his own experiments and performed a few kidney transplants. Unlike his cautious colleagues elsewhere, the South African published little, even about his preparatory heart transplants in dogs, most of which quickly died.

In 1967, he was presented with one Louis Washkansky, a fifty-six-year-old grocer whose failing heart was wracked by just about every ailment the organ is heir to. The plucky former boxer knew he was hanging on by a thread and willingly latched on to Barnard's offer of an unprecedented attempt at rejuvenation. What was needed was a donor, fast. Fate provided that one Sunday afternoon when a drunk driver mowed down twenty-five-year-old Denise Darvall and her mother as they crossed a sunny Cape Town street to buy pastries. The grief-stricken father (whose wife was instantly killed) soon learned that his daughter was brain-dead, with no chance of recovery, and that her heart beat on only with ventilator support. Somehow, he struggled through his anguish and agreed to pass her heart on.

Barnard's autobiography, *One Life,* penned with an accomplished professional writer, tells of his own torment as he approached the operating theater.

> The further I went, the worse it became. With each step, the weight of my doubt grew until it seemed almost unbearable. I wanted to turn back, but there was no turning. Two people—a girl and a man—were now being moved into adjacent theaters. Both of them had living hearts that could not continue to beat for much longer. We were approaching the point where there would be nothing else to do other than cut out both their hearts,

> and place one of them—the girl's—within an empty chest of the man who would otherwise never leave the operating table alive.

Washkansky sat waiting as Barnard approached. Glancing at the nurses, the patient joked, "I kept telling them that I didn't want any Mickey Finns [meaning doped drinks, or anesthetics] until you came to say good-bye."

Barnard frowned querulously. "Good-bye?"

One of the bravest subjects in medical history whimsically rejoined, "Good-bye to the old Washkansky . . . " The patient about to have his chest cored out stared into his potential savior's eyes, caught his breath, and asked innocently, "Aren't you going to give me a new heart?"

"Yes," the illustrious doctor responded.

Washkansky cut to the quick. "So it's out with the old and in with the new . . . *Auld Lang Syne*."

As one team began working on Washkansky, Denise Darvall's life-support system was solemnly turned off. Barnard and his colleagues began a ghoulish vigil by her bedside, watching the vital signs swiftly extinguish. When the EKG line went flat, the physicians moved forward. The operating room filled with tension as the historic procedure commenced, the cutting and cleavering and sawing of tissue and bone that exposed Denise Darvall's heart and lungs. The fourteen specialists around the table were watched by an amphitheater thick with masked spectators. Surreal was the vision of the dead young woman's heart being connected to a heart-lung machine, in order to buy time for the cutting to ensue in the next room.

Nerves failed, fundamental errors ensued, and the leads of a separate heart-lung machine to sustain Washkansky suddenly began spitting blood on the floor. Racing against the clock, Barnard and the technicians corrected the fiasco.

Now the cutting away of Washkansky's heart could commence. Just before giving up the ghost, the organ began falling into pandemonium.

> We could see it with our eyes. The slow roll of the heart, caused by its chambers working together, began to falter and then its

unity dramatically broke apart. The rhythm of contracting ventricles, responding to the pulse of the upper atria, suddenly ceased—causing the muscle fibers to jerk in separate spasms. Each fiber went into its own rhythms—tiny spasms within spasms. What had been a rolling sea became a lake of worms.

The incredible heart of Louis Washkansky had endured long enough to reach the operating table—and go beyond it. With catheters through its upper chamber, it had continued to beat on, even during the three-minute breakdown of the heart-lung machine—refusing to give up though chilled with cold blood. And now . . . the destroyed heart was finally collapsing.

This was a heart for history—equal to the man who carried it, the man who refused to lie down and give up. And now there was little left, other than to watch it die. The fibrillation, tragic spasms heralding the end, became more aggravated. The heart which God had given to Louis Washkansky was approaching its moment of death.

Barnard now departed his patient to make the swift final cuts to free Denise Darvall's heart and transport it back in a metal basin. Only then was Washkansky's heart lifted out of his chest and placed in its own bowl, where incredibly, it shuddered with one last tremor of life. Barnard was awed.

I had never seen a chest without a heart or with such a hole—as though the hole itself was fixed and permanent, while the man, with his chest split open, was merely a temporary object, existing around the hole. And in fact, it was just that—something few men had ever seen: a human being without a heart, held in life by a machine eight feet away.

Barnard and a colleague feverishly sped through the intricate sewing and reconnecting of the woman's heart to the man's, weathering another near-fatal catastrophe before the recipient's new heart engorged with blood. After being suspended in an artificially supported nether zone for three endless hours, the nearly ruined organ began not to beat, but to squirm with life.

"This," Barnard recalled, "was the peak we had struggled to find, climbing over two great barriers: coordinating the death of two hearts, and then joining them." Of the transplanted heart's reviving, he said:

> It shot through the squirming muscles, causing the body of Louis Washkansky to arch upward as though kicked in the back. For a moment, the heart lay paralyzed, without any sign of life. We waited—it seemed like hours—until it slowly began to relax. Then it came, like a bolt of light. There was a sudden contraction of the atria, followed quickly by the ventricles in the obedient response—then the atria, and again the ventricles. Little by little, it began to roll with the lovely rhythm of life, the heartbeat of the world.

At dawn, Barnard at last walked away in exhaustion. Only then did he think to call the hospital's medical superintendent, Dr. Jacobus Burger, and explain his night's work.

"We have just done a heart transplant and thought you should know . . . no, it wasn't dogs. It was human beings . . . two human beings."

Obviously, there was no ethical review board at Groote Schuur then, no intermediary to advise caution. A track star in his youth, Barnard had sprinted straight for the finish line. Word of his breakthrough rocked the world, and the international media raced to Cape Town to file breathless reports. Very few had any sense of the historical context of the quest.

Camera crews from France and Germany, Australia, Japan, the U.K., and the United States pushed down Groote Schuur's corridors and climbed trees to snap shots of Washkansky through the hospital windows. The press reveled in the makings of a fairy tale, a conquering of mortality by an overnight hero. Christiaan Barnard, with his tales of a shoeless boyhood on the South African veldt, made for great copy. It helped that he was Cooley-suave and owned a 225-horsepower speedboat named *Pacemaker*. The media—and women—flocked to him. And the story of Louis

Washkansky, the man with the Frankenstein heart, soon proved to be irresistible as well.

Indeed, for a few days Washkansky struggled back to life to play his role to perfection, cracking jokes and referring to Barnard, not himself, as the brave one. The refugee from Lithuania comforted himself by digging into a book called *Die Rich, Die Happy.* Unfortunately, the idyll was cut short. Washkansky's body turned into a war zone. A lingering preoperative infection infiltrated the lungs of this rebuilt shell of a human being and blossomed into a massive dose of pneumonia. After eighteen days of living with the heart that Denise Darvall and Christiaan Barnard—rather than God—had given him, Louis Washkansky breathed his last. The next day the dejected medical superstar boarded a plane to America to appear with Michael DeBakey, and not Denton Cooley, on a mass-audience television program called *Face the Nation*. Somehow, they still managed to celebrate the brilliant horizons ahead.

As luck would have it, Barnard's next transplant recipient would outlast Washkansky by nearly two years. The Cape Town miracle worker's fame endured. On a whirlwind European tour, he would dine with kings, dance with the Italian movie star Gina Lollobrigida, and consort with the sultry Brigitte Bardot. It was as though medicine had produced its own Mick Jagger.

What followed was a global festival of surgical hubris. Three days after the Washkansky operation, New York's Adrian Kantrowitz—who had rushed out 500 telegrams in a plea for a donor—performed a nine-hour transplant that bought the recipient a grand total of three hours of rejuvenation before death. A month later in California, Norman Shumway jumped into the game. Before giving up the ghost in a matter of fifteen days, his patient would undergo two further massive operations and suffer through the transfusion of 288 pints of blood, nearly *sixty times* what he had in him before the transplant. Down in Texas, Denton Cooley set to work a few months later with his signature high-profile style, performing three transplants in seventy-two hours. Three months later, DeBakey's hotshot young surgical assistant, Ted Dietrich, "harvested" the heart, one lung, and both kidneys from a single donor and scattered them into four recipients as though he proffered the hand of God. A surgical rodeo was in the making. Sure enough, Cooley, never to be outdone, a week later removed the combined heart and lungs from a one-day-old baby and implanted them into a two-month-old infant. From this spectacular derring-do, the baby enjoyed fourteen hours of new life. From Paris to Athens and Bombay—in sixty-six surgical centers around the world, the heart-swapping continued. The story vied with Neil Armstrong's July 1969 walk on the moon for media coverage.

Thou shalt do no harm? The truth was that despite this bedrock tenant of the Hippocratic Oath, this sideshow had been rushed into freakish reality long before the procedural intricacies and miasma of organ-rejection had been solved. As of December 3, 1970, the three-year anniversary of the Miracle in Cape Town, only twenty-three of the 166 patients who had been bequeathed transplanted hearts survived. One by one, the global heart transplant teams bowed out of the game.

Yet, the surgeons were not about to relinquish the limelight. On April 4, 1969, Denton Cooley implanted an artificial plastic heart into a forty-seven-year-old printing estimator from Illinois named Haskell Karp as a bridge to his getting a new heart. The ingenious contraption astonishingly beat on for sixty-three hours while Cooley pleaded on national television for a donor human heart to provide Haskell Karp with a more enduring lease on life. The heart finally arrived; but twenty hours later, Karp was dead. DeBakey accused Cooley of purloining the very device he had been so instrumental in developing. Meanwhile, the disciplinary committee of the local Harris County Medical Society censured Cooley for excessive publicity seeking, the Karp family sued him (unsuccessfully) for $2 million, and the American press began feasting on the bitter and very public rivalry that erupted between the "Texas Titans." It was finally manifest that the heart surgeons were not quite as godlike as they claimed.

A doctor-in-training like Andreas Gruentzig may not yet have grasped this reality, but the heart surgeons' hubris was already presaging their eventual downfall—and thereby setting the scene for Gruentzig's own triumph. What was apparent at the end of the 1960s was that the heart was the center of the medical action, every bit as ripe for discovery as the reaches of outer space.

For the time being, Gruentzig struggled at merely getting his own career started within the brittle hierarchies of academic German medicine. Then again, inching up the ladder was never his style. His beloved Michaela recognized this in Technicolor when she brought Andreas to the craggy Alp of Kitzsteinhorn, "the unicorn," for a ski holiday in 1967. The peak was not far across the Austrian border from her home village of Bad Reichenhall, and close to Adolf Hitler's "Eagle's Nest" refuge in Berchtesgaden. There was little snow on the lower slopes then. So Mikki stuffed Andreas into the new cable car and headed to the peak of the 9,800-foot mountain. The prospect was enough to terrify a novice, and Michaela attempted to demonstrate the parabolas of side-to-side traversing that would safely steer a beginner down in one piece. "All right, now. Your turn!" she beckoned to her dashing boyfriend. "Be careful!"

Andreas laughed at her warnings and launched straight down the vertiginous escarpment, determined to prove that conquering the mountain was but child's play to the likes of him—a would-be medical explorer with a passion for getting to the point. Within seconds, he crashed on a mogul, tumbling head over heels in a stupendous burst of white powder. Michaela, now a widow in her sixties, still laughed girlishly at the memory over lunch in a Zürich café where she once sat countless nights with her husband. But she remains struck by the way Andreas dusted himself off and bombed down the mountainside again and again. Forssmann and Sones, and Vesalius, too, probably would have handled their first skiing lessons just as forcefully.

As the winter of 1967 set in, Andreas Gruentzig departed for a six-month paid fellowship to study epidemiology, or the larger patterns of health and illness, at the University of London School of Hygiene. There he absorbed the thoroughness with which the British handle the business of minding the statistics that are the substance of scientific trials. In the late spring of 1968, Gruentzig returned to Heidelberg, where Michaela was doing postgraduate studies in psychology, to resume working at the clinic. He was now thirty, and it was time to narrow in on a career.

Early in 1969, he sped off in his blue Austin Mini for a six-month assistant doctor's job in Darmstadt, an hour southwest of Frankfurt. At that point, Germany had few training programs in cardiac surgery, so following the swashbuckling lead of a DeBakey or a Cooley was a no-go. But the Max Ratschow Clinic was considered an impeccable center for the study of angiology, a European discipline devoted to diagnosing and treating the entire reach of the cardiovascular system, mostly with drugs.

There he befriended another young student from the dark side of the Iron Curtain, a Romanian of German heritage named Ernst Schneider, who like Gruentzig had outfoxed the local commissars to gain a foothold in the West. Schneider, now an interventional radiologist at the University Hospital in Zürich, is an engaging figure, with a hearty laugh more reminiscent of a Bavarian innkeeper than an expert in vascular disease. The two hit it off, downing beers when rare nights of freedom allowed, sharing hijinks and jokes. A certain department chief's backyard swimming pool began to arouse their mutual curiosity. A dare developed between the two, and one wine-soaked night, Gruentzig, still disdaining boundaries of all kinds, stealthily lead his friends over the escarpment to luxuriate in the warm waters of privilege. He might as well have been setting out a blueprint for the rest of his life. "Don't go there," his superiors would soon proclaim as Gruentzig began to puzzle out his own unique path toward the heart. "Just stop me," was his response.

CHAPTER 5

IN NOVEMBER 1969, Andreas and Michaela Gruentzig packed their possessions onto the roof of their little car and waved goodbye to Heidelberg. A position with long-term prospects had materialized at the University Hospital in Zürich, an institution with a reputation for excellence in treating cardiovascular disease. The lakeside city with its dreamy southern views of the Alps seemed to beckon with opportunity. Numerous German scientific luminaries had launched their careers at that university, including Albert Einstein and a certain Herr Geheimrat Sauerbruch himself. Of course, neither had proclaimed that the Blessed Trinity was a monster with three heads, nor that heart surgery might soon be replaced with little more than a balloon on a spaghetti strand.

By the standards of the staid Swiss, Gruentzig cut a bohemian and slightly defiant figure, with his cascading black hair, Latin complexion, burly mustache, devil-may-care manner, and a neck often wreathed by an ascot. His smile radiated self confidence, and his prepossessing charm deflected attention from his often ill-fitting clothing. The chief of angiology, Alfred Bollinger, remembers his first meeting with the new recruit with lasting amusement.

Glad to meet you. Welcome to Zürich. Hope you have settled in well, Bollinger, a strapping figure still, politely began.

Pleasantries were duly exchanged. Now Bollinger got to the big question: What do you hope to achieve here?

Without a second's hesitation, Gruentzig leaned forward, his dark eyes flashing with his characteristic élan. "I have dedicated my life to vascular disease!"

Bollinger today has the gently aged visage of a man who has enjoyed a full life in the company of his lovely wife, Verena. Retired, he spends his days writing novels and his holidays photographing volcanoes from Italy to Mexico to East Asia. But Alfred Bollinger in 1969 was a staunch if congenial customer, with a keen sense of his place in his institution's hierarchy. And what he heard nearly knocked him off his chair. No Swiss would throw out such a boast at first meeting. To his careful professorial mind, Gruentzig's assertion sounded as over-the-top as Oscar Wilde's proclamation upon arriving for his 1882 whirlwind speaking tour in the United States, "I have nothing to declare but my genius!"

Yet, Bollinger quickly realized that the new assistant was destined to cut a wide swath. Whatever task he asked Gruentzig to do, the thirty-year-old understudy completed with an obvious flair, spending countless hours with patients and filing meticulous follow-up reports, then working into the night to help with Bollinger's research papers. "If I asked Andreas to write an article, I'd have it in ten days. If I asked for revisions I'd have them in two. He was quick and absolutely decisive," the angiologist noted.

Within weeks, Gruentzig was knocking on his boss's door, enthusiastically propounding a novel theory: Patients limping painfully around the angiology ward obviously suffered from a disease that affected not just their choked-off leg vessels but their entire cardiovascular system. Thus, Gruentzig decried, their response to a hammer stimulus to their Achilles tendons should provide a telling index of not only the severity of their walking pain (called "claudication") but also the ability of their hearts to push blood through their extremities.

Perhaps, Bollinger rejoined, wondering what the new man was driving at.

Here, Gruentzig's mind raced ahead. Therefore, he concluded, a measurement of reflex reactions to a stimulus in the ankle might provide an instant window into the severity of their cardiovascular disease.

The idea was beguiling in its simplicity—something that would become a Gruentzig trademark. "After two weeks, Andreas showed up with his reflex-measuring apparatus," Bollinger recalled from his vineyard-surrounded apartment complex overlooking the fjord-like Zürichsee, or lake. From the far wall, a grandfather clock ticked with the orderly precision for which the country is so famous. "Then he starts running up and down the corridor with a patient who was wincing and gasping, 'Oh, this hurts!' Andreas blurts, 'Come in here!' and directs the patient into an examining room where he kneels down and begins tapping on his knee with a hammer."

The memory induced a chuckle from the retired professor. Bollinger then explained how he agreed to join Gruentzig in a more technologically refined study of what remains, thanks to their joint research: a bedrock measure of the severity of walking pain. The recruit's attention next latched onto the diagnostic potential of a then-new technique called Doppler ultrasound, a method that uses sound waves to check for differentials in blood pressure, which reveals the impact of atherosclerotic leg blockages. Gruentzig took to it with a passion.

"He was always very rigorous with these kinds of things in seeking reliability and reproducibility. He was very exact and precise and concerned to make things statistically relevant," Bollinger noted. "I have never worked with anyone in medicine who had the kind of inner fire that he had. It was unbelievable."

So life in Zürich got off to a fine start for Andreas Gruentzig. Michaela soon landed a staff counseling job at the University Hospital. With their combined salaries, the couple were able to live more comfortably than ever before. Andreas bought himself a red motor scooter, which he gunned through traffic on the university district's broad Ramistrasse. A blue beret kept his mane of dark hair from his eyes. On weekends, the couple entertained themselves in Zürich's pleasant cafes and indulged their love of dancing. When the weather was decent, they walked the forested hills nearby or skied the mountains that soared up within an hour's drive. The deprivation of Leipzig receded into memory.

In the summer of 1970, the young Germans were married by a town clerk at Michaela's Bavarian family home in Bad Reichenhall. About fifty guests joined in the festivities afterward with the younger ones sleeping rough. The newlyweds embarked on more adventurous holidays, touring Rome and motoring their blue Mini to a campground in Yugoslavia from which they could swim in the warm Adriatic, tossing aside their clothes as they chose.

A second Zürich year unfolded, and Gruentzig devoured new projects, among them a long-term study of the effectiveness of exercise training in alleviating intermittent walking pain—perhaps not scintillating stuff, except for the manner in which it was pursued. Having absorbed every detail of his statistical training in London, Gruentzig made certain that this work would be a model of precision, exhaustive in its thoroughness. His good looks and joie de vivre worked with a silky smoothness in gaining cooperation not only from aging patients, but also the young female staff at the University Hospital.

One of these happened to be Alfred Bollinger's personal research assistant, a sandy-haired wisp named Maria Schlumpf. In time, this soft-spoken, physically tiny, but ever purposeful individual would play her own important role in the conquest of coronary artery disease. But Alfred Bollinger couldn't help noticing that Maria became swept into his understudy's explorations and grew irritated at her diverted attention. Fortunately, that conflict was resolved. Yet the department chairman was not alone in wondering whether the East German was somehow a touch too charismatic, a measure overreaching. The way the doctor-in-training spun down the hospital halls flashing his captivating grin was especially grating to the new chief of internal medicine, a stern, by-the-book figure named Hans Peter Krayenbühl. The self-important swagger of Germans never carries well with the careful Swiss; to watch one prance about like a movie star only made matters worse, as far as certain superiors were concerned.

Not all of Gruentzig's colleagues were so uptight. Felix Mahler, a resident in internal medicine at the time and now a professor in Bern, remembers wild interdepartmental outings at ski resorts, where flirtation reigned. "We were chasing girls and dancing the whole night. He was a delightfully lively person. We had parties together and it was really great fun, with singing and guitar playing. We were very young."

Mahler was also impressed by his colleague's devotion to his mother Charlotta, who frequently visited from Heidelberg. "She was dark like him, really dark with dark, dark hair. She could easily have had roots in South America herself. But her personality was very German. She wore traditional dresses, flowered folk-type costumes called dirndls. She knew exactly who she was—in spite of her losses. She lost her fatherland, she lost her nation, she lost her husband, and still to me she seemed very centered and proud of herself and both her sons. I always felt that she was his anchor in life."

Rotating to a new position in the radiology department in 1971, Gruentzig kept searching for some special lever of opportunity. "He was a person who always made the best out of what he had in hand. I think that was a really fundamental trait," explained Michaela Gruentzig about her late husband. "He was always determined to make the most out of his life, to find some accomplishment."

How did breakthroughs of genius occur? Werner Forssmann, the daredevil of the heart, found his cathartic inspiration from some engravings in an old French physiology text; Mason Sones discovered his turning point in a misfired puff of dye. Gruentzig, having weathered his initial Swiss apprenticeship, was by now itching for his own opportunity. The notion that haunted his imagination involved something much more profound than tapping on patients' knees. An idea had burned into his mind, a scarcely explored tangent that whispered of uncharted alternatives to the headline-grabbing heroics of heart surgeons like Denton Cooley and Christiaan Barnard.

Shortly before departing for Zürich, Gruentzig had attended a lecture about a peculiar approach to dislodging atherosclerotic blockages in gangrenous legs that went by the odd name of "Dottering." The lecturer happened to be a fellow Osti, Eberhard Zeitler, an engaging speaker who mixed a natural affability with ardent convictions, and who had fled East Germany just before the completion of the Berlin Wall in 1961.

The ticket forward, Zeitler argued, lay in a nonsurgical technique that could be speedily performed with scant injury or pain. This quick-talking, goateed doctor claimed that all one had to do was thread a simple, off-the-shelf catheter—to this point, regarded as merely a diagnostic tool—into a diseased vessel, and then bore ahead by gradually pushing forward a telescoping series of larger follow-up catheters to widen a fresh channel of blood flow. The object was to bring new life to the blood-deprived, or "ischemic," lower extremities, which for diabetics too often led to a rendezvous with an amputation saw. Zeitler illustrated his talk with dramatic before-and-after images, the "befores" showing feet with blackening toes and sickeningly swollen ankles, the "afters" often brought back to the pinkness of health.

Gruentzig was instantly fascinated, even though his boss at the time disparaged the talk as quackery. But the image had set. Gruentzig well knew that he was in no position to try out any experimental Dottering while settling into the University of Zürich's regimented Canton Hospital. So he patiently earned his stripes and sought to explore the technique only after gaining a more secure position in the radiology department. Then Gruentzig jumped and set off for a train journey up the Rhine in August of 1971 to visit Zeitler in his Aggertal Clinic in the rolling hills twenty-five miles east of Cologne. A former spa for patients exhausted by tuberculosis, the clinic's atmosphere was not unlike like that of the languorous Davos tuberculosis sanatorium featured in Thomas Mann's celebrated novel, published in 1924, *The Magic Mountain*. No significant medical events were ever meant to happen there, seeing as the Aggertal's specialty was physical rehabilitation, which made it one unlikely setting for achieving medical greatness.

German physicians are sometimes stereotyped as being ploddingly unimaginative compared to the freewheeling inventiveness common to medical researchers elsewhere. This notion is false. The first pioneer to the angioplasty idea was indeed an American, but he was such a wild eccentric that his vision was laughed out of the U.S. Never mind that Charles Dotter was an erudite student of classical music and fine art, ornithology, stamp collecting, and photography, and a fearless mountain climber. Never mind that his restless mind invented countless new ways to diagnose and treat vascular disease. Dotter, mainly through his own excesses, became branded with the disparaging sobriquet of "Crazy Charlie." Only the Germans grasped the man's genius.

In truth, Dotter was a little odd. A born free thinker, he quickly took the helm of the department of radiology at the University of Oregon in Portland. He initially made a name for himself by his explorations of congenital heart disease, and by helping to develop ultrafast-sequence imaging techniques of the workings of the heart valves. But what obsessed him was the notion that catheters could be handcrafted to treat the furthest reaches of the human body.

That ambition was galvanized by his sabbatical at the Karolinska Institute in Sweden with a doctor named Sven-Ivar Seldinger. Seldinger gained fame by eliminating the carnage that often attended to forcing bulky diagnostic catheters into arteries from an entrance site in the groin. His simple idea was that an introducer needle could be pricked through the skin into the blood vessel's inner channel ("percutaneously," in medical nomenclature) to gain an entry point. Then a long wire guide could be advanced forward, allowing the needle to be removed. Next, a hollow catheter would be slipped over the wire and threaded forward until it penetrated the desired target. Seldinger, whose grandfather had devised a pioneering method for pleating skirts, put it this way, "I can perform the procedure faster than you can write about it. Needle in, wire in, needle out, catheter over wire: that is all."

Dotter swiftly recognized that this technique could have radical implications. But its limitations were also obvious. Even when gently slipped beneath the skin, a catheter doesn't just swim through the sharp bends of the various branch vessels to arrive at its convoluted destination like some foreknowing salmon returning to an ancestral pool. It has to be finessed home, and an errant maneuver can perforate or obstruct a blood vessel while in transit. So Dotter grew obsessed with the quest to engineer an array of catheters tailor-made for each twist of the anatomy. The man tinkered constantly, trying to preform various devices into shapes that would be self-directed toward wherever they were aimed.

The radiologist pursued his every idea with breakneck urgency, in part because he developed an early taste for amphetamines. He whirled through his days with contempt for both boredom and sleep. A mountain climber who scaled sixty-seven peaks higher than 14,000 feet, Dotter was never happy unless facing uncharted vistas. His technicians scarcely knew what to expect when they arrived for work in the morning, their missions sometimes growing so bizarre that they were instructed to fetch guitar strings, piano wires, and Volkswagen steering cables for their next experiments.

Bill Cook, the founder of the multibillion-dollar privately-owned Cook Group medical-device conglomerate, met Charles Dotter at the annual Chicago meeting of the Radiological Society of North America in November of 1963. At that time, the future corporate chief had all the status of a traveling salesman. He earned his keep by manufacturing circulation-punching introducer needles and urinary catheters in his garage and showing his wares out of a small portable booth at medical convention exhibition halls. Pitching his tent in Bloomington, Indiana, he purchased $1,500 worth of catheter extrusion machinery and started out on his own. The still boyishly crew-cut Midwesterner might as well have minted gold. Before long, he would control forty-two companies. These included not only a medical-device empire now producing 50,000 different implements for working within human organs and arteries, but also a charter airline, a musical revue called *Blast!*, a British professional basketball team, and the Star of Indiana Brass & Percussion Corps. In recognition of his munificent philanthropy, an Indiana governor bequeathed him the peculiar title of "Sagamore of the Wabash." That translates into something like "sage chief of the great river."

When proffering his fledgling wares in Chicago in 1963, Bill Cook had no claim to fame. On that day, he noticed a peculiar figure hovering behind him, drilling an intense gaze into his back.

"Here was a short, muscular, bald man with darting eyes—I didn't know who he was, but he made me nervous," Cook recalled. "I turned and asked if I could be of help, and he said 'no'—nothing more—and left. Just before we closed for the day, he returned and asked if he could use my blowtorch and 'borrow' some Teflon tubing. He said he wanted to practice making catheters in his hotel room. Thinking I had a real space cadet on my hands, I said, 'Sure, but may I have your name?' He answered, 'Charles Dotter.' "

The next morning, the beady-eyed doctor returned to the exhibition hall, thrusting out a collection of new catheter designs as though he had just gotten his hands on the clockwork of the spheres. He and Cook talked for hours. Then the bizarre doctor invited his new friend to come to Portland, at the Oregonian's expense.

Today, Bill Cook presides over an 800,000-square-foot glass-and-marble headquarters manned by 3,500 employees. Walking through the vast marbled lobby, it is difficult to reconcile such an empire with the man's humble station when he met Charles Dotter. But Cook proved to be thoroughly congenial and homespun. "Dotter intimidated me, completely and totally. He was probably the most respected radiologist in the field of catheterization. His name was already legend."

Cook agreed to visit Portland immediately, and rented his own plane to cut costs on the trip, since one of his passions was flying. Arriving in Oregon, he was regaled with Dotter's head-spinning concepts for transforming the diagnosis and treatment of blockages in every reach of the human body. It was all a matter of fabricating the right tools, said his host. Cook listened raptly, promising to manufacture whatever was required. Returning to Bloomington, he got busy with his blowtorch and extrusion machines.

Soon Dotter was telephoning to ask if the young entrepreneur wanted to take a vacation in Verner.

"I said, 'Where the hell is Verner?' " Cook recalled.

"He said, 'It's in Utah.' "

That was good enough for Cook and his wife Gayle, who agreed to fly west. "It became friendship almost immediately. If he liked you, Charles Dotter was a very easy man to know. He said, 'You want to climb a mountain?' and we climbed the mountain there."

Cook knew that Dotter's mind was bursting, but neither he nor anyone else had an inkling of what a controversial mark the Oregonian would soon make.

Until 1964, Dotter had merely experimented with his telescoping catheter idea on a few test animals and cadavers. He christened it "transluminal dilatation," which means to make an artery wider, or dilated, by expanding it with devices that would push out from the lumen, or channel, within. Dotter's colleagues shrugged off this concept as another one of the man's weird schemes.

January 16, 1964, started as just another day at the medical center on the heights above Portland, Oregon, known as "Pill Hill." Dotter had performed an arteriogram on the left leg vessels of an eighty-three-year-old diabetic, whose toes were so ridden with gangrene that they were ready to fall off. Laura Shaw's physicians had all insisted that she undergo amputation at the site of her ulcerated ankle, but the woman steadfastly refused. The surgeons decreed that her ruined leg circulation was beyond repair, that there was no other choice. But they sensed that Dotter was keen to try one of his wild alternatives, and passed on a scrawled note saying, "Visualize but do not try to fix."

Dotter wasn't fazed. After all, he was known to kick his Great Danes's turds across his living room floor while in the midst of dinner-party conversation. His X-ray study revealed that blood flow through the woman's femoral artery, the major vessel leading to the lower leg from the groin, was totally choked off by accretions of plaque. Here, Dotter reckoned, awaited the perfect candidate for using his telescoping catheters to push aside the hoary arterial obstruction. Unless something was done immediately, the woman faced amputation.

Dotter set to work. First, he nosed a guide wire into the blockage and followed this with a relatively small-bore catheter. Gaining a purchase, he pulled that one back and then channelled larger catheters forward. Suddenly, he drove all the way through to arterial "daylight." The X-ray pictures demonstrated that the flow of blood into the woman's lower leg had been instantly restored. Within minutes, her foot began to warm. Within a week, Laura Shaw's pain disappeared, and the ulcer healed.

Over the next months, Dotter used his new technique eight more times. "Scheduled amputations—where I got a crack at salvage just as the axe was due to fall," was the way he summed up this period. He described his innovation as "wedging, just as you would put a nail through a piece of cheese."

Charles Dotter was impetuous. In August of 1964, he threw open his laboratory doors to journalists from *Life* magazine, the most widely read weekly periodical in the United States. This was the medical equivalent of "streaking." As the cameras flashed, he performed another procedure with his eyes bulging and lips curled fiendishly beneath a balding pate. Playing the mad scientist to the hilt, the mischievous doctor deliberately adopted a demented-looking mien as he worked. Credit Dotter with comedic timing: this was the same absurdity-loving year that launched the madcap film *Dr. Strangelove or: How I Learned to Stop Worrying and Love the Bomb.*

Splayed across two pages of *Life,* Dotter appeared to be about the weirdest doctor in the universe. Such a garish tabloid profile might have forced more ordinary physicians to jump off the nearest bridge. But not Charles Dotter, even though his peers instantly branded him "Crazy Charlie."

"I am sure he loved it. I am sure he was very proud," laughed Josef Rösch, a Czechoslovakian-born colleague who worked at Dotter's side for two decades.

A secretary recalled, "The physicians said they weren't going to send their patients to someone who looked like that. A lot of patients didn't mind at all how he looked. They thought the idea was stupendous, to get to walk again without surgery. They didn't care what poses he had in *Life.*"

"He was an adventurer, not only in his life but also in medicine," Rösch wistfully remembered one March morning in his conifer-shrouded office. "Only because of this quality could he start angioplasty; nobody else would have done it at that time. Nothing for him should ever proceed traditionally. I don't know how you define 'genius', but I think he was that. His mind was always jumping from one topic to another, always thinking about how something could be improved. . . . He was the most exciting and interesting person I ever knew."

Bill Cook realized that the man was dauntless. "He told me then, 'In twenty-five years dilatation will be an everyday procedure and we will eliminate surgery. He said it to the world, that this is what he was doing and that this is what it was going to grow into."

Being branded "Crazy Charlie" only egged Dotter on to scale new precipices. In the late 1960s, the Oregonian stood like a prophet before major medical meetings proclaiming that the surgeons' hegemony was finished.

To convert skeptics, Dotter created his own little film, which was surely one of the wackiest docudramas in history—it might well have been entitled *Dr. Strangelove* itself. Bill Cook, billionaire in the making, was enlisted to fly around the Cascade Mountains in search of visual metaphors for clearing arteries. Suddenly, Dotter demanded that Cook fly under an arched bridge over the Williamette River. Cook sensibly refused.

The 1968 film started with the stirring "Glory, Glory Hallelujah" refrain of the "Battle Hymn of the Republic," the triumphant Union anthem of the American Civil War and of America's sense of manifest destiny. It next featured a patient in pajamas literally clicking his heels in a display of rejuvenation after being treated by Dotter's supposedly miraculous technique. Another segment included an animation of an American drain-clearing device with Dotter's voice-over declaiming, "There's no reaming or drilling, or anything like Roto-Rooting. Nothing's blasted, nothing's blown out or broken up." To illustrate the point, the movie cut to a vast mushroom cloud following the detonation of a nuclear bomb. This was medicine gone *Monty Python.*

For a careful European scientist like Josef Rösch, Dotter's madcap ways were not easy to fathom. Dotter visited the leading Czech radiologist in 1963 in Karlsbad, and began an affectionate correspondence that included the exchange of rare postage stamps. In 1967, Dotter finally offered his friend a year-long fellowship in Portland—a potential ticket to freedom for an ambitious doctor locked within the confines of the Iron Curtain. Even though this was the so-called "Prague Spring," when the Soviet Union briefly relaxed its grip on Czechoslovakia, getting out of the communist country required some tricky footwork.

Rösch remembers his arrival in Portland vividly. The 300-pound Melvin Judkins, another pioneer in developing pre-shaped catheters to finesse through the circulation, met the father and daughter at the airport. Then Judkins and his wife delivered the weary travelers to an apartment that he had fitted out to their every need, vegetarian style.

Soon, Rösch was caught up in Dotter's hijinks. "I remember his taking me and my daughter on a Friday to see a movie about car racing—*The Champions,* I think it was called [in fact a 1966 promotional short film called *Grand Prix—Challenge of the Champions*]. I loved it, and he loved it, too. The next day we drove together real fast in his GTO [the gaudily named Pontiac *Grand Turismo Omologato*] sports car toward the beach and a Porsche blew past us. Charles looked at me wild-eyed and looked at it, then he shouted 'champions!' "

Here the rangy Czech's soft blue eyes began to moisten with laughter. "Then he hit the gas so hard, it drove us all back into the seats—it was like taking off in a jet! He pushed past the 120-mile-per-hour limit on the speedometer, and we swerved around bends beside the beach with the car fishtailing from one side to the other. The other guy was only going about a hundred miles per hour. When Charles hit 120, we roared past him and didn't slow down for the next seven or eight miles."

The story was inevitably bittersweet, because Rösch's daughter was herself killed in a car crash in 1973. Dotter, in fact a much deeper personality than the caricatures of him ever showed, lovingly helped his colleague through the terrible depression that followed. The two men worked together to pioneer innumerable techniques that no one had ever imagined before. "We were body plumbers," Rösch, the author of hundreds of scientific papers, said. "We opened clogged tubes."

The duo also became such close friends that they scaled seventeen high mountains together. Many of their earliest climbs were conducted with Rösch's daughter. When later arrangements were made for the rest of his family to gain visas out of Czechoslovakia, his son joined in, too. Dotter, for his part, was often accompanied by his mistress of many years, a fellow radiologist named Marsha Bilbao. "In the evening he would say, 'We will see what we do when we awake,' " Rösch explained. Back in the woods, Dotter dressed like a seedy, unshaven tramp.

In March of 1968, Dotter invited Rösch to join a climb on the sheer cliff face of Smith Rock at Mount Hood, the white-domed peak that dominates the Portland skyline from seventy-five miles east. "I didn't like rope climbing, but he said, 'It will be a memory!' There was a high cliff called 'The Thumb' that he had never climbed. He said, 'We will climb up The Thumb and go down' . . . We got up near there in the morning and noon came and I said, 'Okay, let's go!' He said, 'No, we have time—lots of time.' We finally started in the middle of the afternoon and climbed up the back side and were on ropes . . . It was already getting dark."

Dotter insisted that they tackle one more cliff face, but Rösch balked, saying he didn't want to climb the dangerous escarpment. "You have to!" insisted Dotter. Somehow, they made it. Halfway down, a relieved Rösch offered to belay the last few meters to the comfort of a ledge.

"Charles was holding the rope on the other side, and I swung into the rock and broke my ribs. I said, 'Okay, let's go down,' but he couldn't find the way down. He told us, we have to climb back up on ropes what we just went down on ropes, and now in the darkness. It grew freezing. . . . We had to go into a cave and hold each other for warmth."

In the end, the team survived, broken ribs and all.

Manny Robinson, an African-American technician who worked with Dotter for nearly twenty years, saw the radiologist in all his extremes, too—beginning with a scene reminiscent of Dorothy's strange introduction to the Great and Powerful Wizard of Oz. "In the late 1950s I had been working on the eighth floor of the hospital in transportation," Robinson recalled as he settled into a chair beside Josef Rösch. "A friend told me there was a research position for which they were probably going to need some help. I went up to Dr. Dotter's office and the first time I saw him he was sitting at the view box looking at some films. I stuck my head in the room and told him my name and that I had come to see him about the job. He never turned around. But he said, 'Why should I hire you?' just like that. It caught me totally by surprise. And then he says, 'Don't worry about it, all you have to do is tell me that you are an honest man and you've got the job,' and then he turned again without ever seeing me. Two weeks later I came over to join the research team.

"We tied the balloons, we shaped catheters, we made guide wires from guitar strings," Robinson explained. "We used to purchase them from a music supply store called Meier & Frank, then bring them up here and cut off the ends that held the core in, then solder away and put in another core. It was all in his head, and he would come up with these weird-looking instruments every day."

A lifelong resident of downtown Portland, Robinson was uncomfortable even walking in the woods. But one day his boss insisted that Manny, too, should climb Mount Hood. The journey began with a late-night, hundred-mile-per-hour drive into the Cascades in Dotter's red Mercury Cougar. En route to the wilderness, the Cougar was pursued by a siren-screaming police car and summarily bouqueted with a speeding ticket. This deterred Dotter, who collected many of those, not in the slightest. He roared off again, only to suffer a flat tire after turning into a remote back road. Now, the team of adventurers was forced to wait for a rare passing motorist's help, because neither Dotter nor his master technicians knew how to change a punctured tire. Finally, they arrived at 11,200-foot Mount Hood, where the boss supplied the by-now bewildered Manny Robinson with crampons and an ice pick.

"He gave me some instructions about what to do if somebody fell," Robinson recalled with lasting incredulity. "We started out, and it wasn't so bad, but my perception of heights was totally off. . . . We crossed crevasses that you couldn't see the bottom of. We were all roped together. He said, 'If you don't make it, the rest of us will counterweight you.' "

Chest against the mountain, Robinson was frightened for his life and trembling at the tricky work of nursing unseen feet into the right purchase in the sheer rock. "That meant in my mind that you were momentarily away from the ground and nearly dead. We did that several times and we got to a point where it started clouding over. And ice starts sleeting and it is getting real dark near the top. I think the only thing that saved me personally is that by that time I couldn't see the bottom because it was fogged in. And when we finally got to the top, the angle was so steep that I had to crawl on all fours and my sunglasses fell off, so I couldn't see. . . . The plan was to stay there until the next morning."

Robinson drew a breath. "I kind of lost it, being a novice, and I remember he gave me some medication to calm my nerves. To this day I couldn't tell you what it was," Manny guffawed. Rösch, the careful Czech, was fairly slapping his thighs as the story proceeded. It was clear that these two old comrades savored sharing their stories.

"Anyway, the decision changed. We were going to go down, even though by this time it was dark. I cannot tell you how we got down. On the way, there was a pathway at the top of the mountain where you could see this streak in the snow. That was a path left by some medical students from this school who had fallen off—so I am quite nervous, you know?" Robinson noted.

The tale spun forward. "One thing I do remember is we got into a cluster of rocks and what happened is these guys around me seemed to be jumping around like monkeys and they glowed orange and all kinds of funny colors. It was the first time I ever cursed at Dr. Dotter, and I said my few words, among them that, 'You guys are trying to kill me.' I had gotten frostbitten and a little snow-blind because of the sleet in my eyes. So we decided to stay together and we, well, lay in a hole . . . I remember there was a plastic raincoat and we tried to put it over us, but the wind caught it and blew it away. I am not saying Dr. Dotter lost it. But he had a secretary named Eleanor Hasigawa, and while we were lying there he was screaming: 'Eleanor, Eleanor, Eleanor—send the helicopter!' "

Eventually a rescue team arrived to save them. Manny was given a week or two off to recover from frostbite. And then it was back to Pill Hill.

As reckless as the mountain adventure may have been, the former technician still considers Dotter to have been a genius. When not shepherding mad escapades, Dotter worked feverishly at realizing his vision of knifeless surgery. But the surgeons insisted he was nothing but a clown.

So the promise of transluminal dilatation was ignored in the United States. The notion might have died altogether were it not for its continuing pursuit by Eberhard Zeitler in Engelskirchen and Werner Porstmann at the Charité Hospital in East Berlin, erstwhile fiefdom of Ferdinand Sauerbruch. Zeitler himself experimented with using catheter injections of so-called "thrombolytic" clot-dissolving drugs like streptokinase to break up fresh obstructions in diseased leg arteries. Unaware of the *Life* magazine fiasco or the derisive label of "Crazy Charlie," Zeitler began testing out the dilatation idea at the Aggertal Clinic. Impressed, he wrote Dotter to request a meeting. Come soon, was the response.

After a stopover in Los Angeles, Zeitler flew on to Portland—a journey of about 1,000 miles—so that he could observe a scheduled case the next day. The following morning, the wary traveler appeared at Pill Hill at seven A.M. to find his enigmatic host swiveling in his chair and blurting, "I'm sorry, I don't have any proper patient for a procedure today, but I've already arranged tickets for us to fly to San Francisco (500 miles back in the direction from which Zeitler had just flown), and we'll be there in an hour and a half.' "

It quickly became apparent that Dotter had actually arranged this far-distant procedure a week in advance, yet had thoughtlessly failed to explain the situation on the telephone. The still jet-lagged Zeitler found himself hurrying off in a taxi with his host. At the airport, they met Manny Robinson, who was proffering plane tickets, his pockets stuffed with catheters he had crafted himself. "So we flew to San Francisco and went to a hospital near the airport. The patient was waiting on the table there. We washed up, put on operating gowns, and began selecting the proper catheters. Then Dotter did the procedure, which went wonderfully. It was very quick."

Dotter took an X-ray afterward and pronounced the clogged artery below the knee to be newly free flowing. Then he drew on a cigarette. "Once he saw that the dressing was finished, he said, 'Okay, let's see what the next flight is to Portland.' We weren't in this hospital for more than an hour. We got a coffee somewhere and that was all. He said to me, 'You see how fast this can be done without complications or risks?' "

This then was the freewheeling legacy to which Andreas Gruentzig was exposed upon arriving at the Aggertal Clinic. It was a new world of possibility, unlike anything he had witnessed in Switzerland. For the first days in that August of 1971, Gruentzig stood by Zeitler's side, watching a series of Dottering procedures with fascination. He heard his congenial host's musings about the potential for radically improved new catheter designs, perhaps even tipped by some soft balloon material. Clearly, the telescoping Dotter catheters were crude. Yet Zeitler wielded the instruments deftly and demonstrated immediate resurgence of blood flow afterward. In the second week of Gruentzig's visit, the young guest physician was invited to provide hands-on assistance.

The two hit it off. Their evenings were spent lingering in the garden with the talk ranging the horizons of medicine. Neither figure understood that they were forging a friendship that would affect millions, and launch industries that would earn billions.

Chapter 6

In early September, Andreas Gruentzig and his muse, Maria Schlumpf, met in a café on Zurich's Gessnerallee so that he could relate everything he had just experienced. As they watched the evening's strollers pass by, his mind began to race, for Gruentzig had returned to Switzerland charged with enthusiasm. By his estimation, the Dotter method was crying for improvements, which he was convinced he alone could make. If properly realized, why wouldn't the heart itself become the technique's proper home? The question, given his cautious work environment, was where to start. Almost unconsciously, he began to doodle on his place mat. What emerged were his first ideas for a balloon-tipped catheter. Maria Schlumpf would hang onto that bit of café ephemera for the rest of her life, knowing its implications were revolutionary.

A couple of weeks later, Gruentzig took a bigger leap. Zeitler, he learned, would soon appear at a mid-September medical meeting in nearby Lüzern, that pretty medieval city at the edge of the crystalline Alpine lake named the Vierwaldstädtersee. Seizing the opportunity, Gruentzig called to ask his new friend to rush to nearby Zürich with his dilatation catheters, insisting he had found an ideal candidate for the procedure. The seventy-seven-year-old male patient had a short atherosclerotic narrowing, or "stenosis," in the easily accessible top, or proximal, section of his femoral artery below the groin. It was a straight shot, presumably ideal for demonstrating the ease of the dilatation technique. At least on paper.

Things began on cue, with Gruentzig assisting Zeitler while ten radiology colleagues looked on. With the aid of X-ray illumination, the guide wire was feathered into the groin puncture site, then tickled to a firm purchase deep in the shriveled channel within the choke point of disease. On the catheterization laboratory's television-like monitor, any skeptic could see the procedure working plain as day. Zeitler reached into his bag and selected the perfectly-sized devices to push forward. *Wunderbar!* The ugly accretion of plaque was neatly compressed against the arterial wall. Fluoroscopy demonstrated an immediate restoration of flow, and the Doppler ultrasound technique offered confirming measurements of a beautifully restored pulse clear down to the ankles. The first Zürich demonstration of the alternative to surgery Gruentzig intended to perfect seemed blessed by the stars.

But as Zeitler was about to head back toward Lüzern, the patient, wakening from his sedative, started moaning about pain shooting through his lower leg. This spelled trouble. "We realized that we had embolized the entire plaque," Gruentzig recalled in an unfinished monograph, describing a frightening complication that threatened to shut off all blood flow in the lower extremity, possibly requiring amputation. "The radiologists, who were skeptical from the very beginning, now had their proof that the method was of no use whatsoever in human beings."

Fortunately, the problem was solved with medication. Fortunately, too, the would-be medical pioneer somehow coaxed continuing support from two powerful figures: Walter Siegenthaler, the chief of medicine, and Åke Senning, the Swedish chief of cardiac surgery. The German-born Siegenthaler had seen one of Zeitler's presentations in Bonn firsthand, and while not one hundred percent convinced, still believed the technique might hold promise. Senning, the first surgeon to implant a coronary pacemaker, worked on the edge himself and carried much clout.

Gruentzig nonetheless had to charm and claw his way forward by turns. In December, he finally received approval to try some personal "Dottering" on the legs of patients deemed to be overly risky candidates for surgical bypass procedures. Over the next two years, Gruentzig would employ the technique on a total of forty-two such cases, with the tireless Maria Schlumpf compiling exacting before-and-after records. Even the resident doubting Thomases were forced to recognize that there was nothing cavalier about this pursuit. Passionate about scientific methodology, Gruentzig insisted upon rigorous pre-procedural X-ray studies of the patients' dilated arteries, with fresh images taken one day afterward and again three months following each procedure. In fact, Maria Schlumpf would follow the course of these patients for fifteen years. Given his superiors' frosty skepticism, Gruentzig was forced to do most of his "Dottering" at lunchtime or at night.

Even so, resistance could boil up instantly, particularly when complications befell the procedure. The chief of vascular surgery raged at one case that ended up with a welling collection of blood, or "hematoma," at the puncture site. "The technique may have worked, but we ended up having to operate the same as always. This is senseless!" he roared. The acrimony did not augur well for Gruentzig, who had no access to patients without a superior's express referral. The big guns at the tightly controlled in-patient wards of the University Hospital often refused to cooperate.

"I found it was an advantage that my colleagues laughed at Dotter," Gruentzig recalled in an interview with a trade journal. "I laughed along with them. Then I just went about my work quietly."

Colleagues insist that the early friction was bitter and that Gruentzig's foreignness was undoubtedly part of the problem, but that so too was his own demanding personality. "He wasn't diplomatic, and you do not assert yourself in this way in Switzerland. It is in the Swiss character for things to proceed slowly. Things there proceed like the money market, step by step. And people in Swiss universities can be set against foreigners," Zeitler observed.

Zeitler quickly recognized that Gruentzig could alternate between silken diplomacy and contentiousness. "I saw him in discussions changing in a flash from easy and charming to sharp and confrontational. He could go powerfully in either direction."

Bollinger added, "He was also pushy. Pushy people often have problems." Nonetheless, the angiologist remained sympathetic and directed patients Gruentzig's way whenever possible. One of the more colorful was an eighty-four-year-old rabbi from Lugarno, the southern, Italian-speaking Swiss canton. The rabbi was sent to Zürich under Bollinger's care to be treated in the outpatient "polyclinic" wards.

"He came in as a private patient and I saw that he had a bad problem. He was a diabetic, experiencing angina every ten minutes. He had occlusions of different leg arteries, but also a tight stenosis of the popliteal [a lower leg branch] artery. It was obvious that he would have to have his lower leg amputated if we could not do something, so I called Gruentzig to see if there was an alternative."

As it turned out, the rabbi proved to be so obese that merely gaining access to his groin artery threatened to be a risky and difficult undertaking, a human equivalent of Pin the Tail on the Donkey. Gruentzig's first reaction was to beg off, but he finally agreed to try to help. The polyclinic's corridors filled with a collection of relentlessly pacing orthodox Jews in black yarmulkes and long beards. Inside the catheterization laboratory, Gruentzig went to work, and managed to restore a critical point in the leg's impaired circulation, which he hoped would spare an amputation. Outside, the entourage from Lugarno uttered prayerful thanks.

Alas, Bollinger got a call from Gruentzig three hours later saying that the rabbi's foot had gone completely white, indicating a dire failure. The Zeitler protégé cooly improvised, and solved what turned out to be a minor problem. But the technique's limitations remained obvious. The stiff, large-bore Dotter catheters occasionally tore traumatic wounds at their entry sites and could never be brought to bear at twisting segments of the anatomy. The battering-ram aspect of the technique carried a constant potential for dislodging fragments of plaque, with the resulting showers of small particles threatening to cause catastrophic problems in the lower legs, including a shut down in the circulation leading to amputation. Moreover, aggressive handling could drive the devices straight through a vessel's wall. In short, the technique was altogether too clumsy to conceive of ever using in the heart.

So, a far-flung handful of inventors began searching for refinements, one primary idea being to affix something soft and bendable at the catheters' tips. Zeitler and Porstmann in East Berlin had their own ideas about fitting the tip of the device with some kind of balloon material, but their refinements remained crude. Meanwhile, an Italian-born radiologist named Caesar Gianturco fabricated a similar alternative in America but then abandoned it when the first trial procedure went wrong. To all except Andreas Gruentzig, the prospects of finding a gentler approach to fixing arteries looked bleak.

But the East German was never one to be dissuaded. So he experimented night after night, evidently ignoring many responsibilities in the process, including his illegitimate daughter back in Germany, Katrin Hoffman. Over dinner in a Zürich café in September of 2003, she confided that the only attempt her father made at communication consisted of a letter he dispatched when she was five or six. Sadly, her mother never let her see it, and she did not even tell the child that the absent Andreas was her real father until she turned ten. "I kept wondering what I had lost," Katrin, who shares her father's swarthy complexion and nearly coal black eyes, wistfully recalled at the age of thirty-four. "He turned his back on my mother, refused to marry her, walked out, and never looked back."

A student at different times of everything from veterinary medicine to hypnotherapy and yoga, Katrin eventually became consumed by a desire to connect with the father she never knew. While in her twenties, Katrin finally met her grandmother Charlotta, and that brief encounter was intense. "She was a very religious lady, and was at first put off by who I was, and wanted me to just go away. Then she began to warm up and told me that I had good genes from Andreas's side." Her grandmother talked about all her own years of scraping by to raise her sons, of an opportunity that surfaced once to offer lessons on the saxophone, which she herself had never played. "She forced herself to learn to play it well enough so she could give the lessons," Katrin marveled. "She was clearly a strong, proud woman. There was no weakness at all in her eyes. She was observant and strongly focused and altogether sized me up."

Gruentzig kept in close touch with his mother and told her of his arch conviction that an alternative to heart surgery might be found in the shape of a child's balloon. So what if no one yet understood the best way to construct such a device? The Dotter method was but a half-realized dream, a shove of light into darkness. Hadn't the Russians and Americans proved that they could control the orbits of spacecraft hundreds of miles above the earth? Hadn't man walked on the moon? So why should the prospect of bringing pinpoint control to a catheter burrowed a foot or two within the human circulation be considered hopeless? The challenges were only technical, and he was determined to solve them.

One night in early 1972, Andreas invited Maria Schlumpf to share a glass of wine with him and Michaela at the couple's Haldenbach Strasse apartment. The two-bedroom flat, where Michaela still resides, sits above a tranquil lane that steps down to the bustle of central Zürich's old town. From the corner terrace, you can see the city's gilded lights and sometimes hear laughter and music wafting from below. In short, it offered a pleasant place for a social call. But Gruentzig had no such thing in mind.

The time had come, he proclaimed, for some hands-on experimenting. Right now, said Gruentzig. The dutiful Maria soon climbed the dark stairs to the second-floor apartment and was ushered down a long hallway to the kitchen. On the table there awaited not candlelit place settings but a rat's nest of materials for tinkering—garden-variety diagnostic catheters, molds, knives, strips of pliant rubber for forming balloon tips, thread and glue for affixing their seals. Poor Michaela Gruentzig certainly had no idea what a circle of chaos would befall her kitchen for the next two years.

So the work began, initially with ludicrous results. Night after night, the soiree was in fact rejoined, but the contraptions that resulted were nowhere near ready for conquering cardiovascular disease. Air leaked, glue dripped, catheters went limp in pots of boiling water. Fingers burned. Once inflated against any kind of constraining sleeve meant to approximate the resistance of a real-life artery, the balloons lost all physical integrity. Instead of expanding concentrically, the things sought the route of least resistance, bloating at the ends of the constriction in an hourglass configuration. There was little punch at the center where they were meant to force aside accretions of arterial plaque.

Maria's husband, Walter, then working as an engineer, joined the kitchen comedy. Improvising through the nights with Andreas and Michaela, the foursome frequently bellowed with laughter. Then they uncorked some more wine, sliced some Emmental cheese, and tried again. The problem was maddeningly persistent: how to devise a system whereby the balloons pushed outward from one end to the other with uniform, controlled force. Gruentzig proposed that surrounding the balloon tips with a thin reinforcing mesh of nylon might buttress their intended cylindrical shape. Where to find it? He knocked on doors, buttonholed anyone who would listen. But the pieces did not fall into place.

Marko Turina, a young cardiovascular surgeon back then, described his friend's zeal. At the time of the interview, Turina had operated all day after flying overnight back to Zürich from the U.S. But he clearly relished the chance to describe his time with Andreas Gruentzig. "He had the 'sacred fire,' as the French call it. It was what he thought about constantly. I have never seen somebody so centered on a single idea like Andreas was. Never in my life. Everyone was telling him that his idea would never work, and had been tried before, and that he was going to fail, that there were pitfalls at every turn. But the idea was consuming him all the time."

The hard-driving Gruentzig briefly turned to an alternative. Opening blocked arteries from within was but a question of mechanics, his own Dottering procedures had shown. Therefore, a straightforward solution had to lie at hand. Perhaps, he decided, the answer might be found in some kind of finely controlled arterial drill. It was easy enough to run a thin elastic wire through the hollow core of a diagnostic catheter and slip the end through the far tip of a catheter. The near end of the wire, the part meant to extend outside the body, he attached to a drilling machine. The goal was to have the catheter sit tight in the artery while the inner wire spun at 3,000 revolutions per minute, producing a whirling tip. That way, the device could bore ahead like an egg-beater through buildups of plaque, packing the material against the arterial wall.

Between April and July of 1972, sixteen tests on Bernese mountain dogs showed promise, more or less. Experimentally induced soft accretions seemed to be cleanly compacted to the sides of the animals' vessels. Equally reassuring, no arteries were perforated, and no visible scar-like reclosures were apparent during subsequent X-ray studies. However, boring through older atherosclerotic plaque proved difficult, and Gruentzig feared that his new approach could produce showers of dangerous emboli—dislodged arterial debris potentially inducing stroke—under real-life conditions. Reluctantly, he abandoned the idea.

Gruentzig nonetheless found plenty of occasions for celebration, especially when entertaining his older brother Johannes, by then an ophthalmologist practicing in Düsseldorf, and their mother Charlotta, living in Heidelberg. "I remember when his mother used to come here at Christmas, often with Johannes. They would go into the woods to get a tree. This was a childhood remembrance from when they couldn't afford to buy a tree. So they deliberately repeated it here, sort of stealing a tree and sneaking it home," Michaela recalled with a laugh while sitting in the same living room where her husband had decorated his purloined Christmas trees. "Ah ya, I remember this trick."

Certain evenings were also put aside for meals out with Michaela, followed by drinks and dancing. By all appearances, the couple remained deeply in love. Photographs of them laughing over wine glasses show that Michaela certainly had a capacity for mirth, but she also shared an inner privacy and phlegmatic depth that was equal to her husband's. They remained soul mates, friends said. But his wife's patience was being tested at every turn.

In any case, Gruentzig pushed on with his research, returning to his original balloon idea with a conviction that a soft, balloon-tipped catheter still promised a kinder approach than his arterial drill. The problem was to locate the right materials. Gruentzig happened upon a manufacturer of shoelaces, ribbons, and silk stockings, a certain Herr Schärer, owner of a little factory called Niederlenz. The shoelace man heard about Gruentzig's dreams to create balloons encased in braid and grew intrigued. You want some silken mesh to help conquer cardiovascular disease? Take as much as you want free of charge, said Schärer.

Next, Gruentzig grew even luckier. Constantly searching for insights, he walked to the local technical university and asked if they knew any experts on the structural properties of plastics. He was then pointed toward a retired professor named Heinrich Hopff, who had a passionate interest in that arcane subject. In fact, the organic chemist had written a textbook on plastics. The eyes of most young doctors in training would glaze over the moment they encountered such a dusty tome. But not Andreas. He listened closely, then explained his quest. He wanted something thin and malleable, but strong enough not to bloat out of shape when encountering resistance, a material with some kind of powerful structural integrity. Hopff suggested that an increasingly ubiquitous plastic called polyvinyl chloride (PVC) could be the answer. The professor explained that its stretchable long molecules held an impressive capacity for permanent shaping. So resolute was their molecular strength that materials constructed from this compound—for example, bottles of Coca-Cola—returned to their original heat-melded configurations, no matter what abuse they later suffered. Thin distensible sheets of the material affixed to a catheter could therefore potentially maintain a predictable sausage shape from one end to another even when expanded against powerful stresses, Hopff suggested. Gruentzig quickly located a manufacturer of coiled strands of PVC sheathing for insulating electrical wires. Another press on the charm button, and he obtained all the PVC coils he could desire.

It was back to the kitchen table. Out came the heat guns and pots of boiling water, the adhesives, air compressors, surgical thread, and knives. Hundreds of new trials-by-error proceeded, many exploring the optimum conditions for preforming the PVC balloons. Things often grew ridiculous. "Sometimes the [still cooling and setting] balloons all stuck together or onto our fingers so that we had to cut them off with a razor blade. Then the catheters would be hung up by wash pegs for drying on a cord in the kitchen," recalled Maria Schlumpf. Often, the work had to be performed under handheld magnifying glasses.

Maria, like Andreas, often dressed in a white laboratory smock for these occasions, which is not normal dinner dress in Switzerland, or anywhere else. She remembers these scenes as being altogether charming. But then again, the kitchen in question was not hers.

"Sometimes I helped," Michaela Seebrunner-Gruentzig recalled from the very next room years later. During this interview, she wore a plain dress and no makeup, her wavy brown hair cut to almost the exact short length of her youth. A wall clock methodically ticked a few feet away. The living room windows looked out toward the nearby house where James Joyce had labored in similarly relentless fashion on his masterpiece *Ulysses*, which transformed Western literature—although the price in patience that his wife Nora paid was never recorded. "But sometimes it was annoying because he was at it all the time in my kitchen with Maria in the beginning and then her husband Walter. He was close to Maria. It was she who engaged so entirely in his work, she who made it the center of her life: this whole project, his ideas, and his vision."

"Whatever Andreas said was gospel to her," observed cardiologist Bernhard Meier, who joined the University Hospital as a medical resident in the mid 1970s. "He needed support like this because he had enough enemies and she really tried to smooth the way for him. She cleaned out every obstacle she possibly could, and anyone criticizing him got a lot of opposition from her."

Maria Schlumpf to this day is an extraordinarily sensitive figure, slight with soft grayish eyes that can steel when she feels pushed. She will countenance no aspersions upon the memory of Andreas Gruentzig. In describing him, she used German words like *weitsichtig* and *ungeheuer inspiriert,* which mean "visionary" and "unbelievably inspiring." She added, "He would never ask anyone else to do something that he would or could not do himself. He was phenomenally thoughtful."

For two years, the kitchen table experiments took place nearly every night and swallowed most weekends. But somehow Michaela and Maria managed to remain friends, and have kept close through the years. Gruentzig's ready wit undoubtedly helped ease whatever tensions occurred during the endless tinkering. Sometimes, he cooked dinners for the little group, mostly Italian food, including pizzas with exotic toppings.

"It was great. We had lots of fun," recalled Maria Schlumpf, smiling wistfully as she flicked through album upon album of treasured photographs.

Wilhelm Rutishauser, the tall, angular-faced chief of cardiology, had heard all about Gruentzig's work by then, and knew the enterprising young doctor was determined to find a new treatment for not just leg but coronary arteries. In October of 1973, he invited Gruentzig to join him as a resident-in-training in the cardiology department. The stern chief of internal medicine, Hans Peter Krayenbühl, who nursed a visceral aversion to Gruentzig from the beginning, howled that the East German had no proper credentials to make such a leap. An outright nemesis in the years to follow, Krayenbühl viewed the appointment as a travesty against proper Swiss order. What was the supposed cardiology fellow's background? Angiology, internal medicine, statistics, vascular medicine, and on and on—with a specialty in fooling with balloons? And he had never completed board certification in a single discipline, Krayenbühl correctly observed.

Fortunately, Rutishauser took a larger view, having been impressed by Gruentzig's earliest Dottering. "I do not exaggerate—they [patients] came in crippled over and walked out without pain. Knowing that the disease runs from the neck to the big toes, I thought, 'Why should this kind of procedure not work in the coronary arteries?' That was the reason I thought he must pursue his work in cardiology. Obviously, people die from disease in their heart, not their legs." Gruentzig, for his part, had obtained the perfect position for his stretch run toward the heart. Leapfrogging past normal channels and constraints, he had landed exactly where he wanted. Better still, he was now equipped with the wherewithal to realize his vision. So what if Rutishauser demanded that his young assistants start work at 7:15 in the morning and continue for another twelve hours each day? There were always nights and weekends, after all.

Gruentzig pushed onward. His techniques for fabricating balloons steadily improved, so that he soon marveled at Professor Hopff's incredibly resolute PVC material. Even when expanded at pressures of five to eight times the earth's atmosphere (called "bars") in the teeth of squeezing resistance, the kitchen balloons rarely burst, nor did their uniformly cylindrical sausage shape waver. The flabby end-bulging had ceased, and there was no further need for an outer reinforcing sleeve.

"I suddenly realized that the strength of this material was so great that the silk mesh was not necessary. This was a great breakthrough and enabled me to reduce the size of the catheter," he explained in his monograph.

"Suddenly" was an apt description. The night Gruentzig thought he had finally formed the perfect balloon, he phoned Maria Schlumpf at three

A.M. "I must build a catheter right now!" he pronounced, asking her to come to help him at once. With fine thread and epoxy, they mounted the prototype balloons onto the tips of diagnostic catheters. Under these segments, they pricked side holes to divert injections of contrast medium and water from the device's inner channel. That way, the mixture not only forced the balloons to expand from a remote control point but also allowed the operator to guide the procedure by X-ray.

Convinced that he had achieved a workable design, Gruentzig began testing the devices in mechanical models of atherosclerosis that he himself designed, as well as in animals and diseased leg arteries harvested from cadavers. Exactingly thorough, he made certain that the devices worked with fail-safe consistency before considering using them in human beings, no matter how ill. Finally, on February 12, 1974, the moment of truth arrived. A sixty-seven-year-old man by the name of Ott appeared in the Polyclinic, incapacitated by walking pain. The cause, Gruentzig could plainly see, was a short, tight narrowing or "stenosis" in his major iliac leg vessel—by all appearances, the perfect lesion for trying the first balloon dilatation with his device. Bollinger remembered an impressive tenderness that Gruentzig brought to this individual. "In his kind and caring way, Andreas explained the therapeutic principle to the patient during a full hour and obtained his informed consent, long before ethical committees existed at the University Hospital."

Scant documentation remains from this first balloon angioplasty procedure, for no one judged it to be particularly historic at that time. Yet it was. Gradually, under X-ray guidance, Gruentzig expanded his catheter balloon, watching every millimeter of the unfolding procedure on the fluoroscopy monitor with rapt concentration. Within seconds, the diseased arterial channel widened splendidly, and ultrasound measurements at the ankle confirmed that the circulation to the lower leg had begun to course freely. Soon, Herr Ott was strolling the hospital corridors, reporting that his walking pain had altogether vanished.

This success only increased Gruentzig's drive. Whenever an appropriate patient materialized—a great number of them diabetics on the verge of amputation whom surgeons had no interest in handling—he kept applying his new technique. Every case required hours of preliminary work to handcraft a suitable angioplasty catheter. But the results justified the effort. More than eighty percent of the procedures showed measurable improvements in blood flow afterward and the patients' subsequent freedom from excruciating walking pain spoke for itself. At a nearby private clinic, two collaborating doctors named Pierre Levis and Andre Wirtz enthusiastically took to employing the procedure as well.

Though still relegated to a cramped basement office in the hospital, Gruentzig sensed that he was closing in on his ultimate goal. In March of 1974, he gained not only security but a modest increase in salary and some access to the hospital's experimental laboratories, thanks to the intercession of his cardiology chief Wilhelm Rutishauser. Again, Krayenbühl protested loudly. With hard eyes wreathed in thick glasses, the stern, judgmental individual made his contempt for the balloon nonsense known. But there was no stopping Gruentzig now. Under the stewardship of helpful mentors, he learned how to do diagnostic X-ray studies of the coronary arteries, rather than just the legs. A second promotion, to a full staff doctor position termed an Oberarzt—the last he would receive in Zürich—followed within months.

Gruentzig, never modest to begin with, knew his star was rising. At departmental parties, he played his natural charisma to the hilt, surrounding himself with young women and enjoying the wine. On the group's outings to the Alps, he tried to prove that he was the fastest skier, a born downhill racer, despite his relatively scant experience. Let's go for it, said Gruentzig. To his chagrin, a particular young Austrian, born in the mountains, usually won. The evenings were another matter. When the music and dancing started, there was no question as to who captured the most attention. Photographs from that era show Gruentzig mugging at the camera, eyes gleaming in his Omar Sharif way.

Christmas in Switzerland comes with certain uproarious celebrations. One of these involves a St. Nicholas impersonator arriving with a fictitious list of every friend's past sins. Snow flies and the office-party pantomimes begin. Helping the cause in the normally staid University Hospital was a storehouse of cognac kept on hand for recovering patients. Gruentzig and friends charmed the keys off the nurse in control. Then the fun began: the reading of the sin lists, the knocking back of drink. Before long, the rambunctious group was standing on chairs, delivering absurd speeches. The chairs began to crack into bits. More cognac might just fix that, the "thinking" went. Eventually, the bleary-eyed cardiology team decided something must be done to hide the evidence of their bedlam. So the dismembered chairs were delivered to the highest bridge spanning the Limmat River, which flows clean and strong through the center of the city and is surrounded by mighty financial institutions and lavish hotels. The steadfast fixers of the heart, the gatekeepers of human health, looked around, eyed their ruined hospital furniture, and chucked the evidence of their mischief straight into the river.

Gruentzig clearly savored such moments of release. But for all his elation, he knew his technology remained crude. Conventional diagnostic catheters had a set design that he knew would not pass muster with his goals. He therefore constantly fiddled with refinements, such as creating a second channel to allow him to both guide and manipulate his procedures forward in a single pass. Finding a way to inflate a balloon in the midst of an artery without blindsiding and potentially damaging the artery was the greatest challenge.

He frequently sought advice from his friend Eberhard Zeitler. The realization of his dream, they decided, clearly required expert engineering. Together, Zeitler and Gruentzig approached several major German medical-device manufacturers, but their entreaties for assistance fell on deaf ears. Here, the industrialists proclaimed, lay an invitation to throw good money after bad, considering the well-known limitations of Charles Dotter's notions of arterial salvation. Of course, Gruentzig could sweeten the proposition himself by contributing 50,000 deutsche marks for his desired refinements, it was suggested. But the East German refugee did not have such resources, and his collateral, a rented flat and a Vespa motor scooter, was not impressive. A disgusted Zeitler suggested that Gruentzig might be better off exploring contacts in Switzerland, where there was more affinity for perfecting small things.

Where to turn next? Gruentzig talked to anyone who would listen, phoning and knocking on doors. He realized that he must find a way to introduce a second inner channel to his own prototype catheters so that the blood could flow safely forward as the balloon tips expanded. That crucial improvement would allow multiple tasks to be handled at once. If one channel was dedicated solely to inflating the balloon tip, then compressed air could be fed from a foot-controlled pump to do this crucial job. The second innermost channel could work all the while like a conventional diagnostic catheter, with sieve-like openings fore and aft of the balloon allowing both X-ray contrast fluid and naturally flowing blood to circulate freely.

But such a system promised to be hellishly more intricate than any catheter Gruentzig had yet jerry-rigged. So he sought help. This time, the answers came from a brilliant young engineer named Schmid who worked for a small Swiss manufacturer of radiological needles. Gruentzig arranged to chat with the man's boss, Hugo Schneider, to see if he would vouchsafe some staff time for the project. A thoroughly practical sort, Schneider was lukewarm, refusing to finance more than some dabbling. Schmid, however, had no such inoculation against Gruentzig's eloquence. The engineer soon poured his evenings into the quest, free of charge. Welcome to the spirit of the kitchen table, then.

Before long, Schmid devised some ingenious answers to the technical challenges Gruentzig outlined. The biggest was to construct a miniature scoring device, like a tiny carpenter's plane, to inscribe a longitudinal groove on a conventional angiography catheter's outer surface. Then he slipped a long PVC tube over this inner catheter and set to exploring various refinements. It was only a matter of time before Gruentzig was presented with a prototype of his two working channels, one to control the balloon inflations, and the other to provide fail-safe diagnostic functions. When Schneider admonished his employee that the project was intruding on his proper responsibilities, Gruentzig gave Schmid 200 Swiss francs to work at night, sometimes on his kitchen table. The engineer said he would pass it on to his boss to buy peace, and somehow the delicate collaboration continued.

By this point, Gruentzig more or less thought he had it right, and on January 23, 1975, he employed his new system on a patient with a short, circulation-crippling stenosis in the iliac artery of his upper leg. The blockage melted away. The lonely inventor could taste the future.

Chapter 7

In 1975 Andreas Gruentzig behaved like a man who had caught the tail of a dream in his hands. He worked nonstop, racing the few blocks back and forth to the hospital on his motor scooter at alarming speeds. Gruentzig savored life on the edge—the ever-cautious and methodical scientist in him vied with a freewheeling alter ego in thrall to compulsion and even danger.

Visiting a small private airport outside Zürich, he looked admiringly at the winged metaphors for defying human limitations, remembering his father's career as a pilot. Intrigued, he signed up for flight lessons, at first in a Piper Cub, then in a Cessna. Winging over the silvery expanse of the Zürichsee, with the ethereal Alps in the distance, proved mesmerizing. Tranquillity beckoned at the edge of the clouds. Up there, Gruentzig could leave behind all his whispering rivals.

The very prospect of flight made his wife nervous, but Andreas scoffed at Michaela's trepidation. Once he had his license, he would fly her solo over the Alps, darting through mountain passes, scudding to a stop at the airstrip outside Salzburg, close to Mikki's family home across the border. Photographs show Gruentzig standing up from the cockpit on the runway, grinning with self-confidence. He had come a long way from his tremulous break from East Berlin through Checkpoint Charlie.

Still, he sometimes seemed inscrutable. Colleagues puzzled at what drove him to fuss night after night at the tinkering that earned him nothing but opprobrium and a seemingly penitential basement office. And now here he was whirling about in the skies. One day, his chief of cardiology called Andreas aside. Why embrace such risks with all that you have going for you? To this, Andreas smiled, "I like flying because it confirms that I have no fear."

On the ground, everyone knew where Gruentzig said he was heading—inexorably to the heart. Constantly refining his balloon catheters, he achieved ever-increasing precision and compiled exhaustive evidence concerning procedures in the legs. As the months wore on, he employed his sophisticated new double-channel device to tackle increasingly complicated arterial blockages, sometimes in the narrow iliac branches deep below the knees. What he really sought was confirmation that the system, if finely miniaturized, could be safely deployed within the pulsing confines of the coronary arteries. Always at his side stood Maria Schlumpf, meticulously recording the results at one day and again at three months after each procedure, while committing herself to tracking these patients' every turn of health for years. Gruentzig understood that this exacting methodology would be crucial to convincing the world that he stood on the verge of a historic breakthrough.

As he pushed forward, the young doctor tried to anticipate every conceivable objection from his critics. Even if most patients claimed that the balloon procedure relieved their pain, and more than 80 percent experienced measurable improvements in their blood flow afterward, skeptics would inevitably demand a convincing physiological explanation as to what had been achieved. Gruentzig therefore asked the hospital's pathologist, Hans Jörg Leu, to conduct rigorous histological studies of the arteries of a series of animals that had undergone balloon angioplasty. Occasionally, some of his early patients succumbed from one cause or another, even though none ever died as a direct result of his own procedures. The pathologist examined their postmortem arteries as well. The results repeatedly indicated that their atherosclerotic bottlenecks had been indeed neatly compressed—"like footprints upon snow," as Charles Dotter once eloquently put it—without injuring the arterial wall.

Gruentzig was by now overwhelmed. His normal professional responsibilities soaked up twelve hours and more every day, and the balloon procedures, with their demands for building catheters into the night, exhausted nearly every remaining waking moment. His free time evaporated. More and more patients sought out the hoped-for quick fix to their leg pain or threatened amputations. Yet Gruentzig kept insisting that his procedure must be employed with extreme selectivity, to avoid inflicting injury or ruining its credibility.

Marko Turina, the young surgeon, was bowled over by his colleague's caution. "Andreas was extremely careful with patients. The injunction to 'do no harm' was paramount for him. He was ethically and morally beyond reproof. He discussed every single case with his colleagues, and if anything went wrong, he only wanted to know how to do it better next time."

But there were only so many hours in the day. Gruentzig knew he needed help, so he enlisted his friend Felix Mahler, just back from a fellowship in the United States, to begin tackling some of the more routine angioplasty cases. He might as well have invited the man to step on hot coals, so frequently did Gruentzig whipsaw between encouragement and fretful disapproval whenever he let his new protégé try his hand. A colleague recalled, "When Mahler's first procedures weren't completed quickly enough, Gruentzig would push forward and say, 'My dear Felix, I'll take over this.' He was very keen on getting cases done quickly and properly, and if the pupil wasn't as skilled as himself, he would take the catheter out of his hands and finish the job. Sometimes he would also say, 'You can't do this case. It's too difficult for you. I need to do it myself.' "

Mahler winced, but carried on. To any misstep, Gruentzig would snap: "*Das nicht passiert* [that must not happen]!" Mahler is now chief of angiology at the University Hospital of Bern. His hair has gone white, but his blue eyes glinted as he recalled his time of testing back in 1975. "Friendship or not, that was his way. He was quite heartless in that. But he was right, because if you have a new method which is very controversial, complications can kill it. People were watching him, and were envious and actually waiting for a catastrophe."

A young American radiologist named David Kumpe felt otherwise. Kumpe had come to Zürich to perfect his skills in imaging the brain while working with a celebrated Turkish neurosurgeon there. A freewheeler himself, the American had joked to his wife that he wanted to advance his studies "in some place where I can learn German and drink a lot of beer." In June of 1975, he walked down the corridor and looked into the catheterization laboratory as Gruentzig set to work. Kumpe was stunned and thrust out his hand with an American's buoyant affability, saying, "Hi, I'm Dave Kumpe, and I have never been in a place that does angioplasty, and would like to find out about it!"

In a flash, Gruentzig enlisted another recruit. Soon, Kumpe would be scrubbing in as an assistant. The experience changed his life, for this particular protégé would eventually pioneer elaborations of the Gruentzig technique into the no-man's land of the brain. He remembers his new associate's provocative bearing well—especially the way Gruentzig managed to set himself apart by shaping and slanting his scrub cap so that it sat like a rakish beret and by wearing clogs as he worked. A kind of "Cool Acres" Cooley had come to Zürich.

The visiting radiologist was quickly impressed by Gruentzig's dedication, especially after finding him one Saturday performing a second angioplasty on a patient whose leg vessel had closed up following the previous day's procedure. "He had an innate sense of what to do when the whole world was crashing around him," Kumpe observed. "He was better at getting out of trouble than anybody I ever saw. Back then, he had to figure out what to do in all these circumstances for which there were no answers. That was very difficult. That is why he later said, 'If I had an enemy, I would teach him angioplasty.' "

Kumpe sensed that Gruentzig was forever an outcast. An influential doctor in the hospital turned to him and expressed his disgust in no uncertain terms: "Gruentzig is an East German, and East Germans are hungry."

With evidence of the balloon procedure's success in the "peripheral" vessels of the legs growing compelling, Gruentzig pressed harder for outside help in fabricating his catheters. More rejections followed from companies in his native Germany.

So the inventor mined local connections. There was Hugo Schneider, the owner of the four-man radiological needle and dental tool shop where he had found the engineer named Schmid. Schneider's business was housed in a modest two-room ground-floor unit on a back street near Gruentzig's own flat. His reluctance thawed as the demand for balloon procedures spiraled, and he allowed a moody technician to look into the weird new therapeutic idea. The so-called production line was not the smoothest, however, what with the fastidious Gruentzig still insisting on hand-tooling important parts of each balloon device himself. Schneider could kick himself now, since his angioplasty-balloon manufacturing business would eventually be gobbled up for nearly $2.1 billion, long after he himself had bailed out.

The Swiss manufacturer did, however, make a few nimble moves. One was to hire, in May 1975, a quick-witted young man named Werner Niederhauser, who had only recently completed a mechanic's course at the local technical school. "I had no medical experience at all," recalled Niederhauser. "I knew catheters were tubes, and that was about it."

The boss had just let the independent-minded Schmid go, and Niederhauser the neophyte was recruited as a replacement. His primary task was supposed to be fabricating radiological needles. But Andreas Gruentzig soon found a way into the new employee's ear, and showed him how to build his balloon catheters with exacting precision. Ignoring the man's employer, Gruentzig took to directly phoning his "requests" to the catheter builders. "It was very convenient for him just to ring up and say, 'I need another catheter,' " Niederhauser recollected in a Zürich interview a decade ago. "We would make one, and two hours later walk it up to the hospital and hand it to him. He was an exciting guy, but demanding. He could ring you anytime, maybe Saturday night at ten o'clock, and say, 'Look, I need some catheters for Monday morning.' He expected you to come in and make these catheters in the next twenty-four hours so that they would be ready for his first patient on Monday."

Niederhauser drew his breath, as if marveling at the curiosities he once endured. "I liked working with him nonetheless. He was not one of these doctors with a lot of unrealistic fantasy demands. Whatever you did, he understood. You couldn't tell him any bullshit. He recognized straightaway if something was not right, even if it was very technical, because he understood every aspect of the catheter, having built them all himself.

"He knew how to be very motivating. He was what we Swiss call 'sugar bread and whip'—*Zücherbrot und Peitsche.* He would want something by tomorrow, and we'd say, 'Stick it. You can make your own catheter!' But then he would come up to the Schneider office with chocolates or a bottle of beer and in this way get what he wanted. He never forgot the people who did something for him."

If Gruentzig could be sweet, he could also be shrewd, and he well grasped that he might be on to something beyond the capabilities of the little machine shop buried in Zürich's sleepy Clausius Strasse. He resumed trolling for connections with larger industrial fish. If the Germans had snubbed him, why not try the Americans? An obvious choice was Bill Cook's rapidly growing company, so intertwined with the work of Charles Dotter. A contact was made, and in April of 1975 a Danish emissary from Cook Europe named Christopher Simonsgaard found his way to Zürich. Business was conducted in the European style, over dinner, with the Dane's girlfriend sharing the repast. Gruentzig flexed the evidence of his technique's growing promise. Discussions unfolded on ways in which it could be further improved.

By now, word of Gruentzig's novel work was rippling through the world of radiology. So Simonsgaard received the signal "Go" from headquarters and dispatched a letter vouchsafing Cook Europe's desire to produce 2,000 balloon catheters for use in the legs, not one or two on a Saturday night and Sunday afternoon. The agreement would feed 50,000 Swiss francs to Gruentzig's "research fund," with provisions for later payments. Gruentzig's meticulousness did not extend to legal matters, however. "If differences of opinion between the parties should arise, which is conceivable with a change of directorship for example, then a neutral person, an attorney for example, should be given access to the company books. A Swiss trial venue is to be provided," he vaguely stated in a letter of closure.

Simonsgaard shipped a prototype of one marginally refined device. Gruentzig responded with ideas for yet further embellishments, outlined in a detailed drawing without a single patent itemization. In time, he would pay for such naïveté. But Gruentzig was no lamb. Somehow, he quickly forged a visceral dislike for the know-it-all Dane.

Already, Gruentzig had made contact with another American company, a Massachusetts firm called Medi-Tech. This enterprise had started in a former church rectory's basement and was now making its name as a manufacturer of finely engineered radiological devices. The founder, Itzhak Bentov, was a brilliant Czechoslovakian Jew who lost his parents to the Nazi gas chambers. Despite little formal education, Bentov became a wizard at forming plastics, and along the way mastered a speaking knowledge of eleven languages, from German to Arabic to Hebrew. Insatiably curious, he gathered patents on everything from diet spaghetti—later christened "Slenderoni" by the Prince Macaroni firm—to automobile brake shoes, the first disposable hypodermic needle, EKG electrodes, and pacemaker leads. Give the man a problem, and he could solve it. His restless mind worked large and he was even said to possess a miraculous healing touch. In late 1974 he wrote the Maharishi Mahesh Yogi (famous for expanding the consciousness of John Lennon and fellow Beatles) an eleven-page letter describing his explorations of a kind of "transcendometer" intended to track the corporeal and psychic transformations of those undergoing transcendental meditation. What he wanted to explore was "oscillating circuits" in the aorta and various parts of the brain, along with "a pulsating magnetic field" around the head.

The letter, replete with intricate drawings, described vibrating wavelengths in the aorta of meditators' hearts; the brain's response to changing sound frequencies during chanting; aura measurements; drawings of magnetic fields circulating around human heads; and an elaborate theory about the connections of all this psychic energy to the magnetosphere, plasmosphere, and ionosphere surrounding our spinning planet.

> When we stand on the earth, we are in a smooth-layered sea of electrostatic potential, with a value of about 200 volts/meter. Fig. 18. This can be looked at as a very stiff jelly. When our body vibrates rhythmically, it sets up waves in the jelly. We move up and down about 7 times a second, and this is exactly the time required for a pulse to go around the earth (1/7th sec.).

Bentov would die in a fiery Los Angeles commercial airline crash in 1979 while en route to a talk about the transcendent power of meditation. But his medical device business was already being run by a partner named John Abele. Destined to become one of the wealthiest men in New England, Abele was not your average cup of tea, either, having undergone 1,200 injections of penicillin as small child to cure a life-threatening staph infection that had worked into his bones. A seductive conversationalist with a mind steeped in an unusual synthesis of physics and philosophy, he became intrigued by the notion that catheters could be snaked into the heart's arteries to dislodge sudden, heart attack–inducing clots, as opposed to the entrenched plaque Gruentzig wanted to clear aside. That concept had been nursed forward with a nearby doctor who would later become seminal to Gruentzig's quest.

For all their enigmatic qualities, Abele and Bentov assembled some keenly practical expertise. Their biggest achievement was to create highly "steerable" diagnostic catheters that could be snaked more reliably into place than most competitive devices. So Gruentzig invited Abele to visit Zürich in the spring of 1975. He was instantly fascinated. Medi-Tech offered Gruentzig a trip to Massachusetts, his first ever to the United States. On a late June afternoon, Abele met his wide-eyed guest at Boston's Logan Airport and brought him to his still fledgling Watertown facility. They pored over technological issues for a couple of days, and returned to Abele's home at night. There, Gruentzig bridled about the wasteful American lifestyle. "He criticized me for using paper cups . . . He would not use our dishwasher, he just rinsed the plates in the sink and he did that with the minimum amount of water," Abele recalled.

From Massachusetts, Gruentzig traveled with his host and his wife to the resort island of Martha's Vineyard, seven miles off Cape Cod. Windsurfing having become the local rage, Abele asked Gruentzig if he wanted to have a go. The result was akin to his first bombing down the ski slopes on the Kitzsteinhorn. Most people take days to get the hang of windsurfing. Not Andreas. Within about half an hour, he was flying over the waves. "The winds were about twenty knots at the time, so this was incredible," Abele recalled from his chief executive's office of what has become a $6.3 billion-a-year multinational medical device conglomerate called Boston Scientific.

For the time being, Gruentzig's life back in Switzerland remained humble. Now that there was a little money in his pockets, he began scouring the countryside in a newly acquired Volkswagen Beetle in search of a refuge from his weekday pressures. He and Mikki were hoping to have a child. The long fjord of the Zürichsee is split in two by an isthmus, and the territory around the upper reaches casts it own more reasonably priced spell. There in Canton Schwyz, the mountains gain heft and the accents thicken so much that the native German remains nearly medieval.

Eventually, the Gruentzigs bought a little home high on a lane twisting into mountain pastures. There was nothing fancy about that bend called Eggli, where the goats and cows still have bells around their necks to inform shepherds of their straying. Number 11 Eggli sits in a cluster of nondescript cement block houses with a ruined Alpine cable car lying inexplicably in a nearby ditch. By Swiss standards, the enclave is akin to a trailer park. But to Andreas and a soon pregnant Mikki, their "chalet" became an instant refuge. Here the medical pioneer could uncork his Friday night wine, play the flute, and sing along with whatever guests he coaxed into the hills. Now and then, Andreas would persuade his brother Johannes to visit with his young family from Düsseldorf, sometimes with their mother in tow. The silence was total as the sun set. A few miles down the valley, they looked toward the nearest commercial center of Galgenen, a name derived from *galge,* which means "gallows"—it being the local hanging depot during the Reformation wars between the Catholics and Calvinists.

As autumn unfolded, Gruentzig entertained occasional foreign doctors eager to learn about his novel therapy—some from Germany, and a few from England and the U.S. This did not sit well with jealous colleagues. The young medical resident Bernhard Meier joined the angiology division. He quickly noticed the opposing poles crackling around his colleague Andreas. "Some people regarded him as a kind of charlatan. They might acknowledge that he would get lucky sometimes and the patients would improve after his procedures, but they said putting a balloon in an artery is a stupid thing and that he was crazy."

But Gruentzig could not be shaken from his goal. With the obliging surgeon Marko Turina, he set to devising experiments with which to finally explore the balloon procedure's potential in the coronary arteries. Turina set up a methodology by which dogs' coronary arteries were cinched and sutured half-closed to create an approximation of the choking coronary disease that Gruentzig so much wanted to solve. Some of the first attempts to open these constrictions failed abysmally, but the pair pressed on until they got the model right, then began demonstrating just how the technique might work in the human heart.

To circumvent prying eyes within the hospital, Andreas transported his anesthetized test animals back and forth to the research laboratory under the cover of draped hospital trolleys, a subterfuge aimed at preempting fresh rounds of carping from higher-ups. Before long, the cardiology staff set to using the same trick to sneak a black jazz pianist out of the ward for nighttime revelry.

Gruentzig let his hair down again at a departmental party. There, he freed his boss's pet parrot from its cage and strutted before the ladies with the bird perched on his shoulder. But the inventor's days were mostly spent in dead earnest. He labored constantly at better understanding and minimizing whatever complications occurred in his steadily growing numbers of procedures in human leg vessels.

Christmas came with another visit from Andreas's mother, his brother Johannes, and his young family. A fresh tree was purloined, toasts were made, and presents exchanged with Michaela and the rest of the family. Within a few weeks, Mikki would be announcing that another relative was on the way. Life was about to change dramatically. The kitchen would no longer be a place for further balloon building into the night. So Gruentzig pressed his contacts from industry.

He fired fresh letters off to Abele and the Cook company, to see which outfit might best serve his needs. Clearly growing impatient with Schneider's fly-by-the-seat-of-the-pants operation, he had still not filed a solid patent application for his invention. But he had become adept at playing industry contacts against one another, and floated the notion that a worldwide franchise for his devices might be available to whomever treated him best. Meanwhile, he arranged to give a series of lectures in the United States to promulgate his breakthrough.

At the beginning of July of 1976, Gruentzig delivered a talk at Harvard's

Brigham Hospital—which was arguably the most esteemed forum he could have dreamed of. But brows were furrowed. "It was an incredibly successful speech to a hostile audience," John Abele recalled. "It was very, very low-key. It was a classic, because he did the thing that is very important in dealing with a hostile audience—you actually lead the audience to find a hole in your talk. You lead them to that hole. The hole was, when asked how he could know that the technique would work in the heart, to acknowledge that he had done only peripheral angioplasty and had only animal data on the heart so far. But he put his talk together in a way that was impossible to attack."

When the presentation was done, waves of applause rolled forward from the back of the room and esteemed Harvard doctors, the very pillars of cardiology, beamed with admiration. Gruentzig stood at the lectern, savoring the moment of triumph he had waited for all his life. The child of the rubble had just become a player on the world's stage.

But his trials by fire were far from over. In August, the kindly Rutishauser departed for a new job in Geneva, leaving his understudy to report to a new chief—who was none other than Krayenbühl, the censorious, stifling figure reigning over the carefully stratified division of internal medicine. The pedantic senior doctor despised the thought that cardiologists should leap forward into actually performing therapeutic procedures on the heart. And with balloons inflated by some smart-ass German? The idea was enough to make Krayenbühl seethe. Exquisitely tedious measurements of the volumes of blood being pumped through the heart, the sheer glory of hemodynamics—that was his passion.

If words were exchanged before, teeth were clenched now. The tension was not unlike that formerly animating the dialogue between Herr Geheimrat Sauerbruch and Werner Forssmann. To Krayenbühl, Gruentzig *was* in fact the second coming of Forssmann. The more things change, the more they stay the same. Zeitler stopped by and encountered something less than open-armed celebration. "Krayenbühl showed me a wall chart and said, 'He doesn't know how to measure the ventricular volume!' He proceeded to give me a lesson on the method they used to quantify the left ventricular volume of the heart and said, 'I have taught this to him now!' Petty."

By day, Gruentzig did whatever his new taskmaster demanded. But the nights were his own. Even with the new baby, named Sonja Meret, arriving in September, he worked long into the nights, constantly recalibrating the results of his first 220 procedures in the legs, while pondering every nuance of the technique's performance in evolving dog models of human coronary artery disease. Much of this toil was drudgery, but he kept thinking large. "The legs were only my testing ground. From the beginning, I had the heart in mind," Gruentzig later affirmed. "It was my cardiology colleagues who were so strongly opposed. They 'knew' dilatation couldn't work. They 'knew' it couldn't work because it wasn't their idea."

But "show time" soon came. In early November, he departed for Miami to the annual meeting of the American Heart Association, a convention hall extravaganza that now draws 30,000 doctors and medical-industry personnel into a frenzied search for pivotal information. The great business of the convention involves hundreds of presentations of intricate analyses and clinical trials in innumerable lecture halls. Doctors with early reports that are judged "not quite ready for prime time" are consigned to first airing them cheek-by-jowl in a cavernous central exhibition hall. The visitor there confronts rows of eight-foot-high poster boards, each crammed with their separate minutia of radiological images, eye-straining charts, and theories bullet-pointed with experimental evidence in small type. For the newcomer, the experience is not unlike standing in the middle of New York's Grand Central Station at rush hour and attempting to slap restaurant menus into indifferent commuters' hands.

For Gruentzig, the scene was bewildering. But there was no comparable venue with which to make one's name, because the floor show attracted the attention of the absolute thought leaders of his field, along with that of the newly emerging breed of so-called "analysts"—the front men for investment banks that had become keen to funnel capital into medical-device companies.

Gruentzig's hotel beside the giant convention complex was awash in cardiologists from around the world. So were the nearby streets and restaurants—not forty or fifty of them, as Switzerland might muster for a conference in the mountains, but *seven thousand* and more. Gruentzig worked off his adrenaline, slept off his jet lag. The clock ticked, and then his time came. In the bowels of the exhibition hall, he tacked up his modest affidavit.

He was scheduled to deliver a short talk about his laborious work on dog models, documented by dry statistics, followed by a review of his first test procedures on human leg arteries. But tedium wasn't his style. He drove straight ahead and explained that all this work was but the table setting for procedures he would soon be performing in the human heart, and the possible beginning of a new age of cardiovascular therapy. Before him stood an audience of sixty or so craning faces from the world's highest citadels of medicine, almost all strange.

Among them was a Martin Kaltenbach, by now a German master in studying the heart. Another was a tall, lean, red-haired American peering from behind no-nonsense glasses. His name was Spencer King III, an Emory University cardiologist, who spoke with the soft understated drawl of a Southern gentleman. He listened politely, even if his eyes were squinched in consternation. As he wrote later, he didn't think much of what he was witnessing. He certainly had no clue that it would change his life, and that Gruentzig would soon land on his doorstep, crowned as the avatar of a new medical age.

> Andreas stood in the center of a small group. His bushy moustache and ascot telegraphed his European roots even before he began to speak. He clearly was convinced that the method would work. Armed with some knowledge of the pathology of atherosclerosis, I said, "This will never work," and we parted.

One figure in the crowd beheld the presentation with awe. His name was Richard Myler and he was that same engine of boundless optimism who had first inspired Medi-Tech's John Abele to think that catheters might just introduce a new era in the treatment of the human heart. His father had died from a sudden heart attack while playing golf a decade earlier, and Myler had a passion to understand the disease. A prodigious talker, his mind percolated with ideas that he could describe into the night, beautiful and original concepts that would make every listener's head spin. But somehow Myler never converted his imaginary notions into reality.

But there was no more willing catalyst. Abele had already tipped Myler off about Gruentzig's work, even calling him from Zürich to insist that the two should meet. So now the effervescent cardiologist, newly moved to a small hospital south of San Francisco, listened keenly. Up and down nearby corridors droned discourses on fine-tuning dosages of this and measurements of that. But Myler knew he was hearing something much more profound. He heard the strains of a higher music as Gruentzig spoke, and he stood transported.

"My god, it's a balloon!" he blurted to his wife, Sharon, a nurse who shared his ardor for medicine. "We're going to Zürich!"

And with that, Gruentzig had his Saint Paul.

CHAPTER 8

BY 1977, CRITICS BEGAN WARNING that the heart surgeons' vaunted achievements often came with a heavy price. Coronary artery bypass surgery was the biggest cash cow in their field, with approximately 80,000 procedures now being performed every year. But to what end? A controversial study at a number of United States Veterans Administration hospitals had uncovered the procedure's dark underbelly—a death rate of nearly 6 percent, an 18 percent frequency of heart attacks, and no measurable gains in long-term survival when compared to those who merely received anti-anginal drugs. That study, which was felt to be exaggerated, at least found that the supposed wonder therapy was more effective than medication in controlling angina—on the other hand, it obviously required months of recovery. A top American cardiologist named Eugene Braunwald derided bypass surgery's explosive growth as "an insidious problem," adding that "an industry is being built around this operation."

So a debate raged that continues to this day. Meanwhile, heart transplantation, formerly the crowning tiara of the surgical era, was scarcely discussed anymore. The likes of DeBakey could claim what he pleased, but the conquest of coronary artery disease was by no means complete. The time was ripe for a better solution.

After returning from Miami, Gruentzig hungered to find the right patient to test out his balloon alternative in the heart. The impediments remained formidable. His bosses were not impressed by their subordinate's brief ascension into the limelight, and steamed that no cardiologist had ever safely risked more than taking pictures in the coronary arteries. In Zürich, a cardiologist was meant to be a master of exacting calibration—tracking the heart's abnormalities with the precision of a watchmaker and medicating various conditions according to exacting algorithms. But that was supposed to be the end of the line.

Bernhard Meier was intrigued by Gruentzig's zeal to overthrow these limitations, perhaps because his own father had recently succumbed to a heart attack—at the age of fifty-one. Meier, now chief of cardiology at the Canton Hospital in the Swiss capital of Bern, recalled from his office there, "Gruentzig said, 'We just need to find the right patient,' and others, including Siegenthaler, replied, 'That's heretical . . . It's too dangerous and you shouldn't ever do it.' "

Another colleague found an agitated Krayenbühl in the hospital corridor one day, his arms flapping. "He said, 'Here now this crazy Gruentzig says he will now dilate the coronary arteries!' I said, 'Why not?' His voice rasped, like a bark or a small roar and he shook his head."

Gruentzig continued plumbing for further evidence that might satisfy the naysayers. He kept fine-tuning the procedure in coronary arteries extracted from cadavers, which sometimes produced fresh puzzles. Even his trusted collaborator Marko Turina had doubts. "I still thought his idea about the coronary arteries was crazy and might never work."

In truth, there was reason for concern. "Our fear was that the heart could suddenly fibrillate," he said in reference to the sudden wild electrical misfirings that can render a heart useless in seconds. "And secondly, that the vessel would suddenly close after or even during the procedure and an infarction would occur. The third fear was that we had no idea what would happen after blowing up the balloon."

Staff conferences sometimes grew heated, with Gruentzig's bosses arguing that his balloon procedure was a pipe dream, a cavalier recipe for disaster. But Gruentzig kept pressing his case. Hadn't he worked out the smallest details in more than 200 leg procedures? Hadn't he demonstrated his exacting care time and again? Wasn't the surgical gold standard of bypass grafting itself a walk on a tightrope? Would it have ever been started in the face of temerity? Åke Senning, the celebrated chief of cardiovascular surgery, listened and came down on Gruentzig's side. "What's to worry about? If something goes wrong, I will operate," he told the backbiters.

With March, the ides began falling into place. Martin Kaltenbach, considered the master of diagnostic cardiology in Germany, had come to Switzerland for a ski holiday. So Gruentzig, ever deft when fateful cards slipped his way, invited him to stop by his modest "chalet" in the Voralps (fore-Alps) of Canton Schwyz for a joint family get-together. There, he poured forth his charm, keen to forge a plan of action. Kaltenbach was captivated. "I was fascinated by the idea and fascinated by what he had achieved and by his personality."

The discussion deepened, and the Frankfurt cardiologist offered his own theoretical analysis of the technique's promise. Gruentzig listened intently. The fields of snow outside glistened, the contrasting warmth inside the chalet proved altogether *gemütlich*. Seizing the moment, Gruentzig suggested that it would be inspiring to work side by side with his esteemed guest. One procedure might be done in Zürich, the next in Frankfurt, then back and forth in a duet of pioneering collaboration. Such a partnership would demonstrate that the promise of balloon angioplasty was objective and repeatable, and not confined to the claims of a lone individual.

The by-no-means naive visitor realized that this arrangement would also lend Gruentzig desperately needed credibility in silencing the sniping on his hospital corridors. As Kaltenbach recalled from the quiet of his stately Tudor suburban house—once commandeered as an officer's quarters for the invading U.S. Third Army—the play for leverage could not have been clearer. "This agreement was a prerequisite for him to get started in Zürich at all."

But the German cardiologist sensed the potential for a dramatic breakthrough, and was eager to pitch in. He knew a thing or two about passion. His knowledge of the heart was won through the course of thousands of X-ray studies and so many years wearing a shielding lead apron that he now struggles with excruciating back pain. If naturally somewhat reserved, he was ineluctably attracted by Gruentzig's ardent conviction. "I witnessed in my career a truly dramatic evolution," he sighed as he recounted. "The next step clearly was a catheter-based intervention such as Andreas advocated. For me it was an absolute fascination to come closer to this disease."

A deal was cut, and March soon brought Gruentzig other welcome tidings. His mentor, Eberhard Zeitler, organized an international gathering of everyone who had ever performed angioplasty in the legs, be it by the earlier "Dottering" or the new Zürich method. On March 25, about thirty doctors booked into the Atrium Hotel in Nürnberg, located on a site where Hitler stayed when gathering for the displays of Nazi might that so frequently dazzled this erstwhile citadel of the Thousand-Year Reich. The Atrium looks out on an urban woodland where the last holdouts of the S.S. succumbed to Patton's triumphant Third Army. Zeitler, the newly appointed chief of the Nürnberg General Hospital's radiology department, discovered that a star guest was fascinated by the setting for other reasons.

The first morning of the conference, he discovered Charles Dotter in the lobby, attired in his usual turtleneck shirt and casual khaki trousers and clutching a notebook. Zeitler shook his head incredulously as he painted the scene in a much later interview. "He said, 'Can you tell me about this, I was up from four o'clock in the morning and outside in the park until eight o'clock.' He had drawings of five birds he had been observing and whose songs he had memorized and could imitate. He had written their American names beside them and asked, 'Can you tell me what they are called in German?' " Zeitler recalled that Dotter, always craving new challenges, next probed for information on climbing prospects in the Bavarian Alps.

Another figure who chuckles over that gathering is David Kumpe, only recently departed from Zürich for a position back in the United States. Six months earlier, he had presented Gruentzig's pioneering work to the Radiological Society of North America's massive annual meeting in Chicago. Keenly perusing the exhibit had been one Charles Dotter. Arriving in Nürnberg, Kumpe climbed into a taxi with so-called Crazy Charlie. Dotter produced a tiny cylindrical coil of wire mesh, about the size and shape of the springs attached to the ends of push-button pens. He casually suggested that such things would soon be implanted in diseased arteries to hold them open for years. Their impact could be revolutionary, Dotter insisted. (He was right; today, after a number of refinements, this concept, now called "stenting," has been transformed into a $10 billion-a-year industry.)

That first-ever angioplasty conference grew electric when Andreas Gruentzig strolled onto the stage. He boldly proclaimed that he was about to proceed directly into the human heart. Moving with a catlike litheness as he presented his results, Gruentzig disarmed skeptics by plainly describing every complication and worry that troubled him. But he radiated self-confidence all the while.

Watching from the audience was Richard Myler, who had vowed to explore Gruentzig's arc of exploration firsthand. The Californian doctor and his wife, Sharon, were captivated, and after a celebratory dinner, joined Andreas for the trip back to Zürich. There, they were charmed by Michaela. As new parents themselves, they hovered appreciatively over little Sonja and delighted in the emerging Zürich spring. Outdoor cafés were blossoming by the shores of the great lake and in the cobblestone squares beside their grand Hotel Bauer au Lac.

Richard Myler remembers that time as being magical. The days were filled with close observations of Gruentzig's meticulous leg procedures followed by animated discussions and leisurely dinners over the kitchen table of legendary tinkering. The weekend was savored in Eggli. There, cows with bells ringing roamed the high pastures. Promise hung in the air.

But there was also trouble. "We had been there nearly a week and I saw that the chief, Krayenbühl, barely paid attention," Myler explained. "I said, 'Professor, it is a great honor to meet you and see your splendid hospital. I think Andreas's work could be one of the greatest advances of this decade.' I remember Krayenbühl looking at me and ignoring Andreas, even though he was standing right there. Then he said, 'His importance to this department lies in running the clinic. He needs to concentrate above all on the clinic!'

"We walked out of there with Andreas gritting his teeth. I said, 'Andreas, you have to get out of here.' "

As the days passed, Myler proposed that he could offer a much kinder

A crude image of Werner Forssmann performing first heart catheterization on himself, 1929.
(Courtesy Boston Scientific)

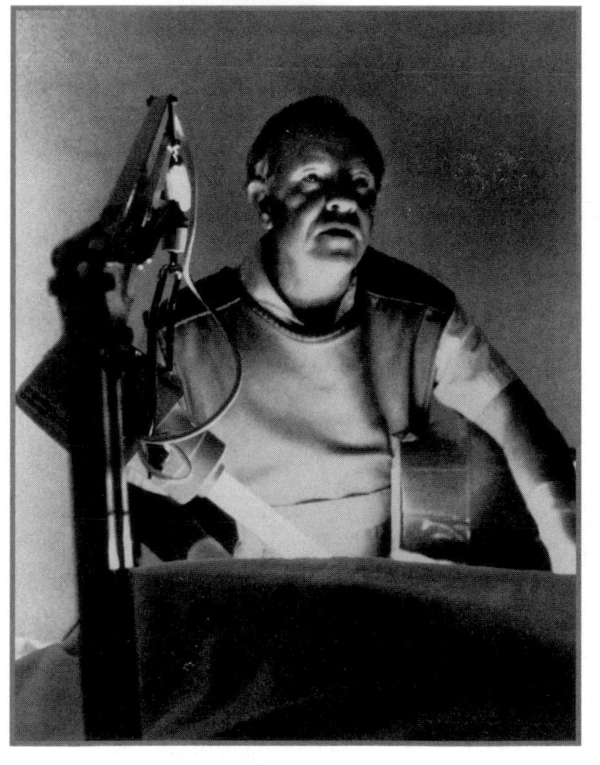

Mason Sones stares from his Cleveland Clinic X-ray pit.
(Yu Kwan Lee, courtesy Cleveland Clinic Foundation)

As famous as a moon astronaut, Christiaan Barnard is besieged by autograph seekers at the 1968 national finalist meeting of young German science students seeking government scholarships. *(Courtesy* Stiftung Jugend *forscht e. V., Germany)*

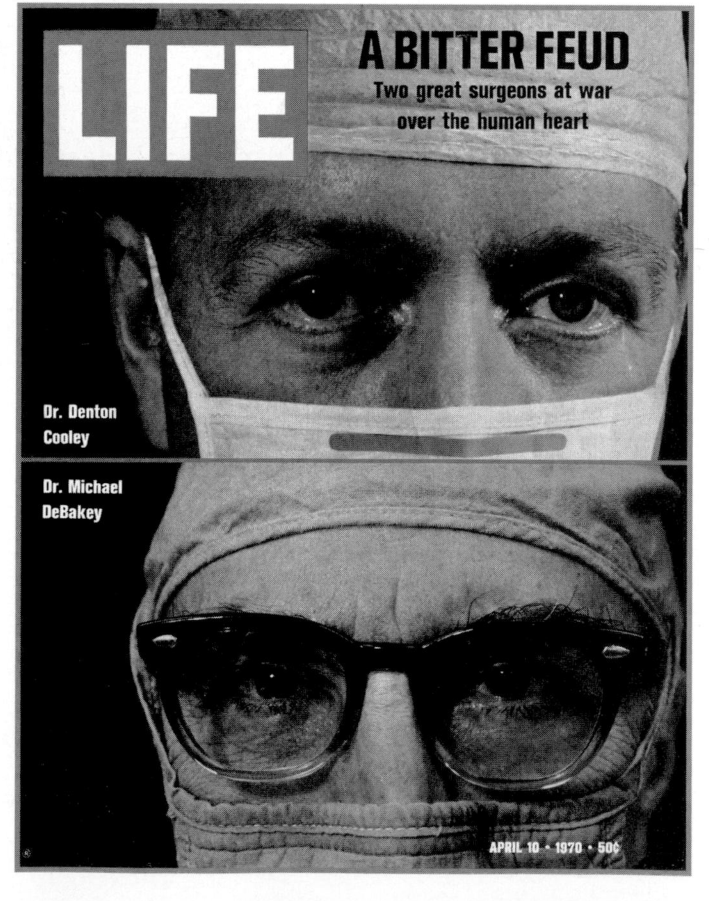

Life magazine turns its eye on the hubris of heart surgeons in an April 10, 1970, cover story. *(Reprinted with permission from Time/Life)*

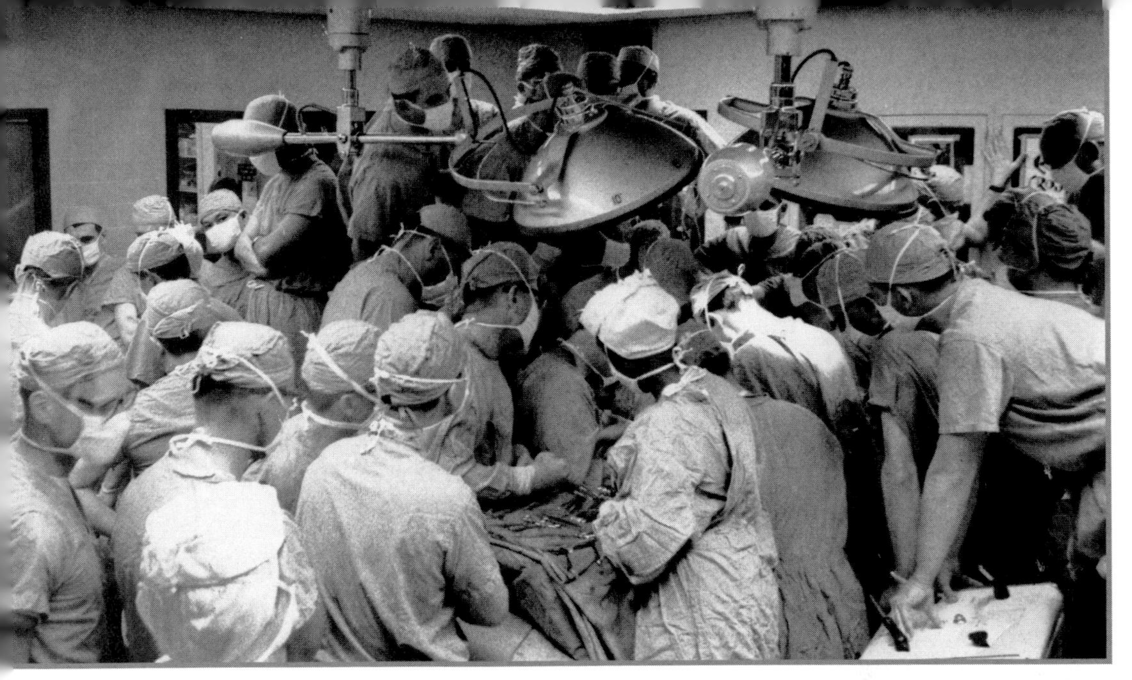

The heart gone hot. Denton Cooley performs bypass surgery in 1970 before eager spectators from seven nations. *(Courtesy Texas Heart Institute)*

Playing the "Crazy Charlie," Dotter hams it up prior to his notorious 1964 *Life* magazine photo shoot. *(Courtesy Dotter Vascular Institute)*

Ever seeking new heights, Charles Dotter prepares to scale a Rocky Mountain precipice. In time he would climb sixty-seven peaks over fourteen-thousand feet. *(Courtesy Dotter Vascular Institute)*

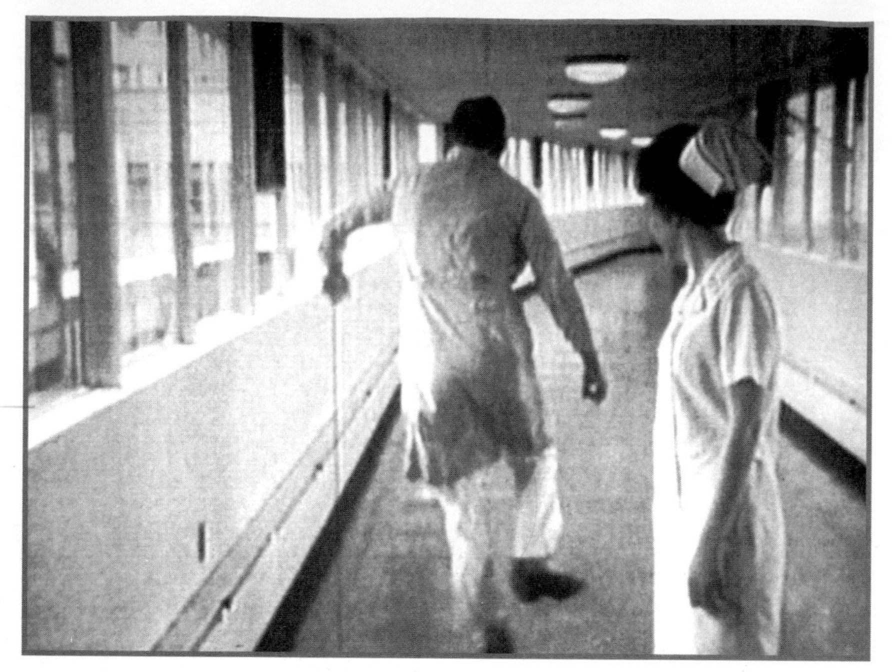

Dotter has one of his "claudication" patients demonstrate instant heel-clicking rejuvenation for his over-the-top short film. *(Courtesy Dotter Vascular Institute)*

Andreas Gruentzig on an early 1970s night out with Michaela and a colleague named Kaspar Rhyner. *(Courtesy Wilhelm Rutishauser)*

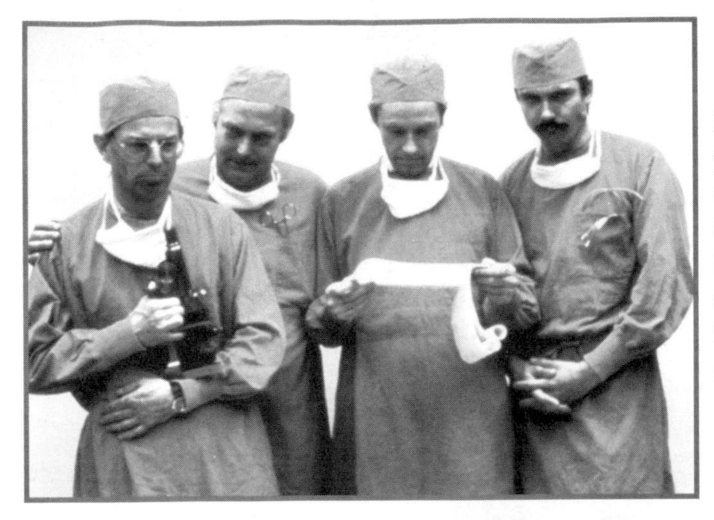

Medical men clowning in pretend-serious guise—*(right to left)* Andreas Gruentzig, Alfred Bollinger, Ürs Brunner, and Hans Jörg Leu pose in Zürich, circa 1972. *(Courtesy Gary Roubin)*

Maria Schlumpf tinkers on a balloon catheter prototype on the Gruentzigs' kitchen table. *(Courtesy Maria Schlumpf)*

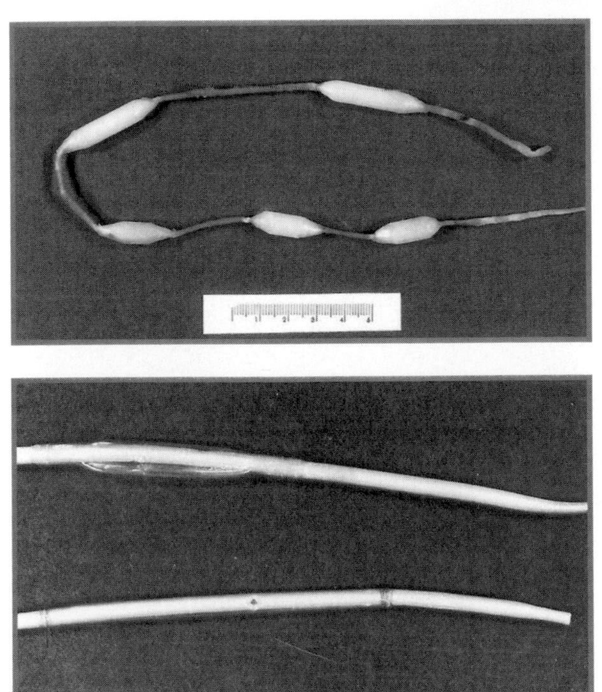

Left, top: A nightmare of early sausage shapes afflicts early attempts at forming angioplasty balloons. *(Courtesy Maria Schlumpf)*

Left, bottom: An early kitchen table balloon catheter handmade by Gruentzig and Maria Schlumpf is shown inflated *(top)*, and deflated *(below)*, as it would be advanced into an artery. The crude device with its long clumsy tip extension and single working channel is not yet fit for use in humans, and bears little resemblance to the streamlined and precisely functioning catheters that would emerge in a few years. *(Courtesy Maria Schlumpf)*

With Rube Goldberg-like origins, the early Gruentzig system required a huge compressed air tank to inflate the tiny heart balloon. *(Courtesy Maria Schlumpf)*

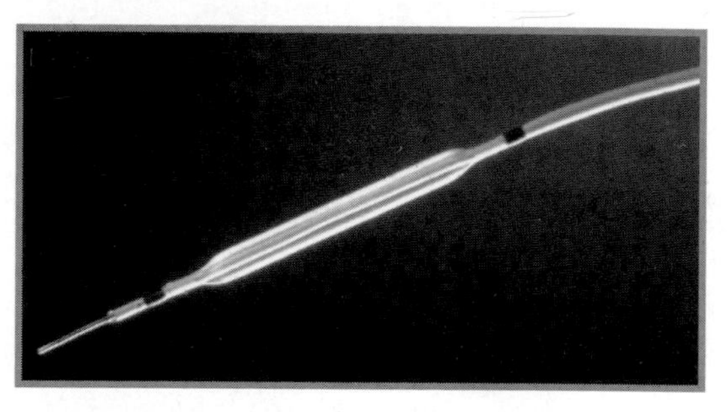

A typical streamlined modern angioplasty catheter, shown inflated as it would be when pushing aside plaque in a coronary artery. *(Courtesy Wilhelm Rutishauser)*

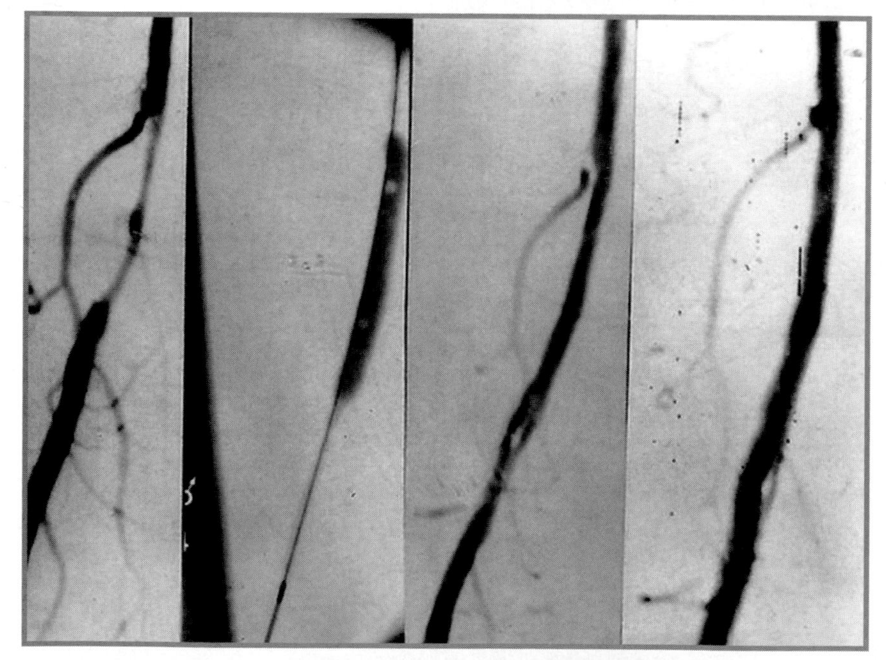

An X-ray image of a successful early balloon angioplasty leg procedure. The blood flow through the top of the artery *(left)* is withered to a filmy strand by advanced atherosclerosis. A ghost image follows *(second from left)* of the angioplasty balloon being inflated within the diseased artery's midst. *Right*, the post-procedure X-ray, shows the artery enjoying a continuous line of rejuvenating blood flow. *(Courtesy David Kumpe)*

A 1975 night out for the balloon makers—Michaela Gruentzig *(left)*, Walter Schlumpf, Maria Schlumpf, and Andreas Gruentzig. *(Courtesy Maria Schlumpf)*

Gruentzig relaxes with Michaela and Walter Schlumpf *(foreground)* over schnapps and coffee at his Eggli, Schwyz, getaway in the late 1970s. *(Courtesy Maria Schlumpf)*

With a bird in hand, a playful young Gruentzig dances with Fraulein Carli, a laboratory technician, at a 1972 party in Wilhelm Rutishauser's apartment. *(Courtesy Wilhelm Rutishauser)*

The 1977 Nürnburg meeting of the minds—Eberhard Zeitler *(left)*, Gruentzig, and Charles Dotter. *(Courtesy of Eberhard Zeitler)*

The 1978 meeting of the so-called International Dilatation Society with James Minor *(left)*, beside Richard Myler, Gruentzig, Lamberto Bentivoglio, Maria Schlumpf *(partially obscured)* and John Simpson *(front)*. *(Courtesy Gary Roubin)*

A 1979 angioplasty teaching course in Zürich shows Andreas Gruentzig seated in the first row, wearing his favored white shoes, with Maria Schlumpf beside him and David Williams seated far left. Immediately behind Williams in a dark suit in the second row is Simon Stertzer, with Richard Myler and then Wilhelm Niederhauser to his right. *(Courtesy Maria Schlumpf)*

environment for testing the procedure in his nonacademic St. Mary's Hospital back in California. There, fifty patients a week underwent coronary angiography, rather than the moribund few whose arteries were studied in conservative Zürich. Myler proffered a vision of America's eager receptivity to new ideas, its impatience with entrenched hierarchies and preset limitations. Arrange an extended visit, he suggested, and we'll test the technique on coronary arteries that have already been exposed in preparation for bypass surgery. Who knows what we may achieve! Gruentzig by now had gathered a few formidable figures around him, but here beckoned one perhaps more providential than the rest: a disciple offering unbridled enthusiasm.

No sooner had Myler departed than a disaster case presented at the University Hospital: A sixty-six-year-old patient with extensive coronary artery disease arrived on the verge of cardiogenic shock, a condition no heart surgeon would touch. Foolishly, Gruentzig agreed to make this his test ground.

Not only was the individual's cardiovascular tree thick with atherosclerosis, but the disease had also bloomed throughout his legs. Worse, the man's blood pressure grew so weak that his failing heart raced erratically in a struggle to compensate. For a trial procedure, this was a minefield. Nonetheless, Gruentzig, surgical cap at a jaunty slant, set to the task. Inevitably, trouble commenced. The man's femoral artery in the groin, the prescribed entry point for the angioplasty guide catheters, was so gnarled by disease that Gruentzig couldn't gain a purchase to advance them but a few inches. Unwilling to accept defeat, he tried an alternative approach to the heart through the left arm, but the prototype catheters refused to follow the twists of this circulatory route for which they were not designed.

"He was desperate to start, because he was ready," Turina recalled. "But Andreas never really performed dilatation." A crestfallen Gruentzig had to withdraw his catheter impotently. His sense of futility escalated when the languishing patient died two days later.

Whispers about the uselessness of his procedure echoed in the hospital corridors. Gruentzig attempted to shrug them aside. "It made me think ten times before making any claims. And I never took their criticisms personally. If I had been in the same position, I would have felt the same way. The idea was crazy. The difference between them and me was I had been doing dilatation on the legs for years. And because of that I knew how my balloon worked. I never would have tried doing the coronaries if I hadn't had all that experience. I wouldn't have had the courage."

Gruentzig resolved then and there to make certain his next patient—if he could ever get another in Zürich—would be much younger and stronger and with his coronary anatomy cherry-picked for the balloon procedure. Medical disaster cases were no place to start. But why, he wondered, had it proved impossible to manipulate his devices in such adverse conditions? Ever methodical, he kept pursuing further technological refinements.

In early May of 1977, Gruentzig flew to San Francisco with Michaela and their nine-month-old daughter Sonja, there to be greeted by his eager new American collaborator, Richard Myler. The following two weeks were spent at Myler's opulent house high on a ridge south of the city. The families again mixed easily—Gruentzig's charm made sure of that. Yet even when feted so hospitably, the visitor could not quite forsake his roots. Germans of his generation nurtured an at once sardonic and affectionate name for Americans: *Wohlstadt's Kinder,* meaning "children of the rich, untroubled state." This was an exact counterpoint to the phrase that still haunted their own past: "children of the rubble or *Trümmerskinder*." More than once, Gruentzig bristled with misgivings about the carefree California lifestyle. The houseguest obviously carried some heavy psychological baggage, a core of inner darkness eddying within the easy charisma he worked so hard to project. As Myler recalled, "He seemed almost anti-American at first . . . He thought Americans were wasteful, extravagant, spoiled, and spent too much money." To teach his hosts a lesson, Gruentzig, once again, often washed the dishes himself—in cold water.

Yet he managed to dance away from such petulant displays with his trademark flashes of warm exuberance and self-deprecating wit. It was *Zücherbrot und Peitsche*—"sugar bread and horse's whip"—again.

Sharon Myler wasn't much annoyed, having decided that Andreas was one of the most brilliant people she had ever met. "He had a very, very strong ego; and he was an artist, he painted, he played piano—he was like a Renaissance man. And the girls, you know, he had a certain magnetism." She turned her head aside for a moment, as if struggling, as many others would, to find a summary description of the man's radiance.

As the Mylers spoke, the Pacific Ocean threw its might at the headland south of Carmel, California, where they make their secluded home. Sharon could not constrain her sense of abiding mystery about Gruentzig's complexities. From evening conversations, she got the impression that Andreas's mother had imbued him with the idea that he was "a golden boy." She also sensed that he shouldered some deep implicit responsibility to look after the more fragile psyche of his older sibling Johannes, he who had first set off alone from Leipzig to spearhead the brothers' flight to freedom. Sharon, like many others, saw in Andreas a figure churning with conflicting and perhaps Oedipal traits. To her, he seemed to be burdened by the need to prove that he was the ideal son, the consummation of his adoring mother's every dream, and a redemptive stand-in for his lost father.

Back then, Richard Myler didn't much worry about psychoanalyzing his new friend. He revered Andreas's sheer zest as they worked side by side, and he grew emotional as he painted the picture. Although recently afflicted by a stroke, the retired doctor had vivid perceptions about those historic days. "We were very close friends; we were like brothers, although our backgrounds were very different. I think there were some people who thought it might be difficult for a German and a Jew to work together. It didn't mean anything to me, and it didn't mean anything to him, either. There was a sense of trust, and there should have been."

Myler, the ready enthusiast, nudged open every door that might help his visitor forward and treated Gruentzig to an openness of the American spirit, contrary to the second-guessing that haunted his progress in Zürich. The cheerful host quickly enlisted the support of St. Mary's cardiac surgeon, Elias Hanna, for a series of exploratory procedures of the exposed coronary arteries of patients whose chests had already been opened for imminent bypass surgery.

The first tests involved no balloons at all, as Myler recalled. "Gruentzig and I were actually putting little tiny catheters in them just to see if we could do it, to see if we knew where we were going." By means of such trial runs, they sampled the ease of first snaking the angioplasty balloon catheter into different arterial branches, thereby evaluating the degrees of resistance to be found in disparate challenges. The anatomy ranged from time-toughened, calcified accretions to pliant, fresh blockages. Surgeons often performed this kind of "diddling" before performing bypass surgery in order to gain a clear notion of what lay ahead. Therefore, Elias Hanna didn't fret, even when Gruentzig and Myler pushed further. The question still loomed: Would any filmy debris cling to these devices once they were withdrawn—an eventuality that would invite potentially massive complications? Lab experiments were all well and good, but the actual procedure would be useless if it dislodged showers of stroke- and heart attack–inducing particles. Myler and Gruentzig saw not a speck of emerging trouble.

Now Myler suggested that they perform actual balloon inflations in the clogged arteries that were destined to be surgically bypassed regardless. "We had short catheters and we knew where the lesions were," he explained. "We could actually see them on the surface. You could feel them. I said, 'When the surgeon opens up the artery to receive the graft, just before he places it in, we will take this little balloon catheter to see if we can dilate it.' " Four of these so-called intra-operative dilatations were performed in the next week. In each case, subsequent tests showed that the circulation was neatly reinvigorated by the balloon procedure. The question dangled: What, then, was the point of the scheduled bypass surgery?

Gruentzig was by now itching to perform a genuine procedure in the heart. Still, Myler hesitated, warning about malpractice litigation should anything go wrong.

Soon, Gruentzig returned to Zürich. There he did what was bidden, while pleading once again for access to an appropriate heart patient. Indifferent shrugs were his reward, and he remained sequestered in his subterranean office that was as cramped as a janitor's closet. By now, Gruentzig had endured six years in the Zürich wilderness, and there was no end in sight.

August came, with no family holiday on a Mediterranean beach. Andreas wanted to visit the Mylers in California again. They flew there, but no suitable patient materialized—those checking into his friend's hospital were either too ridden with advanced cardiovascular disease to be safe candidates, or averse to submitting to such an unproven procedure.

Gruentzig kept busy by presenting lectures about his work, one at nearby Stanford University. In the audience sat a true Californian "golden boy," a young cardiology fellow named John Simpson. Though he would later be accused of being a Judas to The Pioneer, he was instantly intrigued by Andreas Gruentzig. "I thought the concept could be either awesome or just bizarre," Simpson recalled. "It could really be incredibly dangerous, or could be incredibly effective. I told my wife, 'This guy's either going to revolutionize the treatment of coronary disease or he's going to jail.' "

Before long, Simpson arranged to study under Richard Myler's wing, trying to absorb everything there was to learn about Gruentzig's methods. He pleaded for an introduction, without a clue how this would change his life forever.

As things played out, Gruentzig left California forlorn and empty-handed. Back in Switzerland, his caseload piled up in a nightmare of mundane responsibilities—measure this, prescribe that, file report upon tedious report. But suddenly the kaleidoscope shifted. Within ten days, he received word that a potentially ideal patient had materialized.

A thirty-eight-year-old insurance salesman had flat-out refused to undergo heart surgery, even though he had just weathered a terrifying eruption of chest pain. The source of the problem looked to be ideally localized in a short, choking narrowing of atherosclerosis in his left anterior descending artery, the potentially "straightest shot" of all coronary avenues through which to advance a catheter. Though Swiss, "Dolf" Bachmann was one of innumerable German speakers whose parents had blessed him with the namesake of the Führer at the outbreak of World War II. Like many junior Adolfs, he later jettisoned the "A," wanting no part of that legacy. Herr Bachmann seemed robustly sentient now, and he was wide open to discussing Gruentzig's unproven alternative. The two- to three-pack-a-day smoker had come to a crossroads. At first, his bouts of angina often ceased after taking nitrogylcerine tablets. But his pain had continued to intensify until the pills had so little effect that he was swallowing up to fifteen a day. "I ultimately couldn't do anything. With the slightest exertion came pain," he recalled. The next thing he knew, Bachmann was lying on a hospital bed in Baden sweating rivers and groaning with stabbing chest pain.

Bachmann became one frightened man. When the spiraling intensity of his "unstable angina"—often a precursor to a full-blown heart attack—became apparent, he was packed into an ambulance that sped with sirens screaming to Zürich. "There were surreal impressions reeling in my head the whole while," Bachmann said. At the University Hospital, the thirty-eight-year-old underwent immediate diagnostic angiography. The findings were so ominous that urgent bypass surgery was advised. Meanwhile, Bachmann was subjected to the moans of a nearby patient who had just had his chest split open for that very procedure. His young life spun before his eyes, and he was terrified by visions of the saws and knives about to hew into him. When his condition gradually stabilized on its own, doctors agonized about what to do next. The knotted stenosis in his left coronary artery was certainly not going to vanish on its own and had all the earmarks of a future "widow maker" if left untreated. More tests followed. By the next evening, it became apparent that Bachmann might actually be an ideal candidate for the balloon procedure. Andreas Gruentzig was quickly located for a consultation.

Bachmann later described the scene. "Gruentzig wasn't at all overbearing. I developed an immediate trust in his capability. He made it clear that I would be the first man ever to receive this pioneering procedure. At this moment, my diseased heart skipped, making it obvious how dangerous and painfully crippling my situation could become. The severe choice between the scheduled bypass operation and the alternative dilatation probe left little doubt in my mind as to which path was preferable. Gruentzig explained that the method had been heavily experimented on and refined in cadaver arteries. "I said, 'What's the big deal? The difference can't be so great.' "

Gruentzig told Bachmann that a failed procedure might require corrective bypass surgery afterward, but that he was otherwise facing that option from the word go. To Bachmann, the choice was clear. Patient and doctor made their compact. What trepidations went through either's mind the night before has never been recorded. But it is likely that Gruentzig worried his way through the small hours.

At seven A.M. on the 16th of September, Bachmann was wheeled into the catheterization laboratory, sedated but fully conscious—unlike any one of the hundreds of thousands of patients who had already undergone coronary artery bypass surgery. About a dozen anesthesiologists, radiologists, cardiologists, and surgeons stood on the wings. Some had defined roles to play, while others were merely curious about the undertaking.

"Had the world then understood the importance of this procedure, a football stadium could have been filled with spectators," Bernhard Meier observed. Indeed. By a factor of millions, Dolf Bachmann's unfolding experience would prove more fateful to the future course of the heart's healing than the trauma of Louis Washkansky in the supposed "Miracle of Cape Town" a decade before.

As it happened, the first two balloon catheters on hand failed even to inflate during pretesting, and Gruentzig's long-awaited breakthrough threatened to crumble at the start. Fortunately, a third appeared to function perfectly. Now the usual preliminaries could proceed: the pinprick introduction of the guide wire into the groin, the insertion of the lead guiding catheter. Gruentzig delicately advanced this system up through the great aortic trunk vessel and then down into the left ventricle of the heart—a task that can be completed in thirty seconds now. But for Gruentzig it was a grueling forty-five-minute procedure. Nonetheless, Gruentzig found his way to the narrow divide at the opening of the diseased artery. This fact may sound humdrum, but to observers watching the progress on an overhead X-ray viewing screen, the drama was immense. The wide-awake Bachmann himself lay there eyeing the movements of strange probes in the depths of his life-giving arteries—an experience that no patient in the history of heart surgery had ever undergone.

Maria Schlumpf stood spellbound at Andreas's side, assisting. Her recollection is that he betrayed not the slightest sign of trepidation. But then again, she was the man's muse. Surely, Gruentzig was catching his breath. Surely, he would have recalled the abject failure of his aborted procedure half a year earlier, the angry comments about his hubris haunting him ever since. Many eyes bored into his back, the observers well knowing that failure now would consign the great idea of his life into the dustbin of futility, with his reputation consigned to that of a quack. Bachmann himself remarked that the assembled doctors "seemed to be hanging on needles." And no wonder: An absolute prerequisite was that heart surgeons stand in constant readiness to rip open his chest should a disaster unfold.

Werner Niederhauser, the handcraftsman of the balloon catheters, weaved in and out of the room. The progress didn't look altogether seamless to him. "As far as I remember, he had an air bubble moving down the artery and his bosses were standing outside [the glass walled catheterization laboratory] watching him, saying 'Oh my God,' and he was standing there like he had done it one hundred times and there was nothing to it."

Now the balloon catheter was conjured forward through the guide catheter into the left anterior descending artery and pushed forward through the midst of the angina-inducing crush of plaque. The tip slid effortlessly into place. Its every move registered plain as day on the X-ray screen. There was no place to hide. His left hand grasped the inflation syringe that was set to pump up the balloon—and with it, the might of his life's passion. The room grew deadly quiet. Ever so slowly, the balloon in the heart expanded.

Here lay the moment of truth. Would the maneuver trigger a heart attack? Would the balloon split under the pressure, provoking an epithet-spewing rush forward by emergency surgeons, the former toilers behind a barber's pole? And would surgery even have a prayer of success under such dire circumstances? There are countless ways to injure a man when you are brandishing a perhaps not altogether magical wand that has never touched the heart's depths.

If his physical demeanor was cool, Gruentzig was certainly working at a state of hyperalertness now. Fortunately, the EKG held steady; the patient displayed no indications of pain. The pressure in the balloon climbed to twice the earth's atmosphere, then three, and finally five "bars." The sedated but conscious Bachmann watched the dark shape in his artery suddenly grow until it had expanded to three millimeters in width. Every indication suggested that the man's gnarled wedge of atherosclerosis was indeed yielding to the expanding force of the balloon—but how could one be sure? For more than fifteen seconds, the peak inflation pressure was sustained. Slowly—excruciatingly slowly it must have felt to Gruentzig—the beautiful balloon was deflated. Post-procedural X-ray images suggested that the residual hair-thin channel through Bachmann's muddied artery had indeed yawned open. But what did this really mean?

The believers in the audience had to be rubbing their eyes, the skeptics scowling furiously, so peculiar was the entire performance. If the tiny shift in brightness against murk on the X-ray screen signified anything meaningful, Gruentzig had just pulled off one of the most deceptively simple advances in the history of medicine—and indeed in the quest to remake mankind's fate. In a matter of minutes, he had ostensibly achieved the same restoration of vital blood flow that a heart surgeon gains after sawing through a human chest, clamping off the life-giving arteries to the heart, and transferring the very pulse of existence to an extra-corporeal machine. Furthermore, the surgeon must slice away a replacement vessel from the leg and hurriedly stitch it around the imperiled heart vessel while blood gushes forth between his fingers. Then, the human heart must be jump-started back to life, the gaping wound in the chest sewed into some semblance of its former wholeness, the brain waves checked for signs of embolic stroke. Yet to come are weeks of debilitation, depression, impotence, and agonizing pain—not only in the chest but around the severely injured leg vessel.

Gruentzig's nonsurgical procedure was meant to negate all that. Yet the proposition still sounded absurd. A little balloon on the end of a spaghetti strand was going to rewrite the course of heart surgery? Okay. How many dollars were they asking for the Brooklyn Bridge now? The facts were basic enough. Gruentzig had made a needle puncture in the patient's inner thigh and administered local anesthesia. That was the extent of the blood and gore. Three thousand years of bewilderment before the human heart may have just given way to a technique that looked almost mundane—you advanced a hollow tube into the seat of life, pumped air into an ephemeral-looking little balloon, and *voila*!

What had really happened? Gruentzig himself could scarcely be sure. To maximize his chance of success, he inflated the balloon a second time, and then checked to confirm that the force of the circulation had been definitively enhanced, at least as far as he could tell. The pressure gauge did seem to offer a profile of remarkably renewed pumping force. If you were also a believer in the meaning of small shifts of light and darkness on the X-ray screen, Bachmann's imperiled artery appeared to surge with fresh flow.

Yet many observers were less than stunned. Werner Niederhauser, a technician with no training in finding revelations in shadowy X-ray images, was himself fairly indifferent. "I said 'congratulations' and was gone. I didn't realize the importance of what had happened. I was worried whether the material worked, that was all I was interested in. At the time we hadn't seen any pictures of coronary arteries after such a procedure, so we didn't know what to compare them with. We just said, 'Oh, it was a little less narrow than before. What happened to the narrowing?'"

Better-trained eyes perceived their own riddles. Certain observers suggested that the X-ray screen indicated that there might be a subtle second blockage nearby, and perhaps that should be expanded as well. Gruentzig probed that with a fresh injection of dye, then declared that there was no real problem there. Withdrawing the balloon catheter altogether, he executed a final angiographic test. The entire procedure had taken three hours, but the artery looked as clean as a whistle—perhaps Bachmann was actually home free.

Yet few things are ever quite that simple. Indeed, danger signs leapt into sight. A major leg artery appeared to quivering spasmodically—why? Had particles of plaque been dislodged through the circulation, spelling impending calamity? The EKG monitor began to shoot off indications of errant rhythms forming on the right side of the patient's heart. Was a massive heart attack uncoiling?

Improvising swiftly, Gruentzig orchestrated a series of emergency diagnostic tests. Fortunately, he soon learned that he was struggling with chimeras. Every vital sign settled down, and the mood in the University Hospital grew euphoric. Gruentzig later recalled, "Every observer was impressed by the simplicity of the method, and I began to realize that my dream had become true. For a moment, my joy was crushed by the awareness that the patient had developed an EKG abnormality on the right side of his heart. But a test four hours later showed this had normalized. Now we were ready to celebrate."

Bachmann himself remembers Walter Siegenthaler coming up to him afterward, saying, "You are the luckiest man in the world, because thanks to this concept you have escaped the need for surgery."

Gruentzig meanwhile called Myler with the news.

"I've done it!" he told his collaborator in California.

"You've done what, Andreas?"

"I have just performed the first angioplasty on the human heart!"

The patient, Bachmann, was also euphoric. The next morning, he phoned a local newspaper journalist to come by for an exclusive interview. An aghast Gruentzig worried that tabloid headlines about his procedure would create a furor such as Forssmann had endured in Berlin, and he convinced the newsman to hold off until he could get a report published in a proper medical journal.

Bachmann himself looked to the future like a man reborn. His chest pain soon disappeared and before long he cast his nitrogylcerine tablets aside. Upon discharge from the hospital, he quit smoking and sought to eliminate every major source of stress in his life. In time, he divorced his wife, quit his insurance job, and moved to the beautiful mountain canton of Graubünden. There he launched a magazine devoted to a Swiss card game called JASS that he adored. More than twenty-five years later, Bachmann remains a kind of poster boy for the brave new world of rerouting choked arteries without surgery.

After completing his first historic case, Gruentzig decided to proceed quietly, and he returned to his normal responsibilities. A month passed before he joined Martin Kaltenbach in Frankfurt to perform his second procedure on October 18. This one, involving a closure in the vital left main artery, failed miserably. The patient, just like Bachmann, was fairly young and refused to suffer bypass surgery. But he had not one, but two, constrictions in his coronary arteries, as well as more deeply encrusted cardiovascular disease. Gruentzig slipped his balloon catheter through the first easily enough, but immediately realized he could not inflate it fully. The plaque buildup in this case had grown calcified and rigid with age. The patient had to undergo bypass surgery a short time later.

A month afterward, Kaltenbach joined Gruentzig for a third procedure in Zürich, and this one worked like a charm. On November 24, they performed a fourth and also altogether successful procedure back in Frankfurt. A routine was becoming established, and a well-satisfied Gruentzig flew off to Miami for the next meeting of the American Heart Association. This was the moment the erstwhile lonely tinkerer had sought for years.

Mason Sones, the previous bearer of the torch of discovery, wandered around the halls just beforehand, muttering to Richard Myler, "You have to introduce me to that goddamn Kraut." The enigmatic Gruentzig was wearing sandals and a bolo, or string tie, for the occasion.

In any case, fifty or sixty medical eminences from around the world tucked themselves into a small meeting room, where Gruentzig laid forth his story: the tinkering on his kitchen table; the exhaustive testing of his improbable balloon idea in by now nearly 300 diseased leg arteries; his experiments on coronary arteries in dogs and excised human cadavers; and his work with Myler. Then he flashed upon an overhead screen the one-day-before, one-day and one-month-after X-ray images of his first use of the balloon catheter on the heart of Adolf Bachmann. Few in that audience were neophytes at reading arterial X-rays. Most were doctors accustomed to viewing atrociously diseased human arteries every day. They got the picture, fast. They well understood Bachmann's "before" image. The vessel was choked thick. Now Gruentzig flashed before them a picture of the "after" once his balloon had been inflated. Every trained eye could see the dramatic reversal. The pictures were too concrete to dismiss. For decades, the world's cardiologists had been consigned to taking tedious measurements and dispensing pills. They heard much about the wizardry of the likes of Denton Cooley and Christiaan Barnard, while learning firsthand about the trail of carnage many supposed masters of the heart often left in their wake. Suddenly, they could feel the tide turning.

Myler remembers a mood of euphoria sweeping over the lecture hall, an instant gasp of recognition. "There was sudden applause. I'd never heard that in the middle of an abstract presentation before." Gruentzig could barely continue with his talk, so much emotion had welled up from within after all the years of ridicule and deprivation he had endured.

Myler cast his eyes over the crowd, eager to share his sense of revelation. "A man I had known and revered for years came up to me afterward. He was crying. He embraced me. I remember what he said exactly. He said, 'It's a dream come true!' That was Mason Sones."

Sones had every reason to be transported. It was plain that a fantastic voyage into the heart, beginning with Werner Forssmann's crude self-experiments, was nearing the homestretch. Over the last three decades, Sones had born witness to a remarkable arc of discovery, with the advent of John Gibbon's heart-lung machine making way for profound advances in treating the formerly taboo-shrouded organ. He well knew that his own breakthroughs in illuminating the coronary arteries had inspired another quantum leap forward. But this restless seeker had long believed that there had to be a more precise solution to treating cardiovascular disease than the saw and scalpel. Just what it was, he could never quite figure out himself. Now here stood a dashing young German proclaiming that he had finessed away grievous coronary blockages with the most unlikely of tools. Sones could see a revolution coming within that balloon.

Chapter 9

GRUENTZIG'S REPUTATION INSTANTLY SOARED. Doctors who had witnessed his presentation or learned about it secondhand quickly sought the opportunity to learn at his side. Not far behind were those who wanted to capitalize on potential profits in the making. The scene bore similarities to the 1956 movie *The Red Balloon*, in which Parisian street urchins hound a boy with a magical balloon until they succeed in breaking it. In this case it was doctors who were clutching at the strings of a dream.

Out of wealthy private practices, hallowed academic halls, and the number-crunching offices of industry, keenly interested parties materialized overnight. Richard Myler smelled trouble in the gathering stampede. "Andreas had no support from anyone, until the first case was done and presented at the American Heart meeting. And then there were hundreds of flies on the wall, [a few of them] guys that you wouldn't honestly let clip your toenails, and they wanted to do angioplasty, and they wanted to do it after spending only a day with him."

John Abele, the Massachusetts medical-device manufacturer who had wavered between a commercial and mentorship role throughout the procedure's development, sent Andreas a letter on December 2 warning that "various people have already come out of the woodwork asking for prototypes." Abele urged him to begin thinking about minimal requirements for formal training in balloon angioplasty before letting protégés near his procedure; he also advised that it would be necessary to conduct methodical, confirming clinical trials in multiple hospitals. The point, he said, was to avoid a free-for-all that might tarnish Gruentzig's carefully nurtured results.

One of the first acolytes to arrive in Switzerland was a hard-driving specialist named Simon Stertzer from Manhattan's Lenox Hill Hospital, a former Mason Sones understudy who was considered a wizard at diagnostic catheterization of the heart. In time, Stertzer's early mastery of the procedure and keen business sense would gain him substantial holdings in several burgeoning new medical device companies that would earn him tens of millions of dollars. The man was all enthusiasm when he appeared in Zürich just after New Year's in 1978. Gruentzig, for his part, was impressed by his powerful intellect and technical proficiency.

Stertzer watched as Gruentzig deftly handled one easy case, then struggled through a nightmare procedure that nearly ended up with the patient requiring emergency bypass surgery. Yet this didn't diminish the recruit's ardor. "I knew before I went there that this was going to revolutionize the treatment of coronary atherosclerosis," he pronounced a few years later. Never prone to understatement, the expensively dressed and carefully coiffed cardiologist told a 1986 interviewer, "And sure enough the first procedures that I saw were successful . . . [we] marveled at what we had accomplished and wondered about what we could do in the future." The impulse to employ the word *we*, when Stertzer was after all but a mere observer, would haunt the man shortly. A number of colleagues in time accused Stertzer of becoming infatuated with his own sense of prominence. But Gruentzig missed those cues.

Later that month, Myler returned to Europe with his eager young companion, John Simpson, to observe more live cases. The first stop was Frankfurt, to watch Martin Kaltenbach perform another procedure, in which he reduced a 90 percent narrowing in a coronary artery to 70 percent. Today, this would be an unacceptable result, but Simpson said, "To us, it was revolutionary. We declared coronary disease extinct at that moment." He and Myler next visited Gruentzig in Zürich, where a couple of subsequent attempts achieved less than perfect results. Watching alongside were two other new disciples from America named Lamberto Bentivoglio and James Minor. The technique's potential was nonetheless manifest. In a spirit of exhilaration, the fivesome—joined by Maria Schlumpf—capped off their afternoons with long discussions and samplings of the local wine.

Bentivoglio, a native of Rome, possessed the kind of courtly manners that struck an immediate chord with Andreas Gruentzig. "He was bold and imaginative and outspoken," Bentivoglio said of the man who would electrify his own future. "He had such a captivating personality. . . . I remember we went out to dance. He and Michaela were beautiful dancers together, she was very classy. They would dance the tango and use all those complex movements on the dance floor. They were almost like professional dancers."

John Simpson was transported, too. Not yet thirty, he offered ideas for further improving Andreas's catheters. Understandably, his host found this proclamation—later to be realized to the tune of hundreds of millions of dollars—a little cheeky. Gruentzig would not soon forget that affront.

The week's visit ended with a full-blown drinking session in Gruentzig's kitchen, his guests downing bottles of Chianti with Michaela joining in the merriment. As the night drew to a close, Myler pronounced that they should found the first ever "International Dilatation Society." To verify its newly minted reality for the ages, each member contributed a few dollars and all signed their names on the last cork. Gruentzig was summarily decreed president, Myler vice president, Minor treasurer, Bentivoglio and Maria Schlumpf "something else," and Simpson the Keeper of the Cork—which he eventually lost.

Gruentzig was now ready to introduce his breakthrough in America, which he knew was critical to gaining worldwide notice. The avid Myler was an obvious first choice to perform the procedure abroad. The hard-driving Stertzer, on the other hand, was ensconced in all-important Manhattan. Some voices of caution whispered that Stertzer was a maverick, and perhaps not the best choice to anoint as a lieutenant. "Exercise caution," Abele warned. However, Gruentzig was captivated by the power of his enthusiasm and expertise.

On February 4, 1978, the British journal *Lancet*, a seminal publication for doctors around the world, published a letter Gruentzig had written detailing the results of his first procedure. The timing couldn't have been riper. Bypass surgeons were more or less on the defensive, while cardiologists hungered for a better approach. As far as many were concerned, the *Lancet* report announced a transformed future.

Back in Switzerland, the popular press jumped on the story. A local hero, they proclaimed, had come up with an instant cure for "killer heart attacks" —whereas Gruentzig himself asserted no such thing. First out was the newspaper *Tages-Anzeiger,* with a front-page February 7 eye-grabber: MEDICAL-SENSATION: BALLOON TREATMENT AGAINST HEART ATTACKS. The next day the tabloid *Blick's* front page cried DOCTOR ACHIEVES MEDICAL SENSATION, followed by a sub-headline pronouncing, I SAW THE BALLOON IN MY HEART! The glossy *Schweizer Illustrierte* magazine followed on February 13 with a nine-page feature entitled, IS DEATH FROM HEART ATTACK VANQUISHED? THE STRUGGLE FOR THE HEART. To bathe the story in sex appeal, the magazine's cover displayed a topless young lady with a heart painted over her left breast. Then followed another spread of the nubile lass in her underpants contrasted with photos of the dashing Gruentzig and the reinvigorated Dolf Bachmann cavorting about in a swimming pool. This was not exactly the *New England Journal of Medicine*'s approach.

In the fine print, Gruentzig cautioned that it would take months to understand his procedure's strengths and weaknesses, and that its real purpose was to try to control crippling long-term chest pain, rather than evolving heart attacks. But the tabloid coverage set Gruentzig's superiors livid, and made him appear as recklessly self-aggrandizing as a second coming of Werner Forssmann.

Gruentzig caught flak aplenty. Maria Schlumpf remembers Krayenbühl fuming anew. "He was always angry, which was a problem . . . He walked about the hospital floor with patients sitting here and there, barking out, even when they could hear, 'With Gruentzig's procedure, patients will die!' "

The echoes of this bitter disapproval rippled across Switzerland, and it took heated discussions before Gruentzig was allowed to present his findings to a late-winter meeting of the Swiss Society of Cardiology.

In America, Myler and Stertzer were meanwhile jockeying for position.

Trouble awaited them, too. Adversaries in Myler's hospital tried to block him from so much as trying the Gruentzig procedure on their turf. Such a radical intervention should not be attempted, they argued, without full scrutiny by the local Institutional Review Board, a body charged with protecting patients from unproven procedures. For all his outward bonhomie, Myler knew how to play power politics himself. Trolling through his formidable connections, he contacted the top authorities of the Bureau of Devices of the United States Food and Drug Administration (FDA) in Silver Spring, Maryland, essentially the highest arbiter of medical ethics in the land. Myler argued that balloon angioplasty had indeed been tested with unimpeachable thoroughness and, in the end, obtained a verdict that suited his needs perfectly. The Gruentzig technique was ruled similar enough to diagnostic heart catheterization that it need not be labeled, for the time being, as a "prohibited procedure." Ergo, the water torture of a formal application process with years of clinical trials was not required for him to proceed.

So armed, Myler blew past his adversaries. He advised Stertzer, with whom he was in steady contact, that the precedent should work neatly in New York State as well. In an act of apparent showmanship, they simultaneously performed their semi-historic procedures on the very same day, one in Manhattan, the other outside San Francisco. So much for Swiss caution, then.

On March 1, 1978, Myler and Stertzer performed their dueling coronary angioplasties at either end of the American continent. Luckily, considering the flashiness of this debut, neither of these procedures disintegrated into deaths or heart attacks—as was entirely possible. Myler had at least cherry-picked his first patient to exacting specifications: his was another thirty-eight-year-old insurance salesman of generally robust health. Stertzer, on the other hand, chose to tackle a difficult lesion in an older patient—a setting that could have spelled trouble. Nonetheless, Stertzer finessed his way to apparent success. Lamberto Bentivoglio, who was in close contact with Gruentzig at the time, insists that the inventor of the technique soundly demonstrated with Stertzer afterward. He was constantly warning us, "Be safe, don't take chances, because this procedure has not been established, we must give it a good name. And you are responsible as I am for the future of this procedure."

This was all behind the scenes. To the rest of the world, the facts seemed clear: Both of the American patients realized immediate improvements in their circulation, with no need for surgical saw or knife, or the perils of the external heart-lung machine. Even if photographers did not hang from the trees, as they had in Cape Town following Christiaan Barnard's first heart transplant in 1967, word traveled fast.

Interested doctors beat their way to Zürich in droves. There were cardiologists, heart surgeons, radiologists, sundry representatives from the medical device industry—a three-ring circus in the making. William Casarella was one radiologist from the U.S. who made an early trip. Casarella, who would later advance catheter probes into the most intricate recesses of the extra-cardiac circulation, was amazed that his host was not even allowed to make him a photocopy on the University Hospital's equipment.

The co-fabricator of the devices, Werner Niederhauser, watched in amazement, too. "People didn't ask, 'Are you Gruentzig's boss?' They said, 'Are you the one who works with Gruentzig?' So there must have been a lot of jealousy involved. There is the hierarchy you have in the [university] hospital, and he was just jumping steps very fast. This was not done in Switzerland."

Ulrich Sigwart, a fellow German who had met Gruentzig briefly while a medical resident in Zürich a few years previously, arrived at the door. Sigwart was urbane and wry, with an intuitive intelligence that cut to the quick, and the two hit it off. Gruentzig spent hours explaining the procedure to his inquisitive visitor from Bad Oyenhausen.

"We had a good time together, because we were both keen aviators," recalled Sigwart. "I remember Andreas saying, 'Angioplasty has many things in common with flying. As long as things go smoothly, everything looks safe, but once things go sour you get ice on the wings; you get power failure and everything comes apart, and then you have to make the right decisions.' "

Sigwart was dispatched with several balloon catheters and high hopes. He soon hit his own snake pit of hospital politics in northern Germany. His first procedure was hair-raising. The typically unflappable Sigwart's adrenaline was racing, and he could taste the tension in the room as he sought to negotiate the balloon catheter forward. After more than an hour of struggle, he prevailed, and savored the magic of the Gruentzig approach for the first time. But that night he struggled with a nightmare, wakening in sweat as he relived the procedure with the result ending in death. Until recently chief of cardiology at the University Hospital in Geneva, he sighed in an interview there when recounting the earliest days of balloon angioplasty. "It was a high-wire act without a safety net," he said.

That assessment was witheringly apt. Contrary to one of Gruentzig's firmest edicts, the German cardiologist performed seven risky balloon angioplasty procedures in 1978, with nothing more than a whistle in the wind for a fallback. His clinic simply had no cardiovascular surgeons on staff, so Sigwart made arrangements for a surgeon to fly in by helicopter from Hanover—fifty miles distant—at the first sign of any emergency. According to Sigwart, Gruentzig was so eager to get him started that he had no objections to his proceeding under those terms.

Sigwart discovered that he was playing with fire. In three cases he could not even maneuver the balloon catheters into proper position. Another patient succumbed to a fatal heart attack when his dilated artery abruptly reclosed a few weeks later—a little-understood eventuality that would come back to haunt the Gruentzig breakthrough time and again. One colleague considered Sigwart's work so dangerous that he threatened to bring him to court. Regardless, the German pushed on, keeping his balloons primed for subsequent reuse by surrounding them in moist cotton (a frequent economy in a pre-AIDS era). "They were precious things," he remembered.

Another early pilgrim to Zürich was Arnoldo Fiedotin, the fast-talking, Argentine-born cardiologist who joked about "hanging up his balls" in order to forbear training under Mason Sones in Cleveland. He first learned of Gruentzig's fledgling work in the heart during a Chicago taxi ride with two doctors and nearly doubled over with laughter at their story.

"I said to Enrique Leguizamon, 'You're crazy as hell. Who ever heard of sticking something there and squeezing the damn artery? It's going to break to hell.' Enrique said, 'I'll buy you dinner if you go to see this guy.' "

Getting the message, Fiedotin phoned Andreas and asked to visit the very next week.

The stocky Hans Peter Krayenbühl cast his eyes suspiciously on the flood of arriving visitors, convinced more than ever that his subordinate was an attention-grabbing charlatan. Hans Peter and his own boss, Walter Siegenthaler, soon showed that they were not to be crossed lightly. Before he knew it, Gruentzig was informed that his pathetic cramped closet of a basement office was needed for other uses and that he should seek working quarters clear outside the hospital. Ironically, this banishment led to Gruentzig's reestablishing himself in the very building in which James Joyce, the pioneer of the modern novel, struggled through his own Siberia of rejection and indifference in 1918 while writing *Ulysses.*

On the home front, there were other conflicts. At times, the private Michaela found the stream of impromptu dinner guests in her apartment to be a tiresome intrusion. Even worse, she felt that her husband's overnight celebrity was undermining their marriage. "I just had the idea he should be closer to me, and this time was not always easy for me," she confided years later from the same living room where the early supplicants gathered. "I just wanted to have my own life and not always be part of all this. I wanted to be my own person and this became part of our dilemma later on. I was just his wife . . . I had an ambivalent role and I had some resistance to playing this role."

Michaela also fretted that the constant demands on her husband's attentions were distracting him from his responsibilities in raising their young Sonja. "He loved her, I am sure of that. But how much time he could give to her, I don't know," said the mother. A long pause followed those words.

Despite these stresses, the couple packed up on every available weekend to retreat with Sonja to their modest chalet in Canton Schwyz. They struggled to recapture happy times there, watching their toddler cavorting about in the surrounding green fields. Nonetheless, Michaela beheld her husband's rise with anxiety. Practiced as he was at flashing his seductive smile for every gathering group of acolytes, she sometimes worried that Andreas might actually be aching with isolation, maybe even downright depression. Countless photographs from that era show Gruentzig hamming it up among grinning admirers. But did they tell the full story?

"You know, I am a psychologist," Michaela offered. "I have reached a certain time of life when I am returning to think about things from the past.

My view is that he always tended to put himself in the middle and to gather people around him in order to liven things up somehow. But I now think this must have been a way of handling life that he developed. As a child he gave life to his mother and enjoyed the attention. But he didn't really let people get that close to him . . . There weren't many close friends."

Faced with pressure from so many directions, a lesser man might have lost his way. But Gruentzig somehow still held tight to his reins. The cavalcade of pilgrims, he decided, needed to be carefully vetted, lest some fly-by-night operators wreak havoc with his balloons. He therefore told Hans Gleichner, the chief catheter maker, and his understudy, Werner Niederhauser, to withhold the devices from all potential purchasers until they had received prior clearance. Meanwhile, Hugo Schneider, the shop's boss, continued to fret, and insisted that all payments for the catheters must be made in Swiss francs, the only currency he would ever trust. The results of these twin injunctions were sometimes comic.

The high-profile Simon Stertzer took to jetting back and forth to fetch whatever fresh balloons he required for his mounting caseload. Even so, this key American disciple got the Schneider treatment in full: pay Swiss cash on the barrelhead or be gone. Myler and the rest faced the same dumbfounding situation.

The Schneider shop normally posted a clerk at a hallway desk to scrutinize the supplicants' bona fides before allowing them to make a purchase. "We were not allowed to give any balloon catheters to anyone unless Gruentzig said 'Yes,' " Niederhauser recalled. "In the beginning this would be an informal phone call. Later on they had to fill out a paper signed by Gruentzig . . . He would ring the factory and say, 'Doctor This and That is coming, okay? Don't give him any. Tell him you don't have any.' He'd say, 'He's no good. I don't think he should be working with a balloon' . . . I'd say [to the proscribed customers], 'Sorry, you should have rung up before. We don't have any more' . . . You have to understand we were a very small company. We'd just been producing needles, repairing tools."

Even in careful Switzerland, things do not always go according to plan. On one early morning, Mason Sones, the world-renowned father of diagnostic angiography, knocked on Neiderhauser's door. At that moment, however, he was fresh off a transatlantic flight and looking singularly bedraggled as a hard rain pummeled his balding head. Sones announced through the glass that he wanted to buy some balloon catheters—*now.* In quiet and sometimes sniffy Zürich, shops are not always in a hurry to greet their first trade. And in this supposed nursery of a medical transformation, the rumpled Sones looked more like a straggler from the local homeless shelter than a famous doctor.

Niederhauser explained, "He knocked on the window from out on the street and said, 'Can I buy some catheters?' It was nine o'clock in the morning and raining. I said, 'Yes, we have three balloons in stock.' He wanted to pay in cash, in dollars. I said, 'I'm only allowed to accept Swiss francs.' I had no idea who Mason Sones was. His name meant nothing to me. He said, 'Oh, Christ, I have to go down into the city to get money.' He came back at four o'clock in the afternoon completely wet and very drunk. He said, 'Here, you can have your damn Swiss francs!' "

The future medical-device impresario, John Simpson, was also barred from receiving a full package of angioplasty equipment, since a suspicious Gruentzig was not quite ready to anoint the "Keeper of the Cork" to the elect. The young Californian finally set to designing his own version, from which he would make a far greater fortune than Gruentzig ever realized himself.

On April 28, 1978, John Abele wrote Gruentzig to say how impressed he was by a look at Simpson's prototype. "My congratulations to you again on your persistence and patience in handling this whole development under very difficult conditions . . . You are a good tightrope walker," Abele closed.

Tightrope walking? Frictions were already developing among Gruentzig's lieutenants in America. Sharon Myler, for one, was developing a profound distrust for her husband's supposed prime collaborator, Simon Stertzer, who recently gained international notoriety for investing in three Nevada strip clubs for what he claimed to be a move to generate profits to further his cardiovascular research. In a subsequent interview, Sharon Myler said, "Stertzer is a very manipulative person . . . [But Richard] just gave every excuse and said he and Stertzer wanted to work together, and have this collaborative experience.

There was no question that this was a time of pressure for Andreas Gruentzig. He decided to host a late-summer teaching course to gather all his new disciples under one roof, air every difficulty, and provide a springboard for moving forward. After all, no one yet understood the full promise or risks of his technique. Organizing this meeting proved to be an enormous challenge, partially because the University Hospital offered almost no support, other than the time volunteered by Maria Schlumpf.

Gruentzig sometimes worried that he was losing control over his genie. A few months earlier, Felix Mahler, his friend and collaborator in performing angioplasty in the legs, had called Alfred Bollinger from Bern to announce that he had just succeeded in moving the promise of balloon procedure into an entirely different sphere. By dilating a blocked kidney artery, he sought to gain a new foothold in controlling hypertension. Gruentzig, told of the breakthrough secondhand, reacted angrily.

Bollinger feels responsible to this day: "Mahler phoned me and explained what he had achieved. And I said, "That's great, Felix.' Then I told Gruentzig and Andreas got furious and said, 'I intended to do this myself for some time—this is not right!' So what did Andreas do? Ten or fourteen days afterward he did his own renal artery stenosis and proceeded to write very quickly a letter to the editor [of a learned journal to claim first credit] . . .. He was very competitive. I remember this very well, how I met him in the corridor shortly after getting this telephone call from Felix Mahler and thought I would share the excitement . . . I ended up having to apologize to both of them."

That particular reaction may have been excessive, but Gruentzig was being pulled every which way. Entrepreneurs were carefully eyeing the little Schneider company—as well as his increasingly strained relations with the much larger Cook Group, which never got beyond selling his catheters for leg procedures. Rivals from eventual medical-device powerhouses like Cordis, Becton Dickinson, and American Edwards floated overtures to commercialize his advance on a global scale. But they could not find the right recipe to cement a relationship.

Gruentzig's friend John Abele ultimately decided that his Medi-Tech company was in no position to get into the game, since cardiology was not its forte. He also smelled trouble ahead with the U.S. Food and Drug Administration, which can require years of supplication and mountains of clinical evidence before formally approving a new procedure. At the time, the catheter industry garnered but $100 million or so annually and was playing its cards closely. Today, it is generating close to $20 billion in sales each year.

Eventually, a fresh suitor came knocking. This was the newly hired chief of the United States Catheter Instrument (USCI) division of the C.R. Bard health-care corporation, a tall and earnest individual named Dave Prigmore. At first blush, Prigmore came across as an unassuming and utterly pragmatic sort, and a very unlikely midwife for a historic medical breakthrough. Yet, Prigmore was a quick read, and it did not take him long to grasp that USCI needed a breath of new life. But where to find it? USCI's foremost products back then were the catheters Mason Sones had designed ages before, and these were considered mundane things, selling for around $11 a pop. Equally troubling, the FDA had just enjoined the medical-device industry to recast all its instruments as being certified 100-percent sterile. "In 1977," Prigmore recalled, "most catheters were being sold nonsterile in a brown paper bag with no product labeling [about misuse] for as many reuses as you could get out of them. The Medical Device Act of 1976 said you either had to sterilize the product for one-time use or describe how to clean it, and we couldn't do either. So we were about to lose our business. Doctors were using catheters five to ten times. We had to redesign the way we were making our catheters."

Dave Prigmore cast his net about. Only months before, Mason Sones had buttonholed him at the American Heart Association meeting in Miami to proclaim the genius of Gruentzig's breakthrough. But Prigmore at that time was just beginning to make sense of his arcane new world. Nonetheless, he knew how to listen.

The executive had another gift that is not always found at the peaks of corporate power—a curiosity about the nuts and bolts of emerging ideas. "I'm at this point looking for any product for USCI that will take us out of this buggy whip that we're manufacturing. I was just listening to everybody and saying, 'Find me a large rock to jump on because this place is going to fill my boots pretty soon.' " One day, federal marshals would lead him away for arraignment on a questionable charge of conspiracy to defraud the Food and Drug Administration. But in 1978, everyone who met Prigmore found him to be a plain-talking straight shooter.

The best way forward, Prigmore figured, was to rekindle the company's relationships with a variety of leading doctors, none of whom had ever received a dime of USCI royalties for commercializing their inventions. He sent his new point man, John Cvinar, out to visit Mason Sones at the Cleveland Clinic. His task: to work through the redesign of the Sones diagnostic catheter and rebuild good faith with an offer to begin paying the man 3 percent royalties. Cvinar fondly recalls the time. "I was racing all over the world looking for guys who had ideas that might be a next generation or new wave, and when Mason spoke we just ran with it."

From Sones, Cvinar, too, heard an earful about Andreas Gruentzig's potential for revolutionizing cardiology. Prigmore, meanwhile, was getting the same word from Richard Myler. But others were far more skeptical, Prigmore recalled. "The conventional wisdom was that angioplasty couldn't work—you couldn't put anything in the coronary artery without killing the patient. So it wasn't like people were coming around shouting that this is the obvious next thing to do. The vast majority of the world was saying, 'This guy's a lunatic' . . . Even then, he [Gruentzig] thought there were just a small number of surgical patients that might ever get treated."

Nonetheless, languishing USCI desperately needed new products. A marketing colleague therefore flew off to Switzerland and met with Gruentzig. He proposed that USCI should take over balloon catheter sales in North America and perhaps everywhere outside Switzerland. A dubious Gruentzig advised him to go see Hugo Schneider—which proved to be a dead end.

So Prigmore next dispatched letters of entreaty to the recalcitrant Swiss manufacturer. The responses were haphazard. "Schneider was a Swiss who distributed telecommunications equipment behind the Iron Curtain—to Poland, Czechoslovakia, and Yugoslavia. His instruments of power were his teletype machine and Swiss bank accounts. He did business out of like a post-office box in Yugoslavia, all very clandestine stuff . . . He was a wealthy guy with no intellectual interest in the business at all, no desire to build a medical device company or have a success," was the way Prigmore saw it.

However, the buzz about Gruentzig's work kept intensifying, even though only about twenty cases had been performed worldwide by the spring of 1978. "One guy who talked to us in that period who had visited Andreas then called up and said, 'If you guys can make me a catheter like that thing that guy in Zürich's making, the world will never hear of him again. We own the *New England Journal of Medicine* . . . It was absolutely true. I think he turned out to be a skunk in more ways than one." To his credit, Prigmore avoided that back alley, although he evidently obtained a balloon catheter for his engineers to analyze down to the narrowest detail.

In May, Gruentzig flew to Boston to perform a live procedure before a room packed with Harvard doctors. The idea was to showcase his work in that famous citadel. But a near fiasco awaited. The scheduled candidate for the procedure was ill-selected from the start, and the host doctor, Bill Grossman, made things worse by trying to manipulate the catheters himself. As gathered eminences watched, the procedure failed and the artery shut down, requiring the woman to be rushed off to emergency surgery. Appalled and humiliated, Gruentzig silently vowed that never again would he perform a case at this supposedly hallowed institution.

The visit at least led somewhere promising. As the day wound down, Dave Prigmore convinced Gruentzig to join him for a private meeting so that he could press USCI's case once again. "I was impressed by him. I thought he was a very sharp young guy. I was trying to fall in love, there wasn't any question about what I was trying to do," Prigmore recalled. But the conversation soon ground into conflict. "I found him very difficult to deal with. He was mistrusting of America and Americans, mistrusting of USCI. On almost every subject we would try to align ourselves there was no common ground . . . But I had an absolute need and the right people telling me that this was where to go."

In a show of good faith, Prigmore offered to build Gruentzig an improved version of his nettlesome guide catheter, free of charge, in order to demonstrate USCI's manufacturing expertise. He definitely did not promise to improve on the balloon that Gruentzig had designed himself. "We had not been able to make a balloon. We had seen the product, we had the product, we had engineers working on how you might do this. We could not do it. We couldn't figure out how you could make the thing expand and only grow at certain pressures. It was a materials thing. They wouldn't tell us what they were doing, oh no. They were Swiss. Andreas wasn't going to tell us anything, and Hugo wasn't going to tell us anything because we had never met."

Gruentzig reiterated that Prigmore himself needed to meet Hugo Schneider personally, with whom he said he shared a patent on his balloon catheter. So the suitor sent yet another letter to Zürich, requesting the earliest possible visit. Meanwhile, the commercial potential of the technique became clearer, once Simon Stertzer threw a mid-June Manhattan press conference, evidently without much forewarning to Gruentzig.

Suddenly, descriptions of the balloon angioplasty alternative flew through the American media, with Stertzer in several instances presenting himself as a vital party to the advance. The result: On June 15, a New York television station ran an excited report, with these comments from Stertzer: "As of the moment, we have high hopes for it; we have worked on this with two other centers, one in this country, and one in Europe. And in that sense it is still in the developmental stages and has to be considered experimental, so that we should not expect that everybody with angina or with coronary disease could be a subject for this particular treatment at this particular time."

There was that *we* again.

In less than two weeks, Walter Cronkite of the CBS network, the most influential anchorman in American television, gave voice to another enthusiastic report. Meanwhile, the July 3 edition of *Time* magazine ran a glowing feature article on balloon angioplasty that seemed to cast Simon Stertzer as the innovation's key pioneer. Gruentzig was livid, and called Myler to tell him so.

Stertzer later claimed that he patched things over fairly easily, blaming the affair on misquotations. Indeed, he warmly recounted one early meeting with Gruentzig in Manhattan in which The Pioneer hailed his vital contributions to the growth of his technique. "I thanked him profusely one night in New York, telling him that he couldn't imagine what effect my being included in this project had on my career. He stopped me and he grabbed me in his inimitable fashion. He said, 'Don't thank me!' He said, 'I have just as much to thank you for everything because the Swiss Society of Cardiology would have laughed me off the rostrum had I not been able to say my work has been duplicated in the United States in New York and San Francisco.' "

Perhaps.

David Williams, a young cardiologist at Brown University/Rhode Island Hospital in Providence, would hear a different story. Eventually happening upon the *Lancet* report of Gruentzig's first balloon angioplasty in the heart, Williams was immediately fascinated. He himself had experimented for the last year using catheters with perforated tips to introduce a variety of drugs into animal arteries. "We were trying to figure out a way to look at the direct effect of these drugs on the coronary circulation. Nobody else was doing anything like this at the time. We conducted more than a hundred experiments and, even though there were a lot of failures, we realized that we could indeed touch the very edge of these plaques. So we began thinking about ways of actively treating these blockages by catheter. We had begun meeting with bioengineers and pharmacologists to discuss the idea of developing more precise interventions, possibly with lasers or ultrasound therapy." In any case, Williams hopped on a plane to Zürich two days after reading about Gruentzig.

The prospect of the trip filled Williams with excitement. The fight against coronary atherosclerosis was his passion, for Williams was yet another cardiologist who had seen his father die cruelly from heart disease—when he himself was fifteen. But worldly he was not. "I had never been to Europe in my life. I was one of those guys taking pictures out the window of the plane."

Arriving at the hospital the following morning, Williams watched as a technician laboriously struggled to clean the insides of a previously used catheter for the day's scheduled case. Meanwhile, Gruentzig took preliminary X-rays of the patient's arteries, only to discover that they held not one but two coronary blockages—a forbidden indication to proceed at that time. The dejected visitor was led back to Gruentzig's office. But then his host treated him to an extensive review of the successes and problems that attended every balloon angioplasty case he had performed. In time, he railed about Simon Stertzer's publicity seeking. "He waved that *Time* article before me, saying this was intolerable. He was furious. He thought he was betrayed."

The two men grew comfortable, and Williams quickly realized that Gruentzig's X-ray images were more sharply defined than any he had ever seen. "Of course, Andreas had been able to get his hands on some of the most advanced German equipment, but he also knew how to tweak the images better than anyone else. No one else really understood as well as he did what a good image of a coronary artery lesion should look like, and there is no doubt that all of diagnostic angiography progressed a quantum leap thanks to his work."

Afterwards, Gruentzig graciously invited his visitor to come to his apartment. En route, he paused and offered with a smile, 'By the way, will you please call me Andy?" *Andy?* Williams was incredulous, suspecting Gruentzig was working too hard at fitting in with his transatlantic partners. "I said, 'Wait a second, don't do that. You don't need to act like an American; you should just keep your name as Andreas.' "

Inside the apartment, the Rhode Island cardiologist discovered that Michaela and Sonja were entertaining another mother with her young daughter who was celebrating a birthday. The next thing he knew, the adults were all joining hands to sing "Ring Around the Rosy." To Williams, the great medical pioneer was absolutely disarming.

Dave Prigmore, meanwhile, finally had gained his invitation to meet Hugo Schneider and hurried to Zürich. "This is how naive I was. I flew the night flight and got into Zürich in the morning, went to the hotel, and showered. Schneider picked me up at 12:30. It turns out I'm going to lunch at some open-air restaurant. He sits me in the sun—and I remember this clear as day—and orders me veal Zürich style. It arrives in one huge plate and then when you're done they bring the second half . . . It's now 2:30 in the afternoon and we've got to have some fresh fruit with some strong cordial and I am sitting in the sun with no sleep. By this time he had me exactly where he wanted me."

The drowsy Prigmore tried to recollect the details of briefs he had received from Bard's lawyers, warning that the alleged patent behind the Gruentzig idea looked fuzzy and potentially invalid, which would severely jeopardize the returns on any deal. Getting to the bottom of this issue was Prigmore's mission. "What I wanted to know was what was the intellectual property status. I'm trying to negotiate a normal deal. He's saying, "We've got this patent.' "

Finally Schneider consented to bring Prigmore to his attorney's office in the nearby town of Winterthur the next day. "That guy and I got into a horrible fight over who was telling the truth, at which point I thought, 'Well this is never going to happen,' " Prigmore recalled. Although that session ended sourly, further discussions continued, with Schneider making it clear that he would only consider selling off distributorship rights to the U.S. and nothing more. Given the recent explosion of publicity, the Swiss businessman was beginning to fathom the magnitude of the Gruentzig idea.

His American counterpart found Schneider to be irritatingly elusive. "He spoke seven languages or something, so anytime you were in a room [with him] if he were speaking to his colleagues it would be in some language other than English." As the dialogue proceeded, Prigmore, negotiating alone, developed a counter ruse. "I'd bring an empty chair up beside me and when I was having to break in for a minute I'd just talk to the chair, and say [in loud English], 'Don't pay any attention to me, I'm going to talk to my invisible colleague on the chair.' "

Years later, over coffee in a small Zürich hotel, Schneider put a different cast on things, saying of Dave Prigmore, "He was a really nice man and we never had any problems."

Memories of those days perhaps remain a little closer to Prigmore, now living on the beach outside Westerly, Rhode Island. He parted Switzerland with the suggestion he might later outline a tangible offer. He also asked to attend Gruentzig's first teaching course, in order to gain a deeper familiarity with the breakthrough.

The course, which took place August 7–10, 1978, attracted thirty-seven doctors, both radiologists and cardiologists from around the world. The audience included nearly the entire first wave of enthusiasts for the balloon idea—twenty-eight individuals in all—among them Myler, Stertzer, Sigwart, Bentivoglio, Williams, and the keenly watchful designer of a would-be-better catheter, John Simpson. A kinetic charge ran through that small basement auditorium at the University Hospital. From the start, every attendee knew that Gruentzig's cases in the nearby catheterization laboratory would be broadcast live on a television screen—which was something in itself that had never been tried before. As they filed in, they also discovered that the inventive Gruentzig had developed one of the very first laser pointers, just so he could pinpoint the subtleties of the X-ray imagery flashed on overhead projectors.

Prigmore, the would-be deal maker from USCI, was not allowed to share this enthusiasm. No sooner had he and John Cvinar arrived than they were told their presence was unwelcome. Prigmore recalled things this way: "Schneider picked us up at the airport and said 'Dr. Gruentzig has decided he doesn't want you at the demonstration.' I'm in Zürich [expressly for the course]. John Cvinar is with me. *Dr. Gruentzig has decided he doesn't want you at the demonstration?* Well, what about the contract we are about to enter?"

After checking into their hotel, the USCI men walked up to the University Hospital to plead with Gruentzig to relent. Weren't they potentially crucial emissaries for his breakthrough? Didn't they deeply need to better understand the procedure themselves? Gruentzig said, "Forget it."

Shaking their heads in disbelief, the businessmen trudged to a café where they had a "lunch" of pilsner beer. Dejected, they rented a car, rolled down the highway, and soon crossed the border to Germany and toured castles in the Lower Rhineland. Eventually, they made their way back to Zürich and joined several doctors of their acquaintance for dinner, pleading for intercession on their behalf. The next morning, they reappeared at the meeting room's door, as Prigmore recalled. "Finally, before a coffee break he [Gruentzig] announced that John and I were going to be out in the lobby and if anybody wanted to talk to us, they could do so for fifteen minutes. So we're like seagulls and bread on the beach for fifteen minutes and that was the end of it."

Cvinar said that things gradually improved. "As a little time went on we were allowed in the hall. Then with a little more time we were allowed to sit and watch the cases, and then, as a little more time went on, allowed to put some literature on the table—it was like death by a thousand scratches."

Inside the auditorium, the European doctors rapped their knuckles on their desks to applaud every case that went well. Prigmore was altogether impressed by Gruentzig's comportment as well. "I saw him as being enormously brave. He was extremely honest. . . . I thought the fact that he was willing to do those live demonstrations was a tremendous marker of his guts, and his character."

Enthused, the USCI men walked back down the hill for another go-round with Hugo Schneider. Plans were finally made for him to visit them in Massachusetts, accompanied by his tough attorney.

The gathered doctors were taking things in with different emotions—their clinical fascination ranged from elation to dismay. "To tell the truth, a lot of people in the room were discouraged, some were horrified by what they saw," David Williams recalled. "There was a lot of carnage going on in that course. About half of the cases that were performed as we watched crashed and ended up requiring emergency bypass surgery; a couple had heart attacks. We didn't know what to make of it, to tell the truth. You could look at what was unfolding before your eyes as being something that was unprecedented and fraught with promise, or something that was very dangerous. He was really rolling the dice between either raving success or very substantial failure—and it was happening live, before your eyes, which was very daring. Some of the doctors left shaking their heads. Others, knowing these patients all would have had bypass surgery anyway, were encouraged. To me, it was very obvious what the potential of the procedure could be. You could see right in front of you that here was something that could easily take off and be a revolutionary treatment with a broad application."

The magnetism of the idea was its very simplicity. "The angiogram was the gold standard of coronary disease then. If you suspected a problem, you did a cath so that you had a picture of what was wrong. Here you had a technique that would get rid of the narrowing on that angiogram in a totally convincing way, so that the gold standard itself would say, 'The problem is gone.' "

The watching cardiologists, Williams said, sensed that the technique might even transform their own lives. "You would see some very dramatic results immediately, which was of prime interest to people like myself who had spent all their time in internal medicine and were handcuffed compared to cardiac surgeons. We were all internists, not surgeons, and to be able to intervene in a patient like that and recast their coronary arteries was a vision. We were sort of starved for this kind of ability."

At one point, the Rhode Island doctor fell into a conversation with a Harvard cardiologist named Peter Block who had already performed a balloon angioplasty and was still possessed of his prize catheter. Williams related that he had a perfect candidate for the procedure, but despaired about his chances of actually acquiring a catheter before the patient was routed to surgery. "Block actually had a catheter. A single catheter. The idea seemed marvelous. We talked about my using his. This seems so odd now, considering that millions of these catheters have long since been distributed around the world. But when we got our first catheter we used it over and over."

Gruentzig's first teaching course ended on an exultant note. A crescendo of applause erupted whenever he succeeded in manipulating his balloon device past a difficult challenge; even the dismaying surgeries seemed to work out. When all was done, Andreas threw a garden party outside the gray townhouse that housed his lonely office. Among the flower beds, he grilled sausages and helped his guests to their fill of Swiss wine. The sense of an emerging transformation was so palpable that a number of these normally arch-competitive doctors physically embraced their host and each other. A few even shed rejoiceful tears and clamored to be photographed on Gruentzig's red motor scooter.

It was odd the way the twentieth century's hardest drivers into the human circulation nearly all shared an intoxication with fast vehicles—motorcycles, sports cars, and planes—that usually happened to be red. Then again, what was about to happen to medicine was symbolized by a simple notion that had floated over the streets of Paris in a movie from two decades before: a red balloon, the beautiful balloon of human aspiration.

Chapter 10

In some ways, the situation on the ground remained faintly ridiculous. Werner Niederhauser and Hans Gleichner struggled like furious elves to produce a single catheter a day.

John Cvinar, the USCI marketing executive, explained the challenge. "The original catheters were fairly complex instruments, so you had little shafts at the end of the catheter that had to be drawn down, you had little skives that had to be made in them, and all of this was done under microscopes. It was all handwork. There was a metal nose cone that went on the end of the thing, and a little wire that went into one of the lumens and then everything had to work right. So it was all done with jewelers' tools. There would be choices in epoxies and the way they set. That was the problem—it was all hand fabrication done by one individual."

Hugo Schneider acknowledged his slowness to grasp the need for change. "I was naive. I didn't realize that such a big business was developing," he said simply over coffee in a Zürich hotel many years later.

"I don't think anybody at Schneider realized what it was all about or had any idea as to what it could lead to," Niederhauser agreed. "We didn't know whether what he [Gruentzig] was doing was good or bad or why nobody else was doing it, because we thought maybe he was a little crazy."

But the customers kept rapping on the needle maker's shop door. Finally, the message sank in: Advancing the technique needed more resources than Schneider's outfit could provide. One autumn afternoon, Hugo Schneider flew into Boston with his Winterthur lawyer at his side. John Cvinar greeted them at Logan Airport and packed them into his car. As it happened, things got off to an inauspicious start—in fact, with a resounding crash.

Cvinar, who now runs a seed capital business for medical-device start-up companies, recalled during an interview in a Boston hotel, "There was an old beat-up Volkswagen that had no lights on as we came up out of a tunnel. He was sitting on the top of the hill not going anywhere and we ran right up and hit him."

Cvinar's passengers were uninjured and the next morning they met with the USCI group at its suburban Billerica, Massachusetts plant. Before long, they walked away with an agreement closed with a handshake. The deal was that USCI would mass-produce the balloon catheters for North America in return for an up-front fee of a mere $27,000 and 6 percent royalties on each device sold. That was twice the amount USCI paid the great Mason Sones, after all. Schneider, underestimating the situation once again, seems to have raised no significant objections.

To USCI, the cost was laughable, according to Cvinar, who had by now become marketing director. "What it really came down to was, 'What was the risk?' We said, 'What's the front end cost of us doing this? $30,000 or so and some engineering time from a couple of technicians—a couple hundred thousand dollars.' So why would we not do it? Dave and I just looked at each other and said, 'It's so damn cheap, let's just do it.' There was not even a hesitation about it."

To wrap things up, Prigmore soon landed in Switzerland with C.R. Bard's gruff chief corporate counsel, Dick Flink. He was, as Prigmore recollected, "not a fun guy to negotiate with." This time, there would be no stupefying wine-soaked lunches in the sun, no absurd conversations with empty chairs—just hard business. After the usual dickering, the Bard men moved in for the corporate kill. They claimed that they stood ready to rescue the Gruentzig idea from its current manufacturing limbo and transform it into an international bonanza. Sign with us, and every person involved here will get richer, and so on. The Swiss, who should have known better, met with an American version of one of their own avalanches. Schneider insisted on clinging to the European end of the business, but that was about as hard as he played his hand. Before the day was done, Prigmore and Flink walked away with a staggering conquest. They agreed to pay Schneider their previously floated grand total of a few thousand dollars—in Swiss francs, as was insisted—for exclusive distribution rights in North America, Japan, and Australia, along with unfettered access to the balloon catheters' workings. Furthermore, the Americans would be privy to all related technology under development without having to pay a franc more. And here came the kicker—a contingency clause saying the 6 percent royalty payments would be halved if the patent proved to be compromised (which Bard already suspected it might be).

And what would Andreas Gruentzig get for handing over his potentially momentous breakthrough to the American corporate world? Half of the first payment to Schneider—$13,500, according to Prigmore's best guess. Consider: The Medtronic corporation recently paid $1.35 *billion* for exclusive patent and distribution rights to a Midwestern doctor's spinal surgery-abetting devices. But doctor/inventors in the late 1970s had no such expectations, and few specialist lawyers stood ready to work on their behalf. Nonetheless, Gruentzig's arrangement with Schneider was to collect a hefty 20 percent of its gross earnings off his devices—which was not bad.

The wheeling and dealing in the medical device industry was just beginning. In fact, before the Schneider deal was concluded, USCI's Prigmore received a curious visitor, none other than the very Keeper of the Cork of the "International Dilatation Society"—a.k.a. John Simpson. The junior Stanford cardiologist sidled in to show off the alternative angioplasty system he was developing and to perhaps gain some seed funding.

With a fellow cardiologist named Ned Robert, Simpson had been rethinking the Gruentzig concept from the inside out. The two had happened upon an intriguing slippery polymer tubing that was being used as electrical insulation in the U.S. Air Force's Phantom jets. The tubing had amazing properties, Simpson explained. By merely heating the tip, a ready-to-inflate balloon could be seamlessly impregnated there. This approach promised to simplify manufacturing and streamline the entire system by eliminating the cumbersome second heart pressure-monitoring channel, or lumen, in Gruentzig's invention. Simpson claimed his miniaturized and more flexible equipment would slip neatly into formerly impossible-to-reach arteries. Prigmore and his engineers listened cagily, promising nothing. A cat-and-mouse game was about to sneak through the entire industry.

Simpson did visit Switzerland within a couple of months in order to demonstrate his new system to Gruentzig himself. However, the Californian's overture was dismissed with an impatient wave. As things developed, Simpson would have his satisfaction, being based not in Switzerland but beside California's booming Silicon Valley, where investors were beginning to shower money like gold dust upon bright new ideas.

USCI pushed ahead at last, closing its deal with Schneider, and dispatching an engineer to Zürich to study under the one technician who thoroughly understood every detail of Gruentzig's devices, Hans Gleichner (Werner Neiderhauser's boss). The American apprentice was supposed to return with a precise rendering of the technology to enable USCI to gear up its production line. The problem was that Gleichner spoke fractured English and was prone to slamming doors in the face of whoever irritated him. Worse, Gruentzig kept fiddling with new iterations of his devices, his directives often sketched out only intuitively. Adding to the frustration was the suspicion that key technological secrets were being withheld.

In any case, Gleichner, in many ways the only true Keeper of the Cork, was behaving strangely. Under intense pressure to produce ever more balloon catheters, he began envisioning pigeons hanging from his bedroom ceiling. Then, in early 1979, he committed suicide.

Other Byzantine issues surfaced. One was that the Bard legal team discovered that Gruentzig had published an elaborate pre-patent itemization of his invention's workings in a German radiology periodical called *Roto,* and followed with a more up-to-date description in a medical weekly. Dick Flink and colleagues pounced on that intelligence with a vengeance. They judged that the technology had therefore been unveiled in the public domain; ergo, the patent was a dead letter, indefensible in any court of international law. Bard/USCI summarily announced it would cut its royalty payments for each device sold by half, to 3 percent. Even though the Schneider company obediently withdrew its U.S. patent, USCI's victory threatened to be a Pyrrhic one, because Gruentzig himself grew enraged, believing that once again he had been double-crossed by fast-talking Americans. But enough information had been ferreted away for USCI to begin building certain components of the Gruentzig system on its own.

The company was definitely not squeamish. They decided that the price of their first balloon catheter should be $175 a pop—nearly twenty times higher than anything else in their inventory. The field was so abuzz that the cardiologists clamored for the new technology nonetheless.

Indeed, the very government of the United States jumped on the bandwagon. Dr. Michael Mock, a senior physician at the National Institutes of Health (NIH), had attended Gruentzig's first teaching course and quickly decided that his concept merited study as an alternative to coronary artery bypass surgery, which by now represented a colossal drain on the U.S. healthcare budget. The NIH's National Heart, Lung, and Blood Institute (NHLBI), for which Mock provided critical guidance, had sponsored a clinical trial evaluating coronary bypass surgery results from dozens of leading hospitals in the United States, and Mock wanted to conduct a similar study of the balloon angioplasty alternative. In January of 1979, he invited Gruentzig to give a presentation before an advisory committee of top surgeons and cardiologists in Bethesda, Maryland. Suzanne Mullin, an NHLBI research assistant, collected Gruentzig and Maria Schlumpf at Dulles Airport. She soon became impressed. "Andreas was very scientific and humble about presenting his data—insisting that he only wanted to get to the truth about the safety of his procedures. He was very passionate."

Every conceivable issue was aired, she said. Why did angioplasty work wondrously in some cases and fail abysmally in others, leading to the need for emergency bypass surgery? Why did arteries sometimes close abruptly in the midst of procedures and how often did this phenomenon occur? It was also becoming clear that many patients' arteries gradually constricted weeks or months after the procedure, an insidious process labeled "restenosis." How widespread was this problem? Who was best entitled to perform angioplasty safely?

Gruentzig addressed the committee's every concern. A joint decision was made that the angioplasty registry should include consecutive patients from multiple medical centers, and that these patients should be very carefully chosen, to comprise an equivalent sampling of perhaps 5 to 10 percent of relatively easy surgical candidates. With candor and charm—and behind it all his gleaming, dark good looks—Gruentzig once again converted the assembled skeptics.

Mock moved forward, knowing that influential members of the U.S. Congress were keen to find some viable avenue of escape from the ever-mounting expenditures being devoted to cardiac surgery. A cardiologist named Kenneth Kent was appointed director of the study, and a letter was distributed inviting would-be participants to a June planning meeting.

Catheter jockeys of varying expertise so itched to get into the game that Richard Myler warned that the near stampede might be ominous. After presenting the latest angioplasty findings to a packed auditorium at the March 1979 meeting of the American College of Cardiology, he flashed one slide that succinctly advised: "Restraint."

Brown University's David Williams well understood the message, since he was by now struggling with regular trials-by-fire back in Rhode Island. "People hungering to get into this had no concept how difficult these procedures could be. No one had ever done anything in the cath lab like this before, and there were no instructions. I created my own manual, which I kept beside me on a kind of music stand. It wasn't all that much help. You knew each time you started that you could hurt a patient, and it was emotionally stressful. You'd scrub before a case and sometimes think about giving up right after you started."

Restraint? Dave Prigmore remembers the June NIH conference in Bethesda as being a truly bizarre scene. Everyone seemed to have an agenda. USCI's was to leverage the groundswell to gain rapid government clearance for distribution of its balloons, now poised to roll into production. Discussions at the meeting grew heated, with some doctors arguing that only they and USCI should decide who should become involved, and that the federal government should bug off.

The meeting hosted a veritable who's who of the new wave of American cardiology, including not only Gruentzig himself, but also Myler, Stertzer, and the avid John Simpson. Charles Dotter was there, too, although cardiologists by now were doing everything in their power to muscle radiologists like him aside from so much as taking pictures of the human heart. Battling lung cancer and driven to the sidelines at the Cleveland Clinic, Mason Sones watched the proceedings with typically outspoken interest. Sitting beside him was USCI's Prigmore, the technology's gatekeeper, who had by now smoothed things over with Gruentzig.

Somehow, Sones and Prigmore became anointed as the session's possible arbiters. Prigmore still laughs about how that played out. "Things kind of got pointed and I said to Mason, 'Why don't you and I discuss this and come back to report in the morning?' He was just a senior observer . . . All I remember is, I said to Mason, 'Come over to my room at six o'clock or something and we'll have a drink.' We [John Cvinar and Prigmore] ran out and brought some hooch and were sitting there. Chiefs of cardiology are knocking on our motel room door. We were hearing names of people we had never met yet, saying, 'I understand you can get a drink here.' I'd say, 'How did you figure that out?' They'd say, 'Oh, Mason told me.' We ended up with this room full of people we couldn't have bought a chance to get to know at all. We finally thinned them out."

A small coterie coalesced, among them a Florida cardiologist named James Margolis and the erstwhile understudy of Mason Sones, Bob Quint—he who had been knocked unconscious by his mentor more than a decade earlier. "So four or five of us went to this dinner and Mason and I got into this big battle in which I said, 'You doctors have to take control of this thing . . . If you don't get control of it, the FDA's going to take control of it, and you don't want them to decide who's qualified to do this procedure. That's your problem.' The night went on—the night went on and on. We started shouting and finally got thrown out of this restaurant for being too noisy, but we settled on what he was going to say."

As the session concluded, the inebriated Sones consented to be driven back to the hotel. Prigmore recalled, "I remember distinctly dropping Mason off at the front door, and he said, 'I'll see you in the [very loudly] BAR!' I said, 'Yeah.' I went to park the car up against the pool fence, and I finally went back to the bar, and I said, 'Mason, here are your keys; your car is up against the pool fence right there.'

"The next morning, the meeting starts up and there's no Mason. I was in a very awkward position and tried to say something. Finally at 11:30 the double doors to this meeting hall slam open and here comes Mason, shouting 'Where the fuck is my car?' He had blown the whole routine. It was sitting right up against the pool fence."

Ultimately, a consensus formed around important protocols, with Gruentzig dictating a central proviso that every would-be participant in the registry must undergo hands-on training beforehand, with the primary vehicle being attendance at a live demonstration course in Switzerland, another one of which was scheduled for that August.

Back in Zürich, Gruentzig's status remained unchanged. The picky undermining continued and he returned to his peculiar banishment in his off-site office beside the ghost of James Joyce. Meanwhile, his superiors kept boxing him out of any meaningful authority in the hospital. They refused to free more than a couple of hospital admissions a week for potential candidates for balloon angioplasty, even though patients were by now writing, calling, and begging for the procedure from around the world. According to the common structure of European academic hospitals, fees for the balloon procedures went straight to the chiefs, with a wafer-thin cut for the inventor and pioneer of the procedure.

Despite his growing acclaim, Gruentzig was still treated like a serf. A secretary named Katrin Bauben was only allowed to work on Gruentzig's duties after hours, despite the fact that the next teaching course was but weeks away. "I was told to start working for Dr. Gruentzig at five P.M. and not before," she explained. "It was only after everybody went home that I was allowed to start on his correspondence or patient scheduling and reports, and the work on the next course." Frau Bauben nonetheless labored into the night, typically bidding *auf Wiedersehen* to Andreas at ten P.M. She remembers his daughter Sonja frequently play-typing and singing songs at his side in the early evenings, what with her father's time by now being swallowed wholesale night after night. Such were the grinding frustrations of Gruentzig's life, and he began quietly seeking a better job elsewhere.

Invitations for speaking engagements rolled in. The newest was to address an August conclave of American cardiologists and cardiac surgeons on a resort island called Kiawah off the coast of South Carolina. Gruentzig appeared at the expense-paid meeting of the South Atlantic Cardiovascular Society with Michaela and Sonja in tow, perhaps expecting to be wreathed with tribute. After all, a famous California doctor had recently dispatched a letter to Stockholm proposing that he, along with Charles Dotter, should be awarded the Nobel Prize.

No sooner had the Grand Inventor finished discoursing on the triumphs and travails of his first fifty procedures than a sort of Grand Inquisitor stepped forth. This individual was a Hungarian-born North Carolina cardiac surgeon named Francis Robicsek, and he was not only brilliant but an attack dog. Although he had been asked to play the role of formal cross-examiner, no one had expected Robicsek to come forward with both guns blazing. But the man was so determined to speak on behalf of his entire brotherhood of increasingly balloon-hating surgeons that he had reportedly flown to Zürich to prepare a dossier on the shortcomings of the Gruentzig approach.

A kind of public ambush unfolded, with Robicsek dripping sarcasm and slinging barbed questions from the podium. What kind of miracle procedure is this, the interrogator asked, that sends one out of every five patients to emergency surgery with their lives hanging in the balance? What sort of godsend is being offered when these little balloons cannot even be advanced through one out of three of the most cherry-picked coronary narrowings, whereas surgeons can reroute these slam-dunk single blockages with a success rate of nearly 95 percent? Why, nearly half these balloon jobs close up again in a matter of time due to so-called restenosis. Any medication with such a failure rate would be yanked off the market, Robicsek sneered.

Now it was Gruentzig's turn. With aplomb, he methodically addressed each issue that had been thrown in his face, explaining that the numbers were improving rapidly with refinements in the technique and that every one of these patients was scheduled for bypass surgery in the first place, were it not for the alternative of his far more benign procedure. But the ferocity of the surgeon's attack rocked him. "The guy (Robicsek) was so smooth and devastating, it was very funny to everybody except Andreas, and I think he had never confronted anything like this before," Emory University's Spencer King remembered. "He got really steamed and fought back."

Gruentzig may have been unsettled for a moment, but the meeting evolved into an idyll of summer collegiality. The next two afternoons featured tennis, swimming at the beach, cocktails, and oyster roasts before the setting sun. Meanwhile, the visitor was subjected to a charm assault, at which American Southerners are past masters, by a collection of softly accented cardiologists and heart surgeons from Emory, the most influential medical center in the South. At the time, it was impossible to know that his very future was being gathered into a net.

The second live demonstration course on balloon angioplasty began a week or two later in Zürich, in August of 1979, with about sixty European and thirty American doctors in attendance. The latter included Ralph Lach, a private-practice cardiologist at the unsung Mount Carmel Hospital in Columbus, Ohio, an indication of the spreading reach of the Gruentzig gospel. Lach had plowed through his cardiology career with ardor, performing 2,500 diagnostic angiography procedures to this point. He had managed to taste the good life, gaining a substantial house beside a golf course for his wife and brood of young children. Yet this rather ordinary-seeming cardiologist ached from a sense that *something* was profoundly missing. Perhaps the word was drama. Over the last year, he had picked up on the growing buzz about balloon angioplasty and found his way to Zürich to see for himself.

Lach had no inkling as to what lay ahead. A weekend jazz pianist and ardent amateur baseball player, he was no academic and wouldn't rank as an influential figure in his field. He toiled in the fields of obscurity, even though well-rewarded. But he was fed up with referring patients to bypass surgery. No matter what the Robicseks of the world said, surgical intervention dispensed as much woe as triumph, as far as he was concerned. "What we were doing I thought was ridiculous," he recalled one autumn afternoon outside Columbus. "I really suffered for my patients because we were totally in the dark concerning who should go to bypass surgery and who should not. . . . It was the most major of all surgeries, because it always included a big sternatomy [splitting of the chest] and cardiopulmonary bypass and the risk of neurological problems, motor complications, wound infections, and [so on]. It felt bad. Even if we were referring patients for what we thought was the right motivation, we were oftentimes sending people off to a catastrophe. The Gruentzig idea sounded simple and different and I was ready to give it a look."

Here then was another American innocent abroad, tasting his first trip to Europe. He sat himself in the amphitheater beside rows of cardiologists he had never even heard of: among them Jean Marco from France, Myler and Stertzer, Williams, and a team of Wisconsin go-getters named Gerry Dorros and Dick Springer. Nearby were USCI's Prigmore and Cvinar, at last fully welcomed into the fold.

For Gruentzig, it was showtime once again, the chance to perform before a live audience such as had last gathered before nineteenth-century surgeons cutting away before bug-eyed students, or the macabre anatomy work of Andreas Vesalius back in sixteenth-century Padua. Dapper and unflappable, Gruentzig announced that 264 balloon angioplasty cases had been performed around the world to date, and that the success rate lingered at 60 percent. Then he threw more cold water on the would-be stargazers, acknowledging that at least a third of his own successfully treated patients had their arteries wither shut again from restenosis in a few months. Even worse, 10 percent of all patients required immediate emergency surgery once the balloon was withdrawn from their hearts, and up to 2 percent were dying overall, though none at Gruentzig's own hands.

Just reflect, he admonished. Behold the promise as well as the pitfalls. Remember also that any serious examination of bypass surgery reveals nearly equivalent problems, at the cost of exponentially worse travail. So keep your minds open. This is a technique at its most embryonic and fragile stage, and one that will never ripen without exquisite caution. Its future may lie within your hands, Gruentzig advised.

"He was a masterful teacher," Lach remembers. Gruentzig started things by presenting X-ray films from a procedure gone wrong, one that could have threatened brain damage. "The lights went down in this steep auditorium as he showed his first case, and when the right coronary artery was injected with dye there was this little stream of bubbles running and then roaring down the artery and you could hear the gasp in the crowd. This was a gob of air, this was a train, so we watched. This was really a softening up of the crowd, because the patient suffered no ill consequences. Gruentzig just quickly turned and opened his palms and said, 'Let he who is without sin cast the first stone.' "

A moment later, Gruentzig, the born showman, projected a photograph of this same patient lifting his head off the table and smiling happily at the procedure's end. What kind of hocus-pocus was this? The cardiologists in the audience had their heads spun again as they watched patients cheerfully chatting with Gruentzig even as he threaded his balloon catheters into the depths of their hearts. During bypass surgery they would have been stretched out insentient, their hearts stopped in a twilight zone between life and death. And almost no modern surgeon back then would ever have dared televise in real time the unpleasant sights and sounds of his unpredictable trade. But Gruentzig was passionate about holding nothing back, believing that the sheer dint of his scientific honesty would be the key to gaining credibility among a world of skeptics.

Bruno Lorenzetti, an Italian-Swiss cardiologist, thought the very premise of these sessions bordered on the heroic. "Exposing the procedure on a live transmission where mistakes were seen as they occurred and difficulties had to be dealt with instantly, one on one, *that* was his biggest invention."

For his part, Lach sat spellbound as Gruentzig worked at threading a balloon catheter through one particularly forbidding artery. "I remember leaning over to a seatmate of mine, and saying, 'There is no way he is going to get that balloon catheter through that critical stenosis.' But he did. Of course, then I became a believer and saw the balloon as the future."

A leg-artery case was presented by the Romanian-born angiologist Ernst Schneider, who had enjoyed Gruentzig's company while they were fellow postgraduate students in Darmstadt a decade before. Schneider had since moved to Zürich and ultimately assumed responsibility for Gruentzig's peripheral angioplasty procedures. Lach was dumbfounded by his work, because it proceeded as though with a snap of the fingers. "There was this critical iliac stenosis and I saw him just put that balloon in there and blow it up and have a beautiful result . . . It was so completely simple and you could see (by extension) where coronary angioplasty was going and that it would get so much better and so much faster."

Lach approached Gruentzig to share his enthusiasm. The interchange remains fixed in his mind, he said one afternoon in a living room that looks out over a sprawling golf course. "He was one of the most unforgettable people I ever met, full of energy and deep perception of what was going on around him and what his role was."

At the end of the working sessions, the doctors spilled out into the byways and cafés of Zürich to share their elation and concerns. The euphoric Lach chanced upon an itinerant German Dixieland band making merry in a public square. At a glance, he could see that they were improvising madly and lacking key players. The Ohio cardiologist spotted their idle piano and promptly offered to stand in. Suddenly, he was performing "All of Me" for the good burghers of Zürich and his dumbfounded wife.

Gruentzig loved that story. He adored the social aspects of his teaching courses, and treated this batch of participants to a long evening boat tour of the Zürichsee, upon which Swiss wine and delectables were proffered without cease. Relationships were formed, shared excitement bubbled along with many escaping bursts of champagne.

As it came time to depart, Ralph Lach fairly skipped to the airport. "Well, I almost didn't have to get on a plane to fly home, I was so high by that time. I had seen what I thought was near miraculous. I was elated." Charged with excitement, he returned to his hospital and announced before a gathering of thirty or forty doctors in the cafeteria, "Guys, there is a revolution coming. Health care is going to be changed with what's going on with Andreas."

Some enthused, some scowled, but others apparently set to scheming for a toehold of their own. Petty politics danced into the mix. According to Lach, a rival cardiology group impeded the mere filing of his application with the federal authorities to proceed with his first procedure. Even obtaining the requisite equipment proved to be a nightmare. USCI's Cvinar reported that nine out of ten of its supporting systems failed quality inspections. The core balloon equipment kept arriving from Zürich at a trickle's pace. "We'll get it to you as fast as we can," Cvinar promised.

USCI was juggling madly as a medical melee clamored for its attention, while the little Schneider company all but threw up its hands. "There was no quantity at all," Cvinar recalled. "We were getting shipments of twenty-three units a month. It was awful and they couldn't keep up with the volume. People were screaming at us. I am traveling around the country teaching people, trying to get individual units into guys' hands so that they could use them in cases. I was training them so that they could understand, making sure they understand how the system works, because I had spent enough time with Gruentzig by that point. At the end of the day, they (Schneider) just couldn't keep up with what we needed."

Lach called Cvinar every Monday to beg for the equipment. When he finally identified a willing and ideal candidate for the future, he burned up the phone lines. Was there any hope of satisfaction? Cvinar shortly rang back with the good news. "What if I told you are going to have your catheters on December 4th and you can do your case on December 5th?"

This was Christmas come early, said Lach. "I said, 'You wouldn't be fooling me now, would you?' "

A fifty-one-year-old storm-door salesman with excruciating angina was the patient in question. The man was evidently excited at the prospect of becoming the first patient in Ohio to undergo the historic procedure; in any case, he agreed. The procedure was scheduled to occur in a week. Things did not proceed smoothly, once Lach discovered that his rival cardiology group had snuck their own angioplasty candidate onto the catheterization laboratory's rolls ahead of his, even employing a fictitious name in their ruse. All hell broke loose now, with Lach furiously remonstrating that he alone had pursued exacting preparations, having not only enrolled in the Zürich teaching course but also studied with the hotshot Dorros-Springer team in Milwaukee and Simon Stertzer in New York. The hospital administrators suffered through the acrimony and ultimately decreed he should go first.

The historic Ohio day did not dawn auspiciously. Lach had difficulties even opening the packaging, due to a cautious technician having gassed it to ensure sterility. The plastic wrapping had melted and the local Gruentzig protégé was left feverishly trying to dissect his precious catheters from the fried shrink-wrapped packaging with a scalpel. Scrapings strewed beside his feet. At last, Lach was ready to proceed; he caressed the coveted tools in his hands. Balloon angioplasty was poised to make its mark in a new domain. Crowds gathered.

"People stood shoulder to shoulder in the cath lab, beginning with the attending physician and his partner. There were medical students, a bunch of nurses, assistants from the coronary care unit, anesthesiologists, and surgeons and their people. They tell me the crowd stretched about fifteen or twenty feet down the hall. It was just wall-to-wall." In the nearby operating theater, cardiac surgeons stood by to help in case anything went wrong. Excitement hung in the air.

As it happened, the drama quickly devolved into a kind of medical burlesque. A thousand pinpricks of humiliation and grinding frustration awaited this newest Midwestern explorer of the heart. Lach knew he had his work cut out for him, since the blockage in this patient's artery sat at the very mouth, or *ostium,* of the coronary artery, where leverage threatened to be tricky. But hadn't he seen the very maestro in Switzerland handle such a challenge with sublime calm?

Calm? For five and a half hours, Lach groaned through an ordeal for the record books. At first, he simply could not position the guide catheter where it belonged. "You would come to a critical point in terms of trying to rotate the tip of the [guide] catheter into the right coronary ostium where you applied a little more rotation and the catheter wouldn't move. You applied a little more pressure and the catheter moved about a tenth of what you intended, and you applied a little more and 'oops,' now you have buckled the catheter . . . The guide catheters were the consistency of a soda straw," Lach recalled. "We probably made several hundred passes at that right coronary artery."

Finally, Lach had the system lined up to perfection and prayed all would be easy sailing now. Anticipating the moment of long-awaited triumph, he slipped the balloon device through the guide catheter and into the artery—and promptly ran into the cardiologic equivalent of impotence. "It stuck in the lesion, it couldn't go any further—which was typical in those days. There was no power." Desperate to succeed, Lach fussed on, twisting and nudging the external end of the catheter right and left. At last the tip slipped into place. With a squeeze of the external inflation device, the balloon finally began to push against the ugly accretion of plaque. Lach caught his breath as the X-rays displayed the result—the arterial channel was indeed markedly widened! Without saw or knife, the patient's perilous coronary anatomy had been remade. Wiping his brow, the triumphant cardiologist looked around the room for congratulations, just as DeBakey and Cooley were so practiced at doing in Texas. Hadn't he himself watched an entire amphitheater break into applause when Gruentzig finished a tough case back in Zürich? Lach's panorama was a bit more forlorn. There was one spectator left. In the adjacent operating room, the surgeons who had just frittered away five and a half hours, earning little for their standing at the wings, snarled.

CHAPTER 11

IN THE WANING MONTHS OF 1979, Andreas Gruentzig decided to get out of Zürich. Fed up with the incessant pettiness that plagued his life there, he cast about for a position of authority and respect elsewhere. He wanted to shepherd his advance forward in an environment where he could be master of his own fate. But finding his dream position proved to be no simple matter.

Gruentzig's first preference was to return to Germany, equipped with a full academic professorship. But even Martin Kaltenbach, his influential collaborator in Frankfurt, was unable to help him back to his native land. Petition as Gruentzig might, the message rang in loud and clear: He was still regarded as a maverick who was not only short on the requisite credentials but a trifle too flashy for his own good. Resentment lingered over his popular acclaim—it was Sauerbruch versus Forssmann all over again.

Bernhard Meier, the young cardiologist-in-training, remembers the scenario well. "These guys in the high places told him, 'Yeah, you've had your success, you're a nice plumber, you've invented your plumbing technique, but you're not a doctor.' He wanted to prove to them that he was as intelligent and scientific and clever as they were, because it would have given him satisfaction to become a professor in Germany. But the Germans were just like the guys in Zürich: They did not want Andreas because they knew he would be casting too large a shadow over them. There were all these jobs that were available at German universities that Andreas applied for and he was turned down at every single one of them, just because of jealousy. It was ridiculous." The crushing rejections littered Andreas's mailbox for months, and the onset of winter saw the applicant still trudging back and forth through the snow from his office of banishment.

Over in America, it was a different story. Several top institutions expressed keen interest once word of Gruentzig's availability spread. In the publicity-conscious U.S., the prospect of landing a star-in-the-making exerts a nearly gravitational pull. Harvard's Brigham and Women's Hospital started out at the front of the pack, but progress was slow. Meanwhile, California's Stanford University picked up the baton. But here surfaced another problem in the form of a requirement that Gruentzig must first sit, like every other similar recruit from abroad, for the rigorous California examinations for official licensing in internal medicine. To the proud applicant, the very idea felt like a smack several rungs back down the ladder. "Out of the question," Gruentzig said. Richard Myler attempted to step in by beseeching politically connected contacts to create a special exemption for the man he swore would transform the very future of medicine.

At the Cleveland Clinic in Ohio, Mason Sones barraged the top brass with the message that landing Gruentzig would solidify the institution's reputation for years to come. Sones yearned to designate his own heir apparent and kept the negotiations alive as the calendar turned to 1980. But Cleveland, Ohio? Gruentzig's wife Michaela fretted, having little desire to rush off to an America that she perceived as being ridden with noisy capitalism. She cared little about money and less about vain trappings of any kind. In fact, she reportedly had become involved in a Rudolf Steiner-like group which, in the peculiar fringe ideology of the 1970s, preached that careers themselves, along with every vestige of personal possessions, from household goods to the day's clothing selections, were a distraction from all that mattered—the purity of the inner self. What mattered to her in any case was the increasingly formidable task of protecting her own privacy and quietly raising her daughter while her husband turned into the medical equivalent of a rock star. Their lives were by now fraught with friction.

Another major distraction loomed—namely, the advent of Gruentzig's third teaching course. In the middle of January of 1980, doctors arrived in Zürich from around the world. Once more, Gruentzig performed his live cases with élan, led animated discussions, and made every participant feel as though he had walked through a threshold to a new medical world. Photographs from that session show a dozen cardiologists at a time surrounding him with eyes wide. During lectures, Gruentzig pranced about the stage in white shoes. Certainly, no cardiologist had ever beheld such a figure, especially not in their nervous, watch-your-back world—for here was a doctor who would weather every hazard and question before a live audience with aplomb and searing honesty, then laugh and dance into the night.

Gruentzig capped this session with another one of his cathartic entertainments. These events somehow always seemed to work as a kind of initiation rite into his ever-growing yet always select brotherhood. This time, Gruentzig hired his own train for a journey to the Emmental Valley, lush and cowbelled in summer, dreamily snowbound in winter, and emblematic of the heart of Switzerland. The serious business of the meeting over, he plied his visitors with wine as they chugged ever higher into the mountains. He wandered the carriages kissing the wives' hands, clinking the glasses of the men, and dropping frequent notions of the great vistas ahead. *The future is ours,* was his implicit message.

The train finally pulled into the sleepy village of Grosshöchstetten. Out onto the narrow streets, the assembled Gruentzig protégés tottered. Here to not only greet but exalt them were musicians with bellowing mountain horns, decked out in traditional mountain dress. The tipsy doctors proceeded to a central hall where long tables were laid out with the local provender: fine sausages, air-dried beef, and the valley's signature cheese, along with more wine. The repast grew ever more jolly, and finally the mayor stood up to offer a welcoming speech embellished with peculiar perorations—seeing as neither he nor anyone else in his domain had a clue as to what the visitors were really up to—during which he compared balloon angioplasty to Emmental cheese. Both, he grinned, needed to make holes to achieve their perfection. The slapstick simplicity of the man's conceit drew more than a few guffaws, but it added to the levity. After another glass of wine here, a canapé or two there, the giddy sojourners climbed back onto their little train.

On the return journey, Andreas Gruentzig connected one by one with his merry disciples. Before long, he sat beside Atlanta's Spencer King, his host at the South Carolina island retreat the previous summer. A surgical colleague summed up King's personal style this way: "You know how easy he is? He's Mr. Smooth, he isn't going to aggravate anybody. He's going to find out your answer before he asks you the question. That's the kind of gentleman he is."

In other words, the two were a kind of match. King listened attentively as Gruentzig related his dissatisfactions with the situation in Zürich and his growing interest in working in America. The Georgia cardiologist remembered the conversation: "He said, 'You know I've got to get out of this place. I am not happy here. I can use the cath lab only two days a week,' and so on. I said, 'What's your goal? What do you really want to do?' And he said, 'Well, I want to expand the technique. As it goes forward, I see the possibility that it's going to grow to the point where I may lose control of it. And it may destroy the technique if it's not done right . . . I don't think I can do this here . . . I want to teach the technique. I want to shepherd it, and I want to be a professor . . . I am either going to go to Germany or the United States, and the Cleveland Clinic sounds best, because it is a big surgical place and is world famous because of Favaloro and Sones.' "

"Well, you have a point there," nodded King. But one of Gruentzig's notions did not quite add up. Wheels turning and voice softly drawling, King made his move. "I said, 'Well, there's a problem, because the Cleveland Clinic's not really a medical school and you can't be a professor there.' He just stopped. Then he said, 'Really?' I said, 'Yeah.' He said, 'What do you think I should do?' I just threw it out: 'Why don't you come see us and have a look at Emory?' "

Doctor Smooth had made his silken invitation. Gruentzig pondered. What did he really know about Emory or Atlanta, Georgia, or for that matter the entire American South? He related that he was about to fly to Colorado for a medical meeting at the Snowmass ski resort. King suggested that Andreas stop in Atlanta afterward.

So within two weeks Gruentzig beheld an Emory University School of Medicine bursting with new energy and wealth, thanks in part to a $105 million bequest from the Coca-Cola Foundation, which was then the biggest grant ever given to any academic institution in history. That money was delivered in an anonymous bequest from the secretive soft-drink heir, Robert Woodruff, one of the richest men in the South. The eighty-nine-year-old lord of a substantial estate on nearby Tuxedo Road and a 30,000-acre quail-hunting plantation ("Ichauway") in south Georgia, Woodruff had a passion for promoting medical research. Playing golf with his friend President Dwight D. Eisenhower one afternoon in the 1950s, he had sealed the arrangements that would position the new United States Centers for Disease Control beside his beloved Emory University, to which he quietly slipped formidable checks every year—$250 million when all was added up. "Mr. Anonymous," as he was called by a biographer, also saw to it that the American Cancer Society would shift its New York headquarters to a wooded site across the same street.

As he aged, the widowed Woodruff looked to Emory as a kind of spiritual resting place whose interests he would provide for as long as the institution looked after his failing health. A luxurious condominium was constructed within Emory Hospital just for Woodruff, a kind of bunker-in-waiting for any emergencies that might befall him. The place enjoyed such ultra-tight security that the FBI sought—and won—access to it to house the most scandalous pornographer in the U.S., *Hustler* magazine's Larry Flynt, while he recovered from a paralyzing assassination attempt in 1978. Woodruff shrugged aside that controversy. The very next year he shelled out his secret $105 million to Emory Hospital.

So Emory was flush and expanding at a furious pace. The scene nonetheless remained thick with internal politics. To pave the way for Gruentzig's visit, King called on the chief of medicine, a formidable, if courtly, figure named Willis Hurst, who was the author of a world-famous textbook on cardiology. King remembers his chief's initial reaction with lasting amusement.

"What he said to me was, 'Oh, I heard about that guy—he's a prima donna. You don't want to get involved with him.' I said, 'Look, I invited him to come and he's going to be here as my guest, and I want to bring him around and introduce him to people.' " At that point, Hurst acquiesced.

Gruentzig arrived to spend a couple of nights in Spencer King's stately Tudor-style house, which sits on a leafy rise just beside the local golf club. To a man living in an austere third-floor apartment with a three-foot-wide terrace, the scene had to be impressive. King's home is a refuge of mahogany-lined tranquillity, with a winding drive capped by a stone archway. Several Emory colleagues enjoy similarly comfortable residences up and down his wooded road, which is sandwiched among a series of lovely parks designed by Frederick Law Olmsted, the landscape architect who created New York's Central Park. This was not downtown Cleveland. King gave his visitor a tour of the entire Olmsted-designed Druid Hills neighborhood, where sumptuous homes are done up in Georgian or Tudor style, or resemble French manor houses or Italian villas. At last, they passed the exclusive Druid Hills Country Club as they approached the Emory medical campus. That suburban complex is like a small shining city unto itself. Lofty glass and granite edifices are surrounded by flowery plazas and broad walkways, far from the tawdry slums of downtown Atlanta.

Now it came time for introductions to Emory's leading specialists. Knowing that able cardiac surgeons provided the critical safety net for his new procedure, Gruentzig first met with the local practitioners of that craft. One of these was an impressive, square-jawed figure named Joe Craver, a former all-American football player at the University of North Carolina. Because Craver could be Marine-tough if he wanted to, King had set the stage beforehand with a collegial visit. Craver recalled, "He said, 'Joe, we've got this potentially great situation developing here, because angioplasty is going to be big, I think; and would you be interested as a surgeon in supporting him [Gruentzig]? Right now it's really shaky, and there is a lot of need for surgical backup.' "

Craver obliged. "I liked Andreas from the start; I liked him from when we first met on Kiawah Island. I thought he was an engaging man and I thought he might add a great deal of value here." The warmly predisposed surgeon was thus able to play his role to perfection, beginning by outlining the capabilities of Emory's cardiac surgery department. "There were a lot of places that had a stronger reputation," the avid outdoorsman acknowledged one afternoon in an office decorated with stuffed game birds and trophy fish. "But we had long since passed doing a thousand cases a year . . . He wanted to see how we did what we did and whether we were safe, honestly. Because a whole lot of them [angioplasty cases] were still crashing at that time . . . I think he became comfortable by the end of the day that his surgical backup needs would be met."

The next morning, Gruentzig was given a tour of the cardiac catheterization laboratories and introduced to the power brokers of this cagily political Southern university. A key point of call was Willis Hurst, who only days earlier had dismissed the visitor as a prima donna. Spencer King still marvels at what happened next. "Within probably milliseconds of meeting him, Willis was totally seduced by Andreas's charm and became very interested in the whole idea of his coming here."

Another visit was paid to a cardiac surgeon named Charles Hatcher, who was the chief of his department. He also served as director of the Emory Clinic, and was a regular luncheon partner with Coca-Cola's Robert Woodruff. Gruentzig had certainly never met such a type in Zürich. Hatcher is one of those figures of the Old South whose drawling conversational flights leave the head spinning. Deeply connected to the clubby inner circles of Atlanta society, he owns his own quail-hunting retreat in south Georgia, where gentleman friends gather for sporting weekends. He was another charm specialist.

Gruentzig could scarcely have fathomed the quiet machinations proceeding behind the scenes, the talks in the corridors and phone calls to vet the situation with well-placed contacts. For all their outward bonhomie, the Emory kingpins hungered to make their mark on the international stage, and their semi-private clinic had become so profitable that they were suddenly possessed with enough money to recruit whomever they wanted. So the dance of courtship continued.

Finally, the Emory men invited their guest to lunch at the most exclusive venue in Atlanta, the Piedmont Driving Club. The place takes its name from the processions of coaches and fours that used to proceed from its estate-like grounds, and pays homage to the social stratifications of an earlier Georgia. Neither women nor blacks were admitted until recently, and long pedigrees are required even to be considered for membership. The horses and buggies may be gone, but revelers in Jaguars, Mercedeses, and Rolls-Royces are met by its valets. Its balls and gala evenings are black-tie affairs where the elite in booming 1980 Atlanta had much to celebrate. Indeed, James Earl Carter, a native son of Georgia, had just served as President of the United States.

The Emory luncheon party was shepherded into the Piedmont Driving Club's cozy men's grill. On hand were Spencer King, the quail-hunting Charles Hatcher, and Bruce Logue, the chief of cardiology and a central force at the Emory Clinic. Drinks were poured, menus perused, and the talk flowed. Tell us what you need in order to join us, the Georgia doctors asked. Their by-now eager visitor laid out a few concerns—among other things, he worried about gaining a visa to work in the United States and still wanted no part of any laborious American qualifying medical exams. His hosts nodded with understanding. Then Spencer King looked over the wood-paneled room. And here arose a scene straight out of Tom Wolfe's masterful novel, *A Man in Full,* in which he described this very same domain.

"You might have the grandest house in all of Buckhead and the summer place on Sea Island and the biggest private jet and the ranch or two in Wyoming, every toy a man could possibly long for—and yet your failure to make the roster of the Piedmont Driving Club would always be hanging over you, like a reproach," was the way Wolfe characterized the eminence of the place in Atlanta social circles. "The Driving Club was the very sanctum, the very citadel of the White Atlanta Establishment."

Who should be seated at a nearby table (and most certainly making the roster) but the former Attorney General of the United States under President Jimmy Carter, Griffin Bell. Spencer and Griffin happened to be friendly. So Spencer waved. "I said, 'I want you to meet someone.' Griffin came over and I introduced him and said, 'This is Andreas Gruentzig, the guy who invented angioplasty . . . We're trying to recruit him.' So Griffin Bell said, 'If there is anything I can do to help, let me know.' I said, 'Well, I am sure we will have trouble with his visa.' Griffin said, 'I don't know what I can do about that, but I did appoint the head of the Immigration and Naturalization Service of the United States of America, and my law firm does work on these issues.' "

So there you have it, the Georgians more or less said to Andreas Gruentzig: We'll look after you. Think things over closely and y'all give us a call. Returning home, Gruentzig suddenly found that a possibility had developed at Germany's prestigious University of Tübingen. But just when he thought the position was his, that opportunity inexplicably vanished. In March, Gruentzig returned to Atlanta, this time accompanied by Michaela and Sonja.

Spencer King and his fellow cardiologist, John Douglas, were now not only avidly performing balloon angioplasty themselves but keen to make their department a powerhouse of the new technique. They offered an equal joint practice with no higher-ups standing at the sidelines to reap personal profits from their understudies' procedures—as was routine in Europe. The offer was generous in more ways than one, since opening the doors to such a potentially dominating figure as Andreas Gruentzig also carried the risk of upsetting their working equilibrium. Gruentzig nonetheless worried about the details. The administration responded by arranging a joint professorship in radiology and cardiology, thereby providing dual streams of both income and authority, mirroring the arrangement King enjoyed himself. In short, the red carpet was unfurled.

The recruit was ushered in for another session with Willis Hurst, that other key puller of the strings. He indicated he would do anything within reason—and even a bit more—to create a dream position for Andreas. As it turned out, Gruentzig had a list of demands. The white-haired Hurst recalled the negotiating points in his deeply sonorous Georgian accent. "Here are the things he wanted . . . He wanted to be a full professor and I said I would be able to get that done. The next thing was that he didn't want to take the Georgia state license exam . . . Andreas firmly said, 'I will not take the exam. If I have to take the exam I will not come.' "

Here, Hurst smiled his best don't-you-worry smile, explaining that at one time or another he had trained many influential doctors involved with this process. Before long he would have some private talks with them, saying, "I think he's a genius and a 'national treasure.' I compared him to having Einstein here and said, 'We don't want any impediment to landing this treasure out in front of everybody else.' Therefore we were able to get that."

Truth be told, it was not so simple. Just like Stanford and about every other leading academic medical center in the United States, Emory had its own requirement that all foreign-born doctors must pass the state exams in their field before they could qualify for staff privileges. So Willis Hurst, Charles Hatcher, and Spencer King put their heads together and began pressuring the president of Emory University (equipped by Coca-Cola with a 7,500-square-foot mansion) to bend those particularities—just this once, seeing as they were about to make a legal case that a "national treasure" was on the line. Hadn't Gruentzig been nominated for the Nobel Prize?

Hatcher explained that Griffin Bell and his legal colleagues had happened upon historic precedents—exercised for the likes of the German V-2 rocket scientist Werner von Braun—whereby certain individuals considered to be of compelling importance to the American "national interest" were granted instant citizenship. The challenge was to prove that Emory's assertions about the Von Braun-Einstein class of Gruentzig's importance were not spurious. So Hatcher and Hurst began soliciting letters of affirmation from the biggest institutions on the American medical landscape—among them Massachusetts General Hospital, the Mayo Clinic, Stanford, and Johns Hopkins. Working every connection, they convinced the Georgia State Board of Medical Examiners to promise to grant an immediate license to practice medicine if Emory could win its "national treasure" case with the federal government of the United States.

The Emory kingpins played every conceivable card to gain their prize. Hurst recalled, "He wanted a position for his wife Michaela who was a psychologist, so I had to go to the depths of the psychiatry department and convince them that they should recruit her, which thankfully they were willing to do. Then he said he wanted an office with a window." The droll Hurst wove his fingers together and said, "I don't know the background of that, but I would assume [pronounced 'ahsoom' Georgian style] where he was he didn't have a window."

Well, *ahem*. Hurst thought on this one, and came up with a truly magnanimous gesture. Seeing as his own substantial office looked out toward leafy Druid Hills, Hurst offered to divide his personal space by half, just so that Andreas could have the view he cherished. Was there anything else? It turned out that there would eventually be much else. Somehow, every need was met, and the two men began to forge a relationship that was like that of father and son.

Down in Georgia, they know how to get things done. In the middle of the negotiations, the surgeon Joe Craver invited Andreas and his wife and daughter for a "simple" dinner party in his home, the agenda being to demonstrate the Southern hospitality that awaited them should they make the move. About half a dozen doctors and their wives gathered before a candlelit table laid out with silver plate and crystal glasses filled with wine.

Craver quickly tried to introduce a note of humor. He explained that the U.S. government had recently turned a blind eye to an exodus of refugees, some criminal, from Castro's Cuba, called the *Marielitos*. Thousands of them had landed in waves of rickety boats on Florida's shores. Craver slyly suggested, "Consider what Carter did with all the *Marielitos*. Andreas, have you ever considered just wandering onto the beach. You look like a swarthy kind of guy, almost like a Latino yourself. Have you ever considered just wandering out of the surf on Miami Beach and asking for asylum in Spanish (which Gruentzig actually spoke well) and that way you automatically fit in? The only thing you need to do is not get arrested for six months and then you will have made it! You would save Charles Hatcher lots of aggravation and thousands of dollars' worth of Griffin Bell's time.' "

Andreas just laughed and said, "I would prefer to come in the front door and not by the beach."

Food was served, drinks refreshed, conversation looped off in amiable circles. "The Southern charm got poured on as good as it gets, there was no two ways about it," Craver acknowledged. All was happy jocularity until three-and-a-half-year-old Sonja tottered out to the patio where a long swimming pool remained topped by its protective winter cover. Such arrangements being rare in Switzerland, she promptly strode out onto its flimsy midst and collapsed into the water's surface, crying for help. The little girl was promptly fished out, no worse for wear. Yet Michaela, in any case, continued to see something threatening about the American way of life—and in time would be proved right.

The former East German refugee soon accepted a starting Emory salary of about $180,000. This was five times more than he ever earned in affluent Zürich, and eighteen times more than his revered Mason Sones had started with at the Cleveland Clinic—the equivalent today of perhaps half a million dollars. Better still, if the fees from his procedures surpassed the projected $180,000 benchmark, his earnings could ratchet substantially higher. Meanwhile, he would gain everything he had ever dreamed of—fabulous working conditions with his own catheterization laboratory, a gleaming lecture hall, an extensive support staff, the professor's title he coveted, and a rich university eager to promote his achievements around the world.

August came and it was time for Gruentzig to host his last teaching course in Zürich. The meeting promised to be the greatest extravaganza of them all. This session drew participants from every continent, including doctors from Australia, Austria, Belgium, Brazil, Canada, Czechoslovakia, Egypt, France, Germany, Holland, Hungary, India, Iran, Israel, Italy, Japan, New Zealand, Poland, Russia, South Africa, Spain, the United Kingdom, and Yugoslavia, with fresh legions from the United States included among its 210 participants. Knowing that this was his Swiss finale and sensing that his procedure's impact was about to skyrocket, Gruentzig also assembled his most illustrious predecessors—Melvin Judkins, and especially Mason Sones and Charles Dotter. Werner Forssmann had recently died.

The writing was on the wall. Twenty-nine medical centers now stood ready to report on their collective 800 procedures. The technology and the technique were improving so rapidly that the procedure was beginning to achieve success nearly 80 percent of the time. The nightmare complexities of the early days were being resolved one by one.

Once again, Gruentzig performed his live procedures with flair, and he exhibited searching honesty in the subsequent discussions. But this time, the atmosphere was charged. One reason had to do with the host's show-stopping message—he would be leaving soon for Emory and this would be the last Zürich gathering of his "brotherhood of the balloon." This was dark news to some, as USCI's Prigmore discovered when taking his friend Mason Sones out for an evening tour on one of the party boats that ply the Zürichsee. "I sat in the stern of that boat with Mason for a half hour that night while he cried, literally cried, that he had not been successful in getting Andreas to become the next Mason Sones of the Cleveland Clinic. It was an incredibly emotional time for him. And after that, Mason's whole role at the Clinic was over, because that was the last big thing he tried to do."

New wheels of intrigue were turning all the while, as Spencer King found out while pausing for a drink with John Simpson, who was still talking about his as-yet-unperfected refinements of the Gruentzig balloon catheter system. Simpson announced that USCI had finally offered him $25,000 to help him move forward. "I would grab it if I were you," offered King. I am not so sure, said the Californian. His reluctance, it turned out, was well-advised. Simpson went on to found his own angioplasty company; it ultimately evolved into an entity called Guidant that John Abele's Boston Scientific corporation recently purchased for *$27 billion*.

Gruentzig had another ritual celebration in the works. True to form, this affair mixed equal doses of inspiration and seat-of-the-pants improvisation. The notion was to offer a rustic surprise, with selected disciples scheduled to rendezvous on their last evening at a high woodland cabin about fifteen miles outside the city's precincts near a crossroads called Hochwacht. For someone who was such an obsessive perfectionist in medical matters, he approached this gathering with breathtaking nonchalance—scrawling out a plan on a café place mat while drinking wine with Maria Schlumpf. Then he crafted an invitation fraught with oddly spelled words under a photocopied picture of a mountain hut:

> *"Look at this cabin. Do you like it? If yes, we would be pleased to have you with us on a little "Spaghettata-Party" on Tuesday, August 5. The bus will be waiting for you at 6:30 P.M. in the Schmelzbergstrasse and will be leaving to Meilen at 7 P.M.*
> *We'll have some surprises for you.*
> *1. We cook the Spaghettis ourselves. Are there some good cooks???*

2. Wait and listen.

3. Wait and see.

Come!!! Put on some old trousers, an old pair of shoes and take with you a sweater or so.

Looking forward very much,

Andreas Gruentzig

The buses climbed the wooded slopes and discharged the ebullient doctors and their wives at the beginnings of a forested trail. As they let their hair down, these lords of the heart from every continent in the world heard an ethereal music, thanks to Gruentzig's having hired a Romanian "Pan" to pipe for the occasion. Magic reigned—at least at first—since the night was beautiful and wine was available by the crate once they arrived at the cabin. Talk flowered, photographs were snapped, and doctors sat down on the grass beside their charismatic host. The host and Mason Sones gave wine-fueled speeches. Gruentzig lolled with his little daughter in his arms, while his mother and aunt chatted nearby. Eventually, he set himself to cooking cauldrons of spaghetti—on a wood stove in the cabin, which was illuminated by nothing more than kerosene lamps. This proved to be a ghastly concoction. Performing a brand-new procedure on a human heart might be easier than feeding scores of people in such fashion.

Distractions multiplied along with the wine, so when the "spaghettis" began to materialize, the fare ranged from undercooked as cold worms to burnt. Few could eat the stuff, and so they drank. Bernhard Meier recalls, "Everybody had to find a plate and wait. People were drunk and stumbling over everything. Hardly anything was organized. It was like a Boy Scout outing. It was a complete mess."

USCI's Prigmore had similar impressions. "We were there in the dark and Mason is saying, 'Jesus, I hate wine. Can't he afford anything better?' That wine was awful. I mean after the third personal bottle it was okay. But the first bottle of that stuff went down really hard. Mason was just bitching and moaning. So this guy comes out of the woods playing his pipes of Pan or whatever the hell it was."

Lamberto Bentivoglio found the night to be more magical, seeing as he and Gruentzig fell into a deep conversation as darkness descended over the forest. Recently, the retired cardiologist even shed tears at the memory of this somewhat pagan ritual. "This man was there playing the pipes of Pan and Andreas got sort of transfigured and almost lost. He was staring in the distance as if nothing mattered. I think the pipes of Pan brought him back to the ancient times of the heroes of mythology and all that, that's what I think. That was the expression on his face. It was not related to medicine. It was something that transcended everything. . . ." Gruentzig fell into a reverie, saying that the emerging brotherhood was his greatest treasure and suggesting that an inner cadre of these new explorers of the heart might arrange to meet every year on a Grecian island—Michaela and himself having enjoyed a recent holiday sunbathing and racing around on motor scooters in that land. "He said, 'We could all gather there for a few days in summer and swim and dive and lie in the sun all day. And in the late afternoon we would sit together by the sea and talk of the things that really matter and gaze at the stars later and honor friendship and love while sipping wine and basking in the glow of our mutual friendship.' That's the way he was, he saw the big picture."

Ultimately, the darkness was thick and it was time to depart. Theatrical as always, Gruentzig hereupon unveiled bundles of torches that he proceeded to light first for his most exalted guests: Sones, Dotter, and Judkins. Requesting that they help light the torches of their younger colleagues, Gruentzig signaled that each passing of the flame somehow symbolized the transfer of precious knowledge of the heart onward to a new generation. The ceremony further transported his disciples—until it came time to negotiate the steep black trail out of the forest. Dave Prigmore remembers Charles Dotter and his wife nearly crashing into the woods just ahead of him. Dozens of tipsy doctors stumbled and fell as the torch-lit procession continued foot by foot for about half an hour, with the greats of cardiology belting forth silly campfire-style songs as they lurched.

For safety's sake, Mason Sones had been directed down a shorter and more lateral trail to a junction where a few cars had been advanced deep into the woods, mainly for the convenience of the likes of Andreas's elderly mother, aunt Alfreda from Hanover, and young Sonja. Michaela had just turned on the lights of their vehicle when she watched aghast as Sones, the great hero of the heart, fell headlong into a ditch. She pulled him out and drove the muttering Sones back to his supposed hotel—only the one he directed her to was not his. A bedlam of phone-calling eventually located the right one.

Gruentzig, in any case, brushed himself off, and a week later was writing every cardiologist he knew, inviting them to an even larger extravaganza in February of 1981—to be held this time in Atlanta. Michaela could see which way the wind was blowing, what with all the head-turning adulation. She confided to Gail King, who stayed on with her for a week after the August course was done, that she was filled with trepidation. Outwardly, everything looked sublime as they relaxed in the Eggli cottage in the hills. Gruentzig arrived on the weekend and even performed a charming duet—his mother on a newly purchased piano and he on the recorder. Nothing but tenderness did he show to his wife and daughter; nothing but graciousness to his guest, whose husband, Spencer, had already departed for Atlanta. But in a private moment, his beloved Mikki took Gail King aside and confided her fears that moving to America could be the downfall of their marriage. Gail had her own cause to wonder about the man's emotional state when they embarked for the airport in his recently acquired Lancia Italian sports car. "It was early in the morning and it was raining and sleeting," she recounted. "He was going around 200 kilometers per hour around bends. Michaela stayed calm, but she kept saying, 'You will get a speeding ticket!' He said, 'I don't care!' "

So what kind of man was coming to Georgia after all? Gail King wondered.

Chapter 12

"It was like a medical Camelot," was the way one nurse described the aura that Andreas Gruentzig brought to Atlanta. His arrival in late October of 1980 thrilled the Emory staff, who sensed that one of the most important doctors in the world was among them. "He came into our lives like a comet," Spencer King would later eulogize. But even Camelot has its undercurrents, and over the next year Gruentzig's colleagues would witness masterful teaching and powerful scientific integrity mixed with vain petulance and an increasing thirst for self-indulgence, especially with the opposite sex. Doctors and their wives around the conservative Emory campus were alternately mesmerized and dismayed by the hubris of this larger-than-life personality.

Things started cozily enough in Atlanta. Spencer and Gail King graciously invited Andreas to make their home his own as he settled in. King even let his new colleague take command of his 1970 green Volkswagen Beetle (which he keeps in his back driveway to this day). Meanwhile, Gruentzig began shopping for one of the lavish Druid Hills houses that had tempted him on first sight. "It had to have a certain street appeal," Gail King said, remembering houses he dismissed for not being two stories high like her own. Before long, they found what he wanted, a 3,500-square-foot red-brick affair sitting on a pretty rise on a street called West Ponce de Leon. Court records indicate that the price was around $220,000, but the house required extensive remodeling to make it ready for Gruentzig's wife and daughter, waiting across the sea.

In November, the celebrated recruit began settling into his substantial office with the required view, and met his new colleagues. Michael Kutcher, a thirty-two-year-old cardiology fellow at the time, remembers his first impressions. "I was expecting an older gentleman who was going to be fairly buttoned-down, but he had a kind of Italian-cut light gray suit and looked very suave and sophisticated with his black hair and moustache . . . When he walked in he lit up the room, and you could feel there was something special and different about him."

In a few weeks, Kutcher would become Gruentzig's full-time assistant. "It was almost like a dream come true. We were very surprised he chose Emory . . . we were thrilled and excited. When I started working with him it was like working with John F. or Robert Kennedy. They had such excitement and charisma about what they were doing with the country that it was analogous to what Andreas Gruentzig brought to the field of cardiology. He was exciting and youthful and vigorous, and one would almost get caught up in this whirlwind from the word go."

Gruentzig played his cards coolly as he engaged in planning and training sessions, along with personal meetings with every specialist who might be critical to his procedure, among them cardiologists, radiologists, cardiac surgeons, and anesthesiologists. He devoted particular attention to the support staff—the cardiology fellows, laboratory technicians, and nurses—whose roles would also be pivotal. The temporary bachelor regularly treated them to casual restaurant meals in the evenings. Already, Gruentzig was forming a new coterie of affection. It was not hard to gain the females' admiration, not with his dusky good looks and hand-kissing gallantry. As a green-eyed research nurse named Claire Rice put it, "He looked like a combination of Errol Flynn, Clark Gable, and Omar Sharif rolled into one." Another young woman on staff spoke of his "magnificent hands." The flirting started almost immediately.

With his peers, Gruentzig was more circumspect, either enjoying modest get-to-know-you dinners in the homes of various specialists or passing quiet evenings with the Kings. A couple of times, he even cooked pasta himself or made an elaborate salad, the latter an entrée he adored. "That's why I married Michaela, because she makes the best salads," he told Gail King. For a moment, he epitomized the devoted husband.

Gruentzig struck an initial note of steadfast collegiality with Spencer King and John Douglas. In short order, the trio began performing fifteen procedures a week—or about as many as he did in nine months in Zürich following the inception of coronary angioplasty. The entire department was energized by his presence, and his first formal trainee, Jay Holman, set to working almost every night and weekend for his new mentor. "The esprit de corps was just remarkable," Douglas recalled. "I don't think you see it anywhere today, the relationship that we enjoyed between the cardiologists and the surgeons and anesthesiologists. And in a social context, there was nobody warmer and more charming."

William Casarella, the chief of radiology with whom Gruentzig began performing angioplasty in the peripheral circulation (a practice that soon increased his income by $11,000 a month), offered a similar description. "I remember doing some cases with him . . . and thinking, 'Oh my God, he's not going to do *that*,' and he did! And it would work out. For a while, I thought he was lucky to get away with it, but he kept doing it so many times. I began to realize: It wasn't luck. He was incredibly skillful . . . He was a mesmerizing character."

Gruentzig's mornings began with planning conferences with about twenty other faculty members and understudies. Willis Hurst described the scene: "He breezed into the room with a smile. He listened. He usually waited to hear the opinions of others, but then he would rise and walk to the "white board" where he would take a black magic marker and diagram the coronary arteries in such a way that his point was obvious. I suppose every cardiologist can diagram the coronary arteries, but no one can do it as he could. His diagrams were works of art. The lines, which were drawn with lightning speed, seemed extensions of his fingers. He was like Picasso, drawing a single line as no one else could draw it."

To his new colleagues, Gruentzig seemed to weave magic. Many were especially impressed by the compassion with which he treated his patients.

Before procedures, he explained every detail of what he intended to do, outlined the risks in no uncertain terms, and then offered repeated opportunities to bow out. For Gruentzig, coercive double-talking doctors who pushed procedures upon patients were unconscionable. During the actual angioplasty, he offered little jokes to keep the patients calm, even as his catheters feathered into their hearts. Those who seemed shaky afterward were often treated to a tender massage.

Gruentzig cultivated a convivial rapport with staff, telling his chief technician, Joe Brown, "There are no problems without solutions." On certain evenings, he often called them into his office for champagne, deliberately blasting the corks into the soon heavily indented acoustic tiles on his ceiling. The celebrations happened often enough that the tiles had to be replaced.

Despite such initial bonhomie, Gruentzig's transition to Atlanta was scarcely seamless. Collaborators from industry and far-flung disciples rang constantly, wanting one thing or another. Gruentzig, meanwhile, worried that subversive cliques were consolidating within the field he alone had created, and once in a while he heard whispers that specialists from rival disciplines at Emory were complaining about the disruptive burdens of his procedures. But the style in Georgia was so veiled in charm—and so different from the piercing frankness of Germany and Switzerland—that it left him baffled. How could one read this Southern sweetness, he wondered, where a velvet smile did not necessarily mean what it seemed?

Meanwhile, Gruentzig's own quirks were being noted. One was his intense Germanic perfectionism, combined with open perplexity over certain aspects of American life. Even the Kings began to notice peculiarities. A painter happened to be working on their house; Gruentzig proffered second opinions about every slip of his brush. Sampling their dinners, he had his own suggestions about how various dishes might taste if they had been properly executed in Europe.

Another episode involved his first encounter with American-style religion. The Kings were regular Sunday parishioners at the local Methodist church and tried for weeks to interest their house guest in tagging along. Hearing of his great love of choral singing, they finally convinced the nominal Lutheran to join them for a special early December service called "The Lesson of the Nine Carols." Nine carols? Gruentzig couldn't put up with even half of them. Squirming, he quickly signaled his sheer impatience and boredom. "He didn't have the slightest idea what was going on and after about three of these carols he was ready to leave," Gail King recalled. Clearly, Gruentzig had not grasped the conservative mores that ran through the American South.

To help their guest acclimate, Spencer King put Andreas in touch with a fellow German cardiologist named Klaus Rees who had trained at Emory and moreover had a Swiss wife. Their first talks were fairly mundane, with Gruentzig asking about such pragmatic affairs as opening bank accounts and finding an accountant, issues made idiosyncratic since Gruentzig reportedly arrived in America with a substantial satchel of hard cash. But sometimes the conversations drifted into the larger difficulties of adjusting to American life.

Rees empathized with his new friend's difficulties. "He was very much enthused with the medical community and wanted to be integrated in that. But U.S. society is very rigid. When I came over here it was culture shock for me because I saw all those properly attired medical students with their white shirts and ties. In Europe, we went to class fifteen minutes late and with lunch we had a glass of wine; and here they served coffee that we considered an insult to the taste buds. I always had this glorified vision of America as the "Land of the Free," but then I came into a society that was so much more rigid and so much more narrow in its acceptance of people with different ways of living. You had to have a haircut in a certain way, and I had long hair at the time. I flopped into the lecture halls with sandals, and there was a certain conformity that was expected of you . . . I know that struck Andreas."

Rees and Gruentzig had regular phone conversations. The two shared common reservoirs of experience, beginning with the fact that Rees's native Freiburg had been incinerated by Allied bombing, just like Gruentzig's Dresden. "You know, growing up in Germany after the war leaves a special imprint, especially when you are in another country that has been victorious over you. There is something about that situation that you might never really express in clearly defined ways. But definitely, when you grew up in that time, you have stories of hardship that you went through, even hunger—just the basics of life that you were at the very least aware might not all be there tomorrow."

Rees, who has since been remarried to an American Jew and converted to Judaism, next broached another link to Andreas that no German without his credentials could dare express. "We were united through 'guilt by association' of being from Germany, which you very often encounter when you come here, especially within medicine, where there are many Jewish people. Right to my face in my first years they would ask me, 'How can *you* be talking to me?' . . . I feel that kind of background we certainly had in common in many ways, but in an unspoken way."

Gruentzig had plenty of everyday challenges to discuss with Klaus Rees. On the domestic front, there were constant headaches to be handled in purchasing and then remodeling his new residence for his wife and daughter. At work, the pressures rapidly intensified. In addition to seeing patients and training staff, he had to supervise the outfitting of a new catheterization laboratory that was being built to his exacting specifications. Meanwhile, there were heavy preparations in progress for his first American teaching course.

It evidently did not take long for Gruentzig to indulge in familiar ways of seeking relaxation. Almost every associate believes he quickly started bedding a string of young ladies from around the hospital. Heliana Canepa, a self-described new girl Friday at the Schneider company, who would soon transform that haphazard operation into one of the fastest-growing medical businesses in Europe, put it succinctly. "Andreas was a real womanizer and it was easy for him. Every nurse, every woman just"—here the vivacious redhead rolled her green eyes and laughed hilariously—"*swooned.* You couldn't help it. So it was very available. There were some broken hearts at the University Hospital in Zürich when he left."

Before long, the staff became aware that a certain young woman was indeed experiencing near-fainting spells whenever she got close to the debonair German, and that she was getting closer all the time. "It was the strangest thing," John Douglas recalled, "but she came to me and told me how she had a pre-synchopal [near-fainting] spell. It was in a situation where I think Andreas made a pass at her and she got so excited about that she hyperventilated and started to pass out."

On December 21 Gruentzig took possession of his new home on West Ponce de Leon Avenue. Soon, he invited Klaus Rees to one of the parties he began throwing there, pronouncing, "We always have a good time; there will be all these medical students"—clearly meaning *female* medical students. Then came the punch line: "They are *ready*!" Upon arriving, the light-drinking Rees found no Spencer King, no Hatcher or Hurst or Douglas, only a sea of young people letting it all hang out. The scene was too rowdy for Rees's taste, and the cardiologist feels that his reluctance to enter the spirit of this and other bacchanals circumscribed the limits of his relationship with Andreas Gruentzig.

Gail King sensed what was going on. "Some women just threw their clothes off at the sight of him," she said. Having formed a warm relationship with Michaela Gruentzig while in Switzerland, one night she even called her wayward houseguest onto the carpet. She warned Gruentzig to "cool it," that he was making the wrong impression on his extremely watchful Southern university. The admonition was dismissed with a boast about his forging his own rules.

Gruentzig spent Christmas back in Zürich with Michaela and Sonja, now four and a half. There he regaled doctor friends like Ernst Schneider with stories of his American life and photographs of his new house. His "Mutti" Charlotta, and his brother Johannes and his family made their annual pilgrimage. After the holidays, Gruentzig finally transferred his wife and daughter to their scarcely furnished new Atlanta residence. Sonja, adaptable as any young child, reveled in her transformed surroundings, but Michaela betrayed mixed feelings from the start.

Her husband, in any case, felt his domain to be stabilizing, and began making his presence known with increasing authority in the hospital. Michael Kutcher, now a full-time understudy, confronted a situation that was not altogether "Camelot." He discovered he was working with a driven soul who could be brusque one moment and inspirational the next. "He expected everything to be done very precisely. He never chastised or denigrated or yelled at anyone, but he'd fidget if you were doing things that weren't fast enough or not to his liking. He'd take things over, grab the catheter out of your hands . . . But you knew that he was special and that things had to run at a higher level . . . There was a lot of pressure, but it was almost like he brought everybody up to a different level. He brought out the best in people. It was psychologically and physically demanding, but it was also exhilarating—I would tell my wife when I came home that this was the most fascinating experience I ever had in my life."

Inspired by what he had witnessed in Zürich, Bernhard Meier arranged to follow Gruentzig to Emory as a personal medical trainee. There he saw similar flashes of imperiousness. "Like an elephant in a porcelain shop," was the way Meier described Gruentzig's initial impact upon his new world, although it appears that a number of Emory colleagues felt less strongly at first. "Gruentzig was a proud man. He was domineering, he was flashy, and he had angioplasty under his belt. It was difficult to keep up with him, so every now and then he let other people know that basically they were at his mercy."

Meanwhile, the quiet Michaela was left to make sense of an alien environment in which her husband was fawned over like a celebrity, and where she had scant clues as to what to do with herself. What, she must have wondered, had become of the impoverished East German who had ferried her to Switzerland with all their worldly possessions strapped onto the roof of a tiny car? When asked if the demands of acclimation and new wealth had damaged their marriage, she candidly responded. "I don't know. There really developed a sort of dynamic in it. It was a big step to move to America. The way it was in Atlanta, they offered everything—you know, unlimited resources. From a professional point of view, he could do what he wanted to there. But I was not quite prepared to confront this kind of society, full of Southern money and socializing. I was not very comfortable with this. It was against my beliefs. I did not feel good about it. But he tried to find his stance in it."

The full impact of their lives' transformation registered when her husband eventually drove home with the flashiest status symbol he could purchase—a gleaming German BMW. He gunned his motor scooter (which he had shipped over from Zürich) and hot car at breakneck speed. The newly arrived Bernhard Meier said, "He was kind of a reckless driver. He wanted to show off, and I sometimes thought that you can't be Andreas Gruentzig and Sterling Moss [a famous race car driver of that era] at the same time. He wanted to prove that whatever he did he could do at the highest level . . . He clearly drove to the limit."

Meier, whom Gruentzig graciously put up in his new house, began to sense that his mentor was fraught with deeper conflicts than had been apparent in Zürich. The young cardiologist put in hard days at Emory and then returned to a domestic situation that frequently seemed fraught. He could see trouble simmering, especially when Andreas and Michaela discussed the next batch of guests Gruentzig wished to entertain. "They were extremely opposite. She had no interest in appearances. Even the house was like that. There wasn't a picture on the wall, it was barren. He brought her over there and she never wanted even to visit the United States . . . It was a nice house. But she had no intention to furnish it, she had no idea what the style was in the United States and she didn't care. Of course, he was quite busy, and he didn't know what the style was either, so nothing happened. They were completely lost; they just put in whatever they needed, a table and a stove . . . For her, it was absolutely not important. She would say, 'If somebody doesn't feel at home here, then it's a waste of time, because they should be visiting for the human beings, not for the house.' He didn't know what to do. He of course wanted to have an organized type of thing with structured climbing up the ladder and to her that was a horror . . . You could see that this was not going to work out."

Gruentzig had plenty of other matters to occupy his attention. Knowing more than two hundred doctors would be appearing in February for his initial Emory teaching course—the first such showcase of his talents that he organized on the American stage—he threw his passion for detail into perfecting those arrangements. The auditorium for the meeting, like everything else around him, gleamed with the opulence that Robert Woodruff had showered upon Emory. Beyond a half-dozen glass doors, a tiled lobby glistened with light from overhead windows like the entrance to a luxury hotel. Left and right stood bronze statues of ancient medical heroes—among them Aristotle and Hippocrates. They stand before stairs that trip down to a spacious, two-tiered amphitheater with skylights illuminating the stage. Compared to the dreary lecture hall in which Gruentzig inaugurated his demonstration courses in Zürich, this was Broadway.

As the event approached, Gruentzig pretested every aspect of the new technologies that had been gift-wrapped just for him, beginning with his sophisticated catheterization laboratory and a closed-circuit television system for relaying his work in progress to the lecture hall. Two days later, eager doctors from across America and the world filed into the auditorium to watch his wizardry. The eyes of medicine were fixed upon him once again—just as Andreas liked.

He performed his live cases with his trademark self-assurance, and defused tension at every procedural impasse with typical gallantry and wit. Following one case, Andreas slung his legs onto the catheterization table and offered this aside: "The most intensive part of the procedure has to do with the shoes. Without the right shoes you will not have a successful procedure." A close-up showed he was wearing white clogs, which he theatrically lifted onto a table on the stage with a wink to the audience. The message was that even amid all the arch caution he prescribed, confidence and flair were supreme requirements.

Watching with admiration at the front of the room were Maria Schlumpf, who continued to assist Gruentzig in preparing the systematic long-term reports that he felt crucial to his technique's credibility, and his Mutti, wife, and daughter—the soul allies whose presence at every teaching course seemed to be more important to him than those of the most famous doctors in the world. He never ceased to tell his colleagues that Maria, heartbroken at his departure from Zürich, should always be regarded as balloon angioplasty's handmaiden.

Spencer King eyed the proceedings in amazement. "When you have that situation and you have the maestro Andreas performing, you know the excitement level is quite high in that arena. It's like going to watch Michael Jordan play basketball. But I guess the overriding reason was, this was the ticket—to do angioplasty. . . . In those days, you got your little chit that you'd been to the Emory course and you'd been there for four days. You'd seen some angioplasty, walked into your hospital, and your hospital was happy to have you there and said, 'Okay. Hang up your shingle and get to work.' " And get to work these first practitioners did, performing about 1,000 angioplasties before 1981 played out.

Euphoria reigned again as the meeting closed. The venue for the final night's celebration was as dramatic as always—Stone Mountain, the granite dome rising out of the plains twenty miles east of Atlanta, whose chief escarpment boasts gigantic relief sculptures of the three greatest heroes of the Confederacy: Generals Robert E. Lee and Stonewall Jackson, and the president of the Confederate States of America, Jefferson Davis. The early evening segment of the festivities was tame. But this was followed by a dinner under a pavilion where a bluegrass band set to rocking. The other Andreas, the one who loved to throw caution to the wind, stepped forth. Out to the middle of the dance floor he shimmied, and began whirling around every good-looking woman in sight. One of these was Michael Kutcher's young wife. "I'm sorry. Your husband has picked the worst subspecialty for preserving a marriage," Gruentzig said. The message: We are all overworked, overdriven, and oversexed—and set on having our way with the world. The next thing Kutcher's wife knew, Andreas was patting her derriere and speaking suggestively, even though his supposedly beloved Michaela was but a few feet away. The understudy was not pleased.

Around the same time, Marcia Schallehn, a USCI specialist in training new users of the company's technology, enjoyed similar treatment at a sales meeting in Massachusetts—after Gruentzig had flirted with nearly every female sales representative on the staff. At length, he expressed his befuddlement at why American women seemed to be so hidebound in worrying about a man's marital status. "His thing was that in Europe things were very open, you may be married, and it didn't mean anything against your marriage if you played around."

As the first Emory teaching course drew to a close, a number of doctors spent a not entirely festive evening at the Gruentzig's house. There they discovered a lack of both food and furniture. Gruentzig's Zürich colleague Ernst Schneider said of Michaela, "My impression was that she was unhappy. I never asked Andreas and I never spoke about it to him because Andreas was not the guy to share his internal geography with others. But when

I think about what happened that evening, it felt like the trouble was already there. It did not look like a happy house."

David Williams remembered a group of doctors having to sit on the floor while they planned the publication of detailed formal guidelines for performing angioplasty safely. "There was hardly any furniture in this house . . . There wasn't a picture on the wall. Even though it was a beautiful house, it was cold and empty."

The visitors soon went their way and the everyday routines resumed in Atlanta. But the gossip was simmering. Gruentzig's colleagues openly wondered about the state of his marriage. Michaela seemed to them to be disengaged from the start. Despite being married to a man of extraordinary glamour, she refused to wear makeup or dress in a way that flattered her. "I never saw her looking happy," said Joe Craver, the cut-to-the-bones surgeon who was becoming a close friend to Gruentzig. But perhaps Michaela had good reason, and male observers were not seeing her side.

The situation could not have been easy. The counseling job Emory provided her with proved frustrating. In a line of work in which every linguistic nuance is vital, she had to struggle with an uncomfortable foreign tongue. Without proper credentials to gain an American licence to practice as a certified psychoanalyst or even a psychologist—and no intercession by the former overseer of the U.S. Immigration and Naturalization Service, Griffin Bell—she had to sign disclosures for every patient stating that she was a mere understudy to some "real doctor." Meanwhile, her husband's stature rose. Schneider was by now feeding Gruentzig substantial royalties—some coming through indirect USCI payments for shared technology—with the company's representatives landing in Atlanta and bankrolling monthly parties populated by adoring young women.

Conflicts arose over the proper raising of Sonja. Gruentzig accused his wife of lax discipline, but Michaela holds a different perspective, believing her husband to have been much too absent from his parental responsibilities.

Everyone could see the man changing before their eyes. Gruentzig began demanding special favors, telling Willis Hurst and Charles Hatcher that the ten-minute walk from Emory's distant parking lots was wasting his time and that he needed space for his new car right outside the door.

It did not take Gruentzig long to figure out that Charles Hatcher had a direct line to the beneficence of Coca-Cola, because of his monthly lunches with the failing leader of that global enterprise. As Andreas well knew, the biggest story of the moment involved himself. Therefore, he assumed that he enjoyed supreme leverage. Gruentzig's infatuation with this notion took almost comedic turns, according to Michael Kutcher.

"One day, maybe in February or March, he [Gruentzig] was enraged because something about his animal research lab hadn't been delivered. He had a secretary named Hilda, and he said in his lilting German accent, 'Hilda, Call Mr. Woodruff up right now! I want to speak to Mr. Woodruff. These people aren't delivering on their promises!' . . . The irony of this was that Mr. Woodruff was stroked out. I don't know whether he had Alzheimer's or what, but he was an old man oblivious to the world."

Despite such displays of self-importance, Gruentzig was unquestionably under pressure and he worked without cease. He was consumed with shepherding his technique forward with an absolute minimum of harm to patients. Now and then, his innate conservatism ran to excess. Traveling to a late-winter medical meeting in the Swiss ski resort of Davos, he found his erstwhile collaborator Martin Kaltenbach demonstrating a potential minor technological refinement, which consisted of nothing more than crafting a long central wire over which the balloon system might be manipulated with more exacting control. Gruentzig reacted furiously, demanding loudly before the colloquium, "Every advance has its risks along with its promises. What are the risks of your new innovation?" The audience sat stunned at this petulance. But the charming Andreas reappeared in due time. On this same trip, he displayed another signature trait while hurtling down the slopes at a speed that left him splayed out with a broken thumb—he could be accident-prone, thanks to taking foolish risks in his personal life that he abhorred in the practice of medicine.

Gruentzig was by no means paranoiac in worrying that his advance was about to be pushed to limits he felt to be premature and ill-advised. The gung-ho America of the 1980s was no place to keep a lid on the genie of a clearly hot medical technology. Device manufacturers from across the continent gravitated to displays of this therapeutic revolution-in-the-making like moths to a flame. Out in California, John Simpson landed a quarter of a million dollars—with plenty more to come—from a Silicon Valley entrepreneur named Ray Williams to keep refining his rival catheter system. The resultant start-up company, Advanced Cardiovascular Systems (ACS), signed up a variety of technical experts, including an engineer named Will Sampson who jumped ship from USCI with all his knowledge of the secrets of Gruentzig's technology. The new competitors whispered that the founder's system was cumbersome and slow, and that they were nearing completion of a radical simplification. Rubbing salt in the wound was the fact that Simpson and his colleague Ned Robert held a 50 percent stake in the company. In contrast, Gruentzig, who had taught the Schneider company how to make his catheters from scratch, ended up with only a 5 percent stake in that enterprise. His ownership of the Bard Corporation's USCI division, of which he had become the backbone, was zero.

In fact, USCI's Dave Prigmore had his own worries. At medical meetings, he sensed competitive eyes scanning every new twist of the supporting equipment and core balloon catheter his company was honing to higher levels of performance. There was good reason for companies to beware of spying glances in those days; at one 1979 German medical meeting a Cook representative threatened to call the police on a Medi-Tech rep for having stolen a display catheter from his booth, while the Medi-Tech rival hurled the same accusation right back. "I always used to say to the competitors," Prigmore recalled, " 'This is the worst decision I ever made . . . You guys are lucky you didn't get involved in this mess.' Of course, that was partly to keep them from smelling what was going on. The other thing I was doing, I was saying to Bard, 'Don't breathe a word of this to Wall Street,' because I don't want Wall Street to have any idea how good this thing might be, because then money would flow into it, and venture capitalists would be smelling around and people would be saying, 'Wow, this is going to be a billion-dollar business.' "

Gruentzig saw another threat emerging. A new devotee of his technique was hell-bent on exploring its horizons as no one, not even its inventor, had ever dared before. And the word spread that this doctor, Geoffrey Hartzler, was both singularly gifted and unstoppable. Before long, Hartzler, who hailed from the Midwest and wore cowboy boots on the job, was depicted as the procedure's first gunslinger, the Billy the Kid of heart-probing, a faster draw than the Zürich originator of the technique, Andreas Gruentzig. The stereotype was but a cartoon, but it held sway.

Hartzler's catapult to prominence was in fact as unlikely as that of Gruenztig himself. The son of a Mennonite minister in obscure Goshen, Indiana, at age five he saw his mother become paralyzed by polio and then spend the rest of her years breathing with the assistance of an "iron lung" machine. He and his five sisters soldiered on and Hartzler managed to achieve top ranks in his local high school, then excel in chemistry at the state university. In the summer, he drove a cement truck and showed enough pluck that he was offered a career as a cement chemist by the plant's owner. But Hartzler decided that medicine would offer more excitement. In time, he ended up as a specialist in diagnosing errant heart rhythms at the esteemed Mayo Clinic in Rochester, Minnesota, just as new catheter methods were being introduced to chart these abnormalities. Thanks to the Gruenztig example, he eventually learned to cauterize them without invasive surgery. The former truck driver grabbed onto new procedures with a zeal that sometimes took his colleagues aback. "I was known as an 'Invader,' " he frankly recalled from his current Kansas City bolt-hole of a garage office, which also serves as a recording studio and de facto men's club. Now retired from medical practice, he sequesters himself in these peculiar surroundings when not pursuing his wide philanthropic and business interests. Hartzler's biggest joy is to stand with a 1960 Fender electric guitar on his hips and twang away into the night with his friends.

In the spring of 1979, Hartzler's ears filled with stories about the potential wonders of balloon angioplasty. A senior colleague had already been to Zürich and returned with two precious balloon catheter systems, which sat on the Mayo Clinic's shelves for months. Hartzler couldn't take his eyes off them. In October of that year, an apparently ideal candidate for balloon angioplasty appeared while Hartzler's colleague was off to Europe. Despite having never watched a single procedure in progress elsewhere, Hartzler decided to take the leap forward on his own. A trial run on the heart of a dog in the hospital animal laboratory ended in the St. Bernard's death, but Hartzler was another driven soul. And like almost everyone itching to get into the heart game, he faced no Institutional Review Board to tell him what he could or could not do next.

So the preacher's son found himself feathering a balloon catheter into the left coronary artery of one Harry Grummet, a water-tower maintenance man from Waterloo, Iowa. Hartzler and his patient entered into a joint freefall because he was unable to push even the tip of the device past the constriction. The "Invader" could feel his own Waterloo in the making. As things went worse, the Mayo Clinic's chief of medicine—who Hartzler had cagily telephoned only moments before starting the procedure—stormed onto the scene along with a retinue of furious cardiac surgeons. The profound tensions of that episode linger with Hartzler still.

Hartzler recently recalled the scene from his corner garage office. The semi-decrepit neighborhood outside at first seemed more fitting for a "chop shop" where underworld figures scalpel away not cardiac tissue but car registration numbers. The parking garage's outlying car slips gleamed with neon beer signs beside grinning clown masks and a mannequin boasting tasseled nipple pasties. The adjoining office/recording studio had the feeling of a strange sanctuary from the pressures of this world, at least for meeting one of the most formidable figures in the history of American cardiovascular medicine. But Hartzler's recollections were discerning.

Sitting behind a computer-topped desk, he grew emotional as his story unfolded. "I still get choked up about it, because I realized then and I do now how close it came to my career being over. It's really true. I worked well over two hours trying to just push this balloon through and it wouldn't go." The Midwesterner pleaded for five more minutes when his superior commanded that he stop.

It was not only his patient's safety but his own life that was flashing before his eyes. The cement plant in Goshen, Indiana, appeared to be the next stop. "There were a lot of bad vibes in the room by now. But for whatever reasons in those five minutes the balloon passed, and I measured a significant increase in coronary blood pressure . . . And you know, I became a hero.

It was just astounding. I went from being a goat. I would have been fired and lost my job . . . Suddenly I was doing the first angioplasty ever at the Mayo Clinic." Elated, Hartzler reportedly sent a telegram to his traveling colleague (whose catheter he had "borrowed") pronouncing, "The Eagle has landed," apropos of Neil Armstrong's immortal words upon settling the Apollo 11 lunar module on the moon on July 20, 1969.

Colleagues fumed nonetheless. According to Hartzler, "all hell broke loose" when he formally presented the results of the case to them a few days later. Who was he to take such chances with a patient's heart, and to do so on an unauthorized dare? The jury soon weighed in. The Mayo hierarchy decided that the man was trigger-happy, especially when he pursued a burst of new angioplasties in the next two months and then tripled his pace early in 1980. They served notice that his services were no longer required at their esteemed institution. Hartzler fished about and landed in a freewheeling private cardiology practice in Kansas City. One of things he liked about the place was that cowboy boots were still in fashion.

With Gruentzig-like flair, he sent USCI's Marcia Schallehn a dozen roses to ensure that a flow of balloon catheters would follow him to his new domain. Considering the excitement of the time, the gesture was not so extravagant. Goffredo Gensini, a highly respected cardiologist and medical device inventor himself, was simultaneously racing his private plane from his headquarters in Syracuse, New York, to a small airport near USCI's Massachusetts plant so that Schallehn could deliver catheters to the runway.

Hartzler achieved the greater fame. Before long, he would be performing 800 or more balloon angioplasties every year. In June of 1980, he shocked the medical world by employing the technique in an attempt to demolish a heart attack as it emerged—just the therapy that all the president's physicians could not imagine when Dwight Eisenhower suffered his own medical emergency in 1955. It was also just the therapy to stick a thumb in the eye of Andreas Gruentzig, who time and again had warned that the new technique must be used only on the most stable and cherry-picked cases until it was fully refined.

Hartzler was never much inclined to caution. This particular procedure was scheduled under absolutely normal conditions. Yet just as it was about to begin, the patient writhed: His surging chest pain was escalating into a heart attack. The Midwesterner carried on, and prevailed.

Like most stories of overnight breakthroughs, this one has many facets—among them rank acrimony and bitter second-guessing. Some cardiologists maintain that the patient in question had already weathered the full injury of his heart attack and that Hartzler was only mopping up afterward while wrapping himself in a mantle of undeserved triumph. Others say that even if the story is true, the man had no business undertaking such a perilous and untested challenge at this stage. But the fact is that the treatment of evolving heart attacks was about to be transformed, with Hartzler at the lead, regardless of whether Andreas Gruentzig liked it.

In fact, Hartzler's challenge had barely started. Impatient with shortcomings in the Gruentzig device, he began steam-molding better curves into the stiff end of the system's guide catheter so that it would slide more easily into place—one of many innovations that would make him millions. This doctor was on a roll. He ignored the admonition that balloon angioplasty should never be performed on more than a single arterial lesion in one session. He defied the notion that multiple-vessel disease should only be handled by coronary artery bypass surgery. If the new balloon procedure worked in one vessel, why shouldn't it succeed in two or three in sequence? he reckoned. By the autumn of 1980, the "Invader" of the heart began transforming the cautious Gruentzig approach. "Okay, I got aggressive," Hartzler conceded. "I was the hot young Turk."

Critics worried that the man was messianic and somehow determined to single-handedly end the surgical-bypass era. He went much too far, too fast, they alleged, and subjected numerous patients to risks that Gruentzig could never countenance. Before long, Hartzler would leave his own medical audiences gasping before his virtuosity. "Hartzler was very elegant with his hands," Bernhard Meier observed. "Gruentzig was good, he was talented, he was careful. But Hartzler was faster; he was much more aggressive." Gruentzig suspected that his rival's true incentive was commercial—that by doing multiple-lesions for fees of $1,200 per blockage the man could get paid double, or triple, for the same case.

In January of 1981, Simon Stertzer, never short on ego himself, appeared at an Indiana medical meeting alongside Geoff Hartzler. He looked over the upstart's case presentations and heard how he performed the new technique at breakneck speed—as though he were the ultimate wizard of the procedure. Then Stertzer grabbed an aisle microphone and labeled the supposed virtuoso, "The Flash from Kansas City."

The gunslinging between the new kings of the heart was just beginning. And the eager entrepreneurs from California were watching, itching for an all-out duel.

CHAPTER 13

TROUBLE FOLLOWS MONEY, as the saying goes. By the spring of 1981, bouts of increasing tension began to roil the fast-moving angioplasty field along with Gruentzig's own life. Cardiovascular surgeons and anesthesiologists suspected that a mighty engine of profit-reaping was now driving the procedure forward while strewing havoc through their own medical practices.

Emory's Joe Craver described the economic forces at play. "There was a huge emphasis by the national health-care system and insurance carriers to control bypass surgery by promoting angioplasty. It came along at a time when there was a vast population involved in coronary artery bypass surgery, and it was costing a $40,000 to $50,000 hospital bill per case, when you could do this angioplasty therapy for $10,000 . . . So the federal limits on qualifying for reimbursement for angioplasty were very low in order to encourage people to do it, and the payments remained very high. The cardiologists got paid the same amount for a twenty-minute angioplasty as I would for a two-hour bypass operation followed by a patient spending a week in the hospital. The fee for a triple-bypass was around $3,000 and fees for a single-vessel angioplasty were about the same. Then certain cardiologists might do several single-vessels in succession."

The surgeons resented the invasion of their formerly sacrosanct and lucrative turf and despised the proclamations by figures such as Geoffrey

Hartzler that their day in the sun was done—a claim that Gruentzig himself remained loathe to make. Also infuriating to them was the fact that the new wave of "catheter jockeys" were performing occasionally slipshod procedures that required emergency bailout operations until deep into the night. "When you saw somebody doing something poorly because they weren't properly trained and experienced, and when the same condition would be a chip shot if it were done by surgery, that pissed me off," Craver said. "A lot of things the cardiologists would consider 'successes' absolutely would put us out of business. If we had a 45-percent remaining stenosis in a lesion, people would never send us another case."

Gruentzig himself remained ever cautious and insisted on working hand-in-glove with his surgical collaborators. "He always said that this procedure is going to succeed on the strength of the surgical backup it receives in the first five years. If it crashed it would have destroyed him, but the fact is that not one of our first several thousand patients died—*zero* died," the surgeon noted. He told stories of Gruentzig personally paying to fly family members to Atlanta to console them by the bedside whenever a serious angioplasty complication occurred.

Spencer King concedes that the new procedure's boom pressured the everyday functioning of Emory Hospital, what with patients now streaming in from around the world. "Of course, our first years in angioplasty were a windfall. They were paying $1,200 a case in professional fees, when twenty years later we are getting half of that. The surgeons didn't appreciate the complications that might occur on a late Friday afternoon. The cardiac anesthesiologists were very angry that we were cutting into their business by doing an angioplasty that they wanted to assist in doing a surgery on . . . It got heated sometimes, and Andreas almost got into a fistfight with one of them once."

Harsh disputes erupted in far-flung hospitals. "Every angioplasty hospital was pushing the same way because the procedure didn't use a lot of resources and it was relatively cheap. Every hospital involved made a ton of money because they got paid for angioplasty almost as much as surgery, and it was to their advantage to keep the hospital angioplasty commitment full," King said.

Even at congenial Emory, Gruentzig and his closest cardiology colleagues sometimes feuded. Gruentzig was sometimes quick to cry foul over money issues himself, insiders say. The newcomer's oversize personality required more than a little forbearance. Sally Dineen, a research nurse, remembered the haughty maestro mocking a colleague for mimicking the jaunty way he turned scrub caps into ersatz berets. "Dr. Gruentzig walked over, looked him up and down, and said, 'You wear yours for sterility. I wear mine for style.' "

The whole while, Gruentzig bitterly resented Geoffrey Hartzler's rising acclaim, insisting that the man was inviting catastrophe. The conflict grew personal, as Bernhard Meier observed. "Hartzler was impressive and Gruentzig hated that because until then Gruentzig was the only one people took seriously . . . He thought Hartzler was a threat to the method because Hartzler clearly overdid it. He did dangerous cases and people suffered from his overusing the potential. That Gruentzig really hated, because there was a danger that it would really kill the procedure."

These were the best and worst of times for Andreas Gruentzig. His success rates were climbing, royalties were rolling in, and speaking engagements proliferated around the world. Schneider's Niederhauser and USCI's Cvinar paid regular visits to show off vying design concepts, with lavish restaurant dinners and impromptu celebrations following. To help manage the details of his life, Gruentzig hired a personal assistant, a picaresque character called "Sarge." Professional colleagues were baffled by the relationship, because their esteemed associate often seemed to be more comfortable with this rough-edged individual than with them. To his peers, Gruentzig seemed at once methodical and wild.

Young women continued to flock to the rising star, one of them a very attractive Macon, Georgia, debutante named Margaret Anne Thornton. Petite and shapely, she had wavy light brown hair and soft hazel eyes that could make an older man melt. Nearing the end of her undergraduate medical studies, the twenty-four-year-old volunteered to assist in a special research project under Andreas Gruentzig's wing. The two definitely looked each other over closely. "That was the day that sunshine came into my life," Gruentzig later said to her mother.

It did not take long for Margaret Anne Thornton to make her attributes known. She was a Southern belle par excellence—smart and even driven, but self-deprecating and sweetly flattering, while blessed with a vivacious laugh, impeccable manners, and a seductive flirtatiousness. The former sorority queen dressed with flair and innate sex appeal. In short, Margaret Anne Thornton knew how to use her gifts to maximum effect. "She was a foxy girl," said Joe Craver. In numerous respects, the young medical student was the polar opposite to the then thirty-six-year-old Michaela Gruentzig. At the first Emory teaching course in February of 1981, Gruentzig discovered her love of dancing and spun the lithe young Margaret Anne, all 102 pounds of her, around the Stone Mountain dance floor with delight.

Gruentzig's colleagues watched as Margaret Anne drew ever closer to her mentor, even melting into his arms for a comforting hug when in distress. Other opportunities for contact soon emerged. Gail King asserted, "She wasn't discreet at all about running around with Andreas. She was really fascinated with him. There would be parties and she would be all over him."

One day, the visiting Mylers got their own surprise when Andreas brought them to his house and explained that since Michaela was away for a few weeks, he had a medical student staying with him instead. "This girl came in and she was just this Southern belle fawning all over him, and I said to myself, 'I don't think this is going to be good,'" Sharon Myler said.

Gruentzig was scarcely keeping the affair secret, as Klaus Rees found out over dinner one night at a restaurant in Atlanta's trendy Buckhead district. The famous doctor and his young love snuggled together breathlessly. Love, with all its foolishness, was in the air. The couple later flew off for a holiday in Bermuda, where they checked into the Surfside Motel. But this was not their first getaway.

Early in the summer, Margaret Anne's mother, Mary Jane Thornton, and her husband, Earl, had headed off with their daughter for a family holiday on Sea Island, five hours south of Atlanta and just a couple of miles off the coast. The parents drove two separate cars to provide extra freedom upon arriving at their rented cottage. In the lead vehicle, Margaret Anne endured a bit of maternal questioning. "We got halfway down to the island

and I said, 'Well, tell me about your work with Dr. Gruentzig,' and she said, 'Oh, Mama, he is jus' wonderful. He is so brilliant with patients and he is so brilliant with his work,' Mary Jane Thornton recalled.

"I said, 'He is an older man.'

" 'No Mama, you wouldn't believe the age he is . . . He is very active and he is a *wonderful* dancer.'

"I said, 'Oh, you've danced with him?'

"Well, over at his house a few times.' "

Here, Mary Jane Thornton giggled at the recollection. "And I said, 'Oh, *really*?'

" 'And Mama, he's coming down this weekend, if that's all right.'

"I said, 'WHAT?' And I realized this was kind of serious, and I turned on my blinking lights, and as soon as we could, Earl and I pulled off the highway." When her husband approached Mary Jane Thornton's window, she said, "Earl, I think Margaret Anne's fallen in love with Dr. Gruentzig."

Friday evening, the family drove to the small local airport to greet the arriving medical star. Margaret Anne's parents immediately sensed Gruentzig's true age of forty-two. Margaret Anne ran into his arms and he hugged her and held her, and I said, 'Earl, it is beyond our control, it is out of our hands.' . . .We were just stunned, to be honest with you."

Mrs. Thornton found Gruentzig to be far less eye-stopping than others said, and was put off by his burly moustache. Her husband Earl disliked his thick accent. The Thorntons also decided that he was dressed like a rube, considering he was visiting an exclusive resort where American presidents and European royalty paid frequent visits, along with some of the richest figures in the United States. The center of the island's social life was an elegant and grand seaside hotel called The Cloister. "This is one of the ritziest hotels anywhere," Earl Thornton pointed out, "and he didn't even have a coat, he didn't have a necktie. He had on baggy khakis and a plaid shirt or what have you. That's all he had . . . We were going to take Andreas to dinner at The Cloister but Andreas couldn't get in the door like he was dressed . . . and we got in the car and bought him a coat and tie, a pair of britches, and a belt."

On the beach, a graver problem surfaced, at least to the Thorntons' way of thinking. "He did bring his bathing suit and it was a bikini. Back in those days, you weren't seen on Sea Island beach in a bikini, that wasn't done," Mary Jane explained. "And so I went up and bought him a proper bathing suit and gave it to him and told him I would appreciate it if he put that on."

Talk about making a hit with the potential in-laws! Meanwhile, it turned out that John Hurst, son of the Emory chief of medicine Willis Hurst, had spotted Andreas on the strand.

Gail King recalls, "John Hurst came back from his vacation at Sea Island with his family and he says, 'Oh, I met Andreas and I saw him with his wife.' I said, 'He had his wife down there?' John said, 'Yeah, he had this real young, real cute girl with him. I said, '*Hello*! This is not his wife.' He said, 'Well, her mother introduced her as his wife.' I mean the whole thing was just like a soap opera."

Spencer King, for his part, was surprised by how much Gruentzig gushed about The Cloister itself. "He was kind of blown away. The Cloister is kind of the closest thing to an old-world palace and is almost European. Even though the people may be ordinary it seems bigger than life. The deference is obvious. You could think you are in the nineteenth century with all the scraping and bowing."

Meanwhile, the "soap opera" hit Act II. Late in the summer, Michaela Gruentzig invited Gail by to help with an afternoon cookout in honor of her departure back to Switzerland, with Spencer to come by a bit later. The Kings by now half understood the situation. "I knew that he was seeing Margaret Anne," Gail explained of what became a bizarrely ironic afternoon, "that they were, you know, 'dating.' But I didn't realize that Michaela knew, and so when they decided to separate she talked to us and said they decided it was better if she went back home."

Before the barbecue, Andreas invited Gail King to sit with his wife and daughter and watch a home movie of a recent trip he had taken with Sonja—evidently in Bermuda. The viewers sprawled out on the floor for what was evidently a movie filmed by none other than Michaela herself. "So he was showing the video and the two of them [Andreas and Sonja] were playing and having a wonderful time, but I kept seeing this other person in the background. So I said, 'Michaela, is that you?'

"She said, 'Oh no, I didn't go.'

"I said, 'Is it the babysitter?'

" 'No,' she said. I couldn't believe I was so stupid . . . And Andreas said, 'Oh no, that's Margaret Anne.' Then he showed all the rest of it with Margaret Anne in the video and she looked like she was about nineteen. She was darling looking, but she looked like a little teenager, flitting around in that tiny little bikini. And there was his wife and Sonja and we were all watching the thing together, and they [Andreas and Margaret Anne] were hugging and kissing and playing and frolicking in the water . . . I thought, you know, 'I would just kill him if it were me!' "

Yet Gail King somehow sensed that Gruentzig was torn. "I know he loved her," she said of Michaela, the betrayed wife. "It's so hard to say why, just the way he was tender with her. Even while she [Michaela] was living here, he was obviously very much in love with Margaret Anne but that didn't preclude his still loving his wife. It's just that it was a whole different thing, that there was this very young, very exciting woman on hand." In any case, he enjoyed a September getaway with Margaret Anne to Charleston, South Carolina, and soon afterward his wife and daughter were gone.

Following the severance, Gruentzig entered a rudderless phase, according to several associates. Only with reluctance did he join several colleagues at a teaching course at the fabled Mauna Kailua Resort in Hawaii. Upon arriving, he discovered a waiting paradise at the opulent former Rock Resort, beside a spectacular crescent beach with the 14,000-foot Mauna Loa volcano lofting up behind. Gruentzig gave his presentation and then more or less proceeded to run wild. In tamer moments, he rode horses and went scuba diving with Joe Craver.

Meanwhile, other seductions lured from the sparkling white beach. "He would come down to the beach after the morning sessions," Craver explained, "and he was naturally tan. He had this gypsy type of tan and he had this mini-, mini-bikini, and he would go walking on the beach and then start flirting with girls, and get flirted back at. He would come up to me with my wife there and say, 'These women are unbelievably flirtatious, they won't leave me alone.' I would say, 'Andreas, go out and have a good time.' There was nobody more flirtable than Gruentzig. He shagged a few people. He would show up in the evening after having a massage and wearing a white suit and a Panama hat and no shirt, just a jacket and white trousers and these little white sandals. He looked like somebody right out of a Humphrey Bogart movie, just charming the ladies right and left."

Spencer King remembers one strange incident under the setting Hawaiian sun. "I was sitting there on the pavilion with Gail and our son, Spencer . . . We were outside at this restaurant and it was a beautiful setting. Andreas was having dinner with us and two or three other people and the waitress kept coming around. She was the opposite of attractive but Andreas was physically all over her, he virtually manhandled her right there." According to John Douglas, Gruentzig ostentatiously tried to hand the waitress his room key, in plain sight of his colleagues and boss.

Another evening featured a cocktail party on the hotel's sweeping lawns above the sea. The actress Anne Blythe, moderately famous for her film roles thirty or so years earlier, was another guest at the resort. The potential for fiery romance wasn't quite the same as Christiaan Barnard faced with Brigitte Bardot, but it was good enough for Gruentzig that night. "There were maybe twenty-five people and Andreas showed up in a white Panama-like suit with a big straw hat," Spencer King recalled. "It didn't matter that Anne Blythe was, like, twenty years older than Andreas. He went into full performance mode. The comment I remember Willis [Hurst] and I made to each other was, 'He is trying to get somebody to slap his hand, he is just acting out like a teenager' . . . There was no way he was going to pick up the waitress or Anne Blythe, but he made his display. He was like a public show-off. He wanted us to view him right out front as somebody who didn't give a damn."

In any event, the Emory doctors had a much graver concern on their hands one afternoon when a surfer on his honeymoon was smashed off his "boogie board" by a monster wave. Bathers screamed that a young man was floating limply in the surf. Gruentzig and Joe Craver churned through the water to try to help, then hauled him onto the beach and alternated with Spencer King at furious attempts at CPR, but the honeymooner's neck was broken and he died.

The stresses and temptations in Gruentzig's life kept growing. By late 1981, the commercial potential of balloon angioplasty was manifest. Dan Lematrie, then a stock market analyst for Cowen and Company in Boston, was one of the earliest financial wizards to take notice. "It was patently clear early on that angioplasty had the capacity to really alter the competitive landscape in cardiology . . . The reality was that back in the early 1980s, this was viewed as *the* growth opportunity in medical devices—pure and simple. There was nothing like this out there, and Bard was at the forefront . . . It was very clear that Bard was sitting on an extraordinary opportunity," Lematrie said.

Bard/USCI's competitors began playing hardball, making overtures to Andreas Gruentzig at every turn. Meanwhile, Bill Cook, chairman of the rapidly growing Cook Group in Indiana, received a startling telex from him on November 6, 1981. In it, Gruentzig curtly asserted that the Cook catheters that were still being manufactured for use in the leg circulation did not meet his quality requirements:

> THE PRESENTLY USED CATHETERS AND MATERIALS ARE DIFFERENT FROM THE ONES I HAD DEVELOPED AND ARE BEING PRODUCED WITHOUT MY APPROVAL. BECAUSE MY NAME IS NOW BEING COMPROMISED PLEASE STOP USING MY NAME FOR THIS PRODUCT IMMEDIATELY AS WELL AS FOR ANY OTHER CATHETERS AND INSTRUMENTS SUCH AS PRESSURE GAUGES. I EXPECT YOUR CONFIRMATION WITHIN ONE WEEK.
> SINCERELY, ANDREAS R. GRUENTZIG

Bill Cook's response was short and amicable, but skirmishes between rival device companies broke out into virtual warfare. Bard's top executives in Murray Hill, New Jersey, set out to squash the still modest Schneider company, their supposed partner in advancing balloon angioplasty.

Hugo Schneider had arrived from Zürich with a single colleague and pitched his little booth at a remote corner of the March 1982 American Heart Association's sprawling commercial hall, seeing as he couldn't even sell his company's catheters in the United States. Bard's chief, George Maloney, abruptly demanded a complete reworking of the joint licensing agreement to grant USCI rights to balloon catheter sales throughout the entire world, save little Switzerland. By now, Schneider's commercial viability had become utterly dependent upon USCI's back-feeding various core ingredients, which they threatened to stop providing.

Hugo Schneider recollected, "He pressured me saying, 'You must give us rights to the whole world, except Switzerland—you can hold on to Switzerland—otherwise we won't deliver the material.' This was blackmail. It was blackmail! I couldn't believe this."

He soon would, because Bard made good on its threat. Just as European demand soared, the hapless Schneider company found that it was incapable of building a single further catheter in any viable fashion. Heliana Canepa, by now handling distribution, ordering, production oversight, and sales and marketing, remembers the predicament vividly. "Our production people, well, Jürgen Hoffman, came and said, 'I have nothing, and we can't produce any more balloon catheters.' "

Unwilling to capitulate to Bard's pressure, Schneider and Canepa located another American company that might help produce the needed materials. After making arrangements for an exploratory visit, the duo landed at Kennedy Airport and hired a car. They carried with them a satchel stuffed with the preferred Swiss paperwork for cutting important deals—cash. But they soon discovered that their potential new upstate New York supplier was a Bard subcontractor, and the proposed deal fizzled on the spot.

C.R. Bard executives had time on their hands and let the weeks play out. Sensing they had Hugo Schneider right where they wanted him, they suggested a meeting in London, where Dave Prigmore might entertain more benign negotiations. There, the USCI executive turned the screw. "They wanted the entire world rights except Switzerland, the same as Maloney had demanded before," Hugo Schneider said. When he resisted, Prigmore suggested that Schneider sell his company outright.

According to Heliana Canepa, her boss was sorely tempted. "Mr. Schneider came back and said, 'There is meat now on the bone.' I was alarmed, and asked why. He said, 'Well, one million [dollars].' I said, 'That's not meat, that's *stealing*! That's stealing!' I said, 'No way, NO WAY!' I was completely against submitting to an acquisition because I knew that this business could really boom. I was the one who was against even talking about acquisitions, but of course Mr. Schneider was over fifty. To him in his heart this whole thing was his second leg in business . . . And of course we were in a very disadvantageous position."

Seeing there was no other way forward, the small but rapidly growing Schneider company bit the bullet and bought its own expensive equipment to streamline catheter production without interference from the bullying North Americans. Before long, however, Bard/USCI came up with another indispensable component of their system in the form of a subtly responsive, "steerable" guiding wire over which to slip angioplasty catheters deftly into place, a piece of supporting equipment that had finally been refined to a level that Gruentzig could accept. It proved to be another cudgel with which to beat the competition. When Hugo Schneider declined Bard's latest acquisition offers, the American corporate executives reached back into their bag of brass knuckles and refused to provide any of these subcomponents either, forcing the Swiss to create their own version. These skirmishes were but child's play compared to what would happen next.

Chapter 14

Things were not about to slow down for Andreas Gruentzig, not in the year ahead, nor ever again. Soon after the shenanigans in Hawaii, he decided to make a fresh start with Margaret Anne by purchasing a new house just for the two of them. When a fine Druid Hills residence a few blocks away hit the market in December of 1981, he instantly snapped it up at cost, declining to haggle whatsoever. The quietly tasteful, single-story brick West Ponce de Leon dwelling had a columned entrance leading to a handsome spread of finely appointed rooms. Better yet, there was an expansive back garden with abundant space for installing a pool and cabana, which would be perfect for entertaining little Sonja on her visits, or the stream of medical luminaries who came to call.

Balloon angioplasty quadrupled its growth in 1982, and its charismatic torchbearer grew steadily more famous—and richer. His own caseload at Emory doubled and a series of new catheterization laboratories were constructed to handle the flood of incoming patients seeking an alternative to invasive heart surgery. These were heady days—except, of course, for those spent negotiating his impending divorce. Hospitals and medical societies around the world proffered lavish speaking invitations. Palm Springs, Heidelberg, Rio de Janeiro, Cape Town, Zürich, and Davos—there was no telling where Gruentzig might fly off to next, usually with young Margaret

Anne at his side. The former child of the rubble held the world in his hands.

Bernhard Meier observed, "Margaret Anne was of course exactly what he needed. He needed a coach and she was absolutely his coach to teach him how to perform in his current role in the United States. She knew how the house had to look and what he needed to do in terms of parties and invitations."

Joe Craver and his wife were often on the guest list, even though the surgeon nursed occasional misgivings about Andreas's young partner. "A lot of people didn't like her, but I kind of tolerated her and that helped me bond with Andreas . . . He would have colleagues from around the U.S. or business people come in and have a dinner and invite my wife and me. He had huge parties, lavish parties that were just lovely with no expense spared—caviar, flowers, caterers, lovely food." As it happened, the increasingly dollar-savvy Gruentzig often had the Schneider company (still leapfrogging in size despite Bard's strong-arm tactics) foot the bill. He also gained a 5-percent ownership of the enterprise's newly issued stock. What with any number of medical-device makers seeking to acquire the former garage shop, the arrangement promised to be providential.

Visiting Atlanta with her husband, Sharon Myler wondered about the transformations taking hold of her former houseguest who had insisted on personally washing her dinner dishes in cold water. Still a soul mate of the banished Michaela, Sharon pinned the new extravagance on Margaret Anne Thornton. "She dressed him up. I mean, this is a guy who would nearly wear two differently colored socks, he never cared about clothing. Suddenly, he began turning into someone who would be seeing his [medical] residents in his house in a silk smoking jacket."

Perhaps such characterizations of Margaret Anne were uncharitable. One of the young woman's lecturers in medical school later remarked that in her experience of teaching 2,000 medical students, the Macon native shined with singular appeal. "Not only was she pretty and bright, but she had a lovely glow of happiness, cheerfulness, self-confidence, poise, and warmth all the time, and she had a wonderful sense of humor . . . She had a sparkle and enthusiasm that made ordinary days much sunnier." And the radiologist-in-training had achieved such high marks through every segment of her studies that she turned down three coveted alternatives to continuing her medical residency at Emory. So she was clearly a winner, and not just by dint of personality alone.

Colleagues nonetheless gossiped, perhaps not grasping that Gruentzig was headed in a direction he himself desired. The inspired seeker was being manipulated into the tricky seductions of Southern social climbing, whispered numerous doctors and their watchful wives. Sarge, Gruentzig's swarthy personal assistant, took to openly declaiming that his boss was being led down the garden path. Gruentzig was not exactly fully tamed himself, as he proved by purchasing a .38 caliber Smith & Wesson pistol and continuing to bound toward the many available women who crossed his path.

He confided to Richard Myler that he would no longer entertain any speaking engagements without assurances that he could fly and stay first class en route. "He said, 'If they are taking you away from your family and your workplace, you should be as comfortable as possible and they should pay for it.' He was lecturing me as if I were four years younger than he, rather than the reverse." Gruentzig's colleagues at the German Society of Cardiology invited the maestro to give a summer talk. Thumbing his nose at the very dons who had shunned him in his years of need, he demanded deluxe treatment and nonetheless showed up a day late. His displeased hosts surmised that the man was intoxicated by his own success.

Singular triumph has always brought special privileges and Gruentzig was perhaps entitled to think he deserved these. Visiting Paris with Margaret Anne for a medical meeting, he booked into the Ritz Hotel along with Joe Craver and his wife. Long gone were the days of singing in Heidelberg beer halls for chump change or motoring in a beat-up car with VIVA ESPAÑA haphazardly scrawled on its side. According to Mary Jane Thornton, Gruentzig hit the boulevards with a view to extending his wardrobe. "He was so envious," she claimed, "because Joe Craver had such good-looking suits and clothes . . . So Margaret Anne took him to Yves Saint Laurent and got him a tuxedo . . . He got back to the hotel and the first thing he did was call Joe Craver. He was so pleased." The shopping spree included the purchase of fine French antiques and an expensive jacket for Earl Thornton's upcoming birthday.

In March, Gruentzig had the proudest acquisition of his life flown in from Germany: a gleaming red Porsche. The car was hot, and he drove it at extraordinary speeds. The young cardiologist Steve Clements remembers the impression the Porsche made. "One time I was driving down this little road called Ridgewood. You come to a stop sign and there is a hill veering up ahead. Suddenly, this red blur went 'chumph!' around me like that. It was Andreas running through the stop sign and shooting up the hill."

He whisked Margaret Anne off to more getaways in Bermuda and the Caribbean, along with visits to the fabled Cloister hotel on Sea Island. No couple at the formal dances there sparkled more than Andreas Gruentzig in his silver tuxedo and the adoring Margaret Anne Thornton in her filmy chiffon gowns. Some guests compared them to Rhett Butler and Scarlett O'Hara of *Gone With the Wind*. The Cloister's staff simply called the love pair, "the prince and the princess," according to the concierge, Huston Visage.

Despite his affection for the high life, Gruentzig never wavered from his passionate stewardship of his technique. During a Massachusetts brainstorming session with USCI, Gruentzig outlined the potential breakthroughs he foresaw: Nearly every one proved amazingly prescient. Soon, he promised, balloon angioplasty patients would be processed through the hospital in a single day. He vowed that the then still scarcely comprehended modality of laser therapy was about to further revolutionize the treatment of cardiovascular disease—and a few years later was proven right again. But what of the vexing tendency of so many of his current angioplasty procedures to fail in a matter of months and require a second intervention? The solution to this so-called Achilles' heel of restenosis, he asserted, could well lie in implantable scaffolding devices that would essentially shore up ruined coronary arteries for years upon end—two million such devices, called "stents" (which Charles Dotter originally envisioned but did not pursue) are now inserted annually around the world. These little gadgets, which resemble nothing more than the springs attached to the ends of push pens, now generate roughly $10 billion in sales every year.

Gruentzig kept a fretful eye on his emerging rivals. By the beginning of 1982, the Silicon Valley–bankrolled Advanced Cardiovascular Systems, led by John Simpson, was talking up the imminent release of its first products. The so-called Flash from Kansas City, Geoffrey Hartzler, unnerved him even more. Whether Gruentzig liked it or not, the situation in the jumped-up 1980s was in flux. The days had long passed since Mason Sones succeeded in deliberately hushing up his first visualizations of the coronary arteries and John Gibbon labored through two decades of quiet experimentation in perfecting his heart-lung machine. Ambitious doctors and eager-eyed entrepreneurs alike sensed fantastic opportunities loose in the wind, and they saw no reason that any one individual, no matter how brilliant, should continue to control its every nuance.

Hartzler had by now burst into prominence. The Midwesterner's prowess in swiftly accessing the deepest recesses of the coronary arteries brought gasps from many cardiologists—but not Gruentzig. He felt that the man was a charlatan who minimized the profound risks of the procedure. When word arrived that a patient had just died following one of Geoffrey Hartzler's procedures as a "guest operator" abroad, Gruentzig raged.

Also infuriating, after all of Gruentzig's lean years of kitchen-table experimentation, was the fact that Hartzler was fiddling with his prized system, reshaping guide catheters to his specifications while experimenting with exquisitely flexible wires that might slip the stiff Gruentzig balloon catheters into place more reliably than those anyone else was making. Dave Prigmore and his increasingly edgy fellow executives at USCI watched the developments closely. They worried that the company's de facto helmsman was stubbornly ignoring a coming sea change. Their hegemony over an exploding market took its first blow in early March when Simpson's company received FDA approval to market its system, which promised to fairly float balloon catheters into their desired territory over a gently manipulated guide wire. The threat remained minimal for a while, since Simpson's critical differentiating element—the guide wire over which the miniaturized equipment was supposed to trick into place—was initially a fiasco. The thing waffled and contorted every which way, and Gruentzig made a show of its limitations in several public demonstrations. But the inventive Hartzler had plenty of ideas up his sleeve for improving upon it.

USCI's executives sensed that no good could come from Gruentzig's disdain for the slightest deviation from his methodology, including his insistence on laborious blood pressure measurements before and after every treatment. Hartzler soon proclaimed these to be pointless—a shot of post-procedural X-ray dye was generally all the confirmation of success angioplasty needed. Get on with it, he urged. As a rival, the man proved to be formidable, for he was not only fearless but brilliantly intuitive. In the face of the most dire situations, Hartzler kept his cool and fashioned solutions on the fly. He also burned with conviction and pronounced that the delicacy of balloon angioplasty, rather than the blade work of surgery, was poised to become the front-line treatment for coronary artery disease—and not five or ten years down the road, but almost overnight. "The only indication for bypass surgery is failed angioplasty," he often proudly proclaimed, to the conservative Gruentzig's dismay.

A perhaps still hotter issue was the idea that heart attacks might at last be stopped in their track. The trigger mechanism had finally been narrowed down to the sudden formation of a clot that plugged up a coronary artery and totally shut off the life-renewing flow of oxygen-rich blood to the heart. Functioning like tourniquets, these catastrophes erupt without warning, causing the cells of the surrounding heart muscle to undergo spreading circles of asphyxiation by the millionfold. Leave the vortex of destruction uninterrupted for a few hours, and entire swaths of the heart are rendered into dead or "ischemic" tissue that can no longer contribute to driving the circulation onward. The result can be rapid death or a dire weakening of the heart's pumping force. Worse, heart-attack survivors are often left with a permanent susceptibility to misfirings of the heart's intricate circuitry, which is a death trap of another kind.

Pinning down the relatively simple but to that point murky mechanics of an emerging "myocardial infarction" was an epic detective story in its own right. Suffice it to say that once the syndrome was finally understood, pioneering doctors from Russia and Germany set to experimenting with drugs to break up the fiendish clots that are its trigger. Well aware of Gruentzig's success, the first heart-attack fighters employed hollow catheters to infuse a clot-dissolving drug called streptokinase into the very nexus of the circulatory crisis. The idea, spearheaded by Peter Rentrop of Germany and explored in depth by colleagues like David Williams and later by a young Cleveland Clinic doctor named Eric Topol, was to dissolve the clots in an emergency procedure and leave it to cardiac surgery or angioplasty to eliminate the chronic buildups of plaque where the filmy clots invariably lodge. The early results of this new "clot-busting" intervention looked so promising that leading cardiologists rapidly made plans for a major clinical trial of the therapy.

The impulsive Geoff Hartzler proclaimed that his academic colleagues were ignoring the obvious, since he had already successfully used balloon angioplasty to attack heart attacks in the same early minutes of peril. He vowed that bold deployment of the balloon might do a better job—with more or less *instant* results—than a slow trickle of drugs. The man was a loose cannon, the Gruentzig camp admonished. The very boots on his feet, the caricature went, pronounced that he was a "cowboy" trifling with mortal risks.

Hartzler remains embittered by that charge. The electric-switch plates in his present recording studio/garage headquarters may have clown faces painted on them, but there was no caprice in what he said next. "The thought that I was as irresponsible as a 'cowboy,' as that term is applied to medicine, was appalling to me, and I would categorically deny that. My activities might have been cutting-edge and unique, but 'unsafe' was not something I was willing to accept."

A "cowboy" is the last label a visitor would ascribe to the man now. An enigma seems more accurate. The ordeal of being raised by a mother confined to an iron-lung machine perhaps drove him to try to cheat death with an inordinate zeal. From his odd office, he currently manages numerous multimillion-dollar businesses, consults on the development of new heart devices, and dispenses a substantial flow of philanthropic funding to his charitable interests—when he's not strumming his Fender guitar.

A decade ago, Hartzler abruptly stopped performing any further angioplasties and went his own private way. For one thing, the toll of performing endless procedures while draped in lead aprons had ruined his back, ultimately requiring five major orthopedic operations, while his marriage fell by the wayside as well. And after performing nearly 6,000 angioplasties, he had little left to prove. He nonetheless still broods over his conflicts with Andreas Gruentzig. "If you respect that his intent was to ensure the long-term survival and viability of the procedure . . . it would make some sense to have it be restrained and have people just pick the most ideal of patients—that proximal, concentric, single-vessel, non-calcified lesions and so on would assure the best results," Hartzler allowed. "But to do so would deprive the overwhelming majority of patients from benefiting from the procedure."

The fact is that Hartzler, in the role of rival prophet, was wracked with nearly the same pressures as Gruentzig himself. "The long-distance referrals came from everywhere, every day from some different country, from Europe for a while, and then I had a host of people coming from Asia, Japan, and Australia, just on and on and on. And they didn't come because they were easy [cases]."

In the early 1980s, allegations snowballed that, for all his talent, Geoffrey Hartzler was losing too many patients, perhaps well over a score of them, the records indicate, before the methodical Gruentzig had lost a single one himself. Hartzler, who then drove a Harley-Davidson motorcycle, acknowledged that "about fifty-five to sixty" individuals had succumbed at his hands over the course of fifteen years. "That's a lot . . . But what I was taking care of were horrible patients, the sickest of the sick—many of them unstable and some in shock. They were referred to me because surgeons refused to do anything with them." He made no claims as to how many lives he had extended by his heroics, although any fair observer would concede that this was "a lot," too. The consensus is that his pioneering work profoundly extended Gruentzig's breakthrough. But when prodded to elaborate on why he dropped out of the profession, Hartzler's voice grew tremulous. "It was because I couldn't stand any more death," he said.

Geoffrey Hartzler absorbed many withering criticisms in his day. Marketing executives and doctors in the fold of USCI, the company with the most to lose by his ascendancy, lost no time in spinning the message that he was a supremely cavalier risk-taker. But closed-door meetings told a different story. Time and again, Gruentzig was beseeched to meet with his supposed nemesis and search for some common ground. Hartzler himself was eager to collaborate and implored USCI to transform his concepts into prototypes that could be properly tested. But an embittered Gruentzig decreed that the Midwesterner's every idea was innately contaminated, and shouldn't even be explored. Dave Prigmore reached his wit's end. "Andreas fought and fought. We were struggling to figure out how we were going to leave behind this bullet-nosed thing, which he had invented, and move on to over-the-wire catheters. We tried to get Hartzler and Andreas to join forces. I struggled to try to bring these two personalities together . . . Gruentzig wasn't going to do it and went to his death with that attitude." He wouldn't even be seen in the same room with his rival, Prigmore added.

In time, Hartzler was driven into the arms of John Simpson, whose company was barreling ahead with refinements to its original technology. The alignment USCI dreaded was at hand. "ACS glommed on to me," Hartzler said. "I was a kid. My motive in none of this stuff was money, any of those things, it was just to have products that worked. They [ACS] were clearly very rapidly responsive. I'd call them, give them an idea, and within a week they would have a prototype in my hands."

John Cvinar, USCI's director of marketing, slipped into a medical meeting Simpson held in California and saw a changed future looming. He reported back that the nimble new system held the potential to simplify balloon angioplasty so much that it might extend the technique's applications to tens of thousands of new patients. Moreover, since time saved means money earned, the quicker new equipment promised to lure droves of formerly fence-sitting cardiologists onto the angioplasty bandwagon. A stupendous market shift beckoned.

The USCI brass was not dense. The company's officers instructed their engineers to begin copying and hopefully bettering their new competitor's technology. Meanwhile, they had to sweet-talk away Gruentzig's carping complaints about this work. Prigmore explained, "Gruentzig was a tremendous challenge to me . . . How do you manage this character who's very strong-willed, very bright, very brave, but very stubborn?"

Back in Zürich, the rapidly growing Schneider company was also fretting. The Swiss outfit could see that the technology was leapfrogging, whether Gruentzig liked it or not, and instructed its technicians to begin devising their own guide wires. Heliana Canepa remembers Andreas Gruentzig coming down like a ton of bricks on the company's every attempt to improvise off his core concept. "We would get a new idea and he would kill it right away," the Swiss businesswoman laughed as she inhaled on a cigarette. "He'd say, 'You have my name and under my name you cannot do anything that I don't agree to.' We could have copied Simpson right away, but he didn't want it. He was very cautious. He was better than any quality inspector you have today . . . He didn't want any failures. He said, 'In the heart you must be able to steer it' . . . So we followed his instructions, of course. Yeah, it was like that. You just listened when he spoke."

As Dave Prigmore asked, how do you manage such a figure?

Well, you ply The Reluctant Party with favors. USCI offered an expense-paid trip to the Waldorf-Astoria in New York, from which base Margaret Anne might just get in a bit of shopping. Prigmore also invited Richard Myler and Simon Stertzer by to help lay the issues forth—namely, that Gruentzig's thunder would be stilled unless he agreed to supervise a new engineering blitz to further finesse his entire system. By the weekend's close, they got their consent. A few months later, USCI hit the angioplasty market with its own me-too miniaturized and "steerable" catheter, floating into place over a guide wire, which they called "The Mini." The angioplasty wars had commenced.

Ralph Lach, the Ohio cardiologist whose career had been reborn once already, well remembers the transformations that ensued once more miniaturized catheters could be easily slipped into place over neatly controllable wires. "Suddenly, we went from a 60-percent success rate overall to a 91- to 92-percent rate once we had steerability. One of the early reasons for failure was that you could not reach the lesion, or once reaching the lesion you could not advance through it. But with the guide wire advancing very distally as an anchor then you could advance the catheter relatively quickly. . . . The sort of little hockey stick on the end of the wire allowed us to steer into this vessel instead of another one; it could rotate 360 degrees."

A little twist to a wire here, a smaller balloon there, and suddenly angioplasty had become a formidably sophisticated technology, and one ready to give cardiac surgeons a run for their seemingly limitless earnings.

Meanwhile, Gruentzig's fame continued to grow. The popularity of his teaching courses soared, now that the full potential of his technique was apparent. Ever more stylish parties capped the courses' serious clinical work and inspired teaching. At one, Gruentzig had his new pool checkered with red balloons—every one of them an emblem of triumph. Another featured a roasted pig lying in a bed of fruit and with an apple stuffed in its mouth, the allusion this time being to a traditional German feast. A born hostess, Margaret Anne hired the finest caterers for another backyard dinner for 200 in tents outfitted with linen tablecloths and chandeliers; a dance orchestra capped off that summer night's entertainment. Playing the role of consummate hostess and dutifully world-traveling wife could not have been easy for a woman so young, caught up in the struggle to resume her own medical career.

Mostly, Gruentzig's free time was spent in homely pleasures. Little Sonja returned for the summer and delighted in the wonder of the backyard swimming pool. Margaret Anne took to teaching her tennis and leading the then-chubby child through aerobic exercise and dance classes. At night, she tried to draw nearer by bathing the six-year-old and singing her lullabies. But it was impossible to forget that the girl was not hers. At times, Margaret Anne complained that the child was laxly disciplined and petulant. Gruentzig himself would make occasional displays of reasserting his parental authority, despite being all but absent from Sonja's life for nine months of the year. The mood was usually buoyant, nonetheless, on Sundays when the threesome drove down to Macon to spend the afternoon with Margaret Anne's parents, who had by now become quite enamored of the dashing German's exuberant style.

Soon after Sonja departed in the autumn, Gruentzig decided that he was ready to buy another house—this time a weekend cottage on luxurious Sea Island. After a quick reconnaissance, he asked Margaret Anne's mother if she knew anyone involved with the private company that trades all of Sea Island's real estate. Things in Georgia working as they do, a choice connection was made for a closer tour. Gruentzig promptly wrote a check for a $60,000 down payment—enough to cover a $600,000 transaction—and advised the mother and daughter to get the job done on their own and buy whatever seemed right. "This is the way he thought, everything must be taken care of at once. We found the house that she [Margaret Anne] thought he would like because it was on the beach . . . He had said, 'There is no point in buying anything that is not on the beach,' " Mary Jane Thornton recalled. She told her daughter that they mustn't sign over the blank check without prior legal consultation. But they called Gruentzig and heard otherwise. "He said, 'Buy it, buy it, buy it.' "

The so-called cottage sat at the end of one of Sea Island's dreamy, shady, private lanes called Guale Street and stared out over low dunes to a three-mile-long private beach. Shimmering a few miles beyond was Jekyll Island, a playground of nineteenth-century tycoons including the Rockefellers, Morgans, Astors, Goulds, and Vanderbilts, since largely turned into a national park. Among Georgia's so-called Golden Isles, Sea Island has assumed the greater prominence. It was, as Spencer King said, "where the high cotton met." One modern cottage on this resort island boasts an interior waterfall and helicopter landing pad on its roof, and the Spanish moss–dripping "live oaks" of the central avenue shade one stunning Tuscan- or Tudor-style villa or mansion after another. In addition to a string of visiting presidents—Coolidge, Hoover, Eisenhower, Carter, and both Bushes among them—movie stars and European royalty have enjoyed the hospitality of the island's sublime estates.

Gruentzig's cottage was not quite so lavish. But it offered finer amenities than he might have ever dreamed of in the darkness of his Zürich apartment—four spacious bedrooms on two levels and the beach at his door. Palm trees spread their fronds on the expansive front lawn. And having a beachfront address on Sea Island—well, one would have to spend about $4 million for the same property today. Around the clock, a private security force patrols every whimsically named lane—"Tomalatoe" gives way to "Aguizila" —to protect against intrusions from the outside. Spectacular yachts tilt back and forth offshore. In the evenings, the crash of the surf is the only backdrop to the clinking of cocktails on verandas checkered with elegantly dressed guests. A few hundred feet up the shore from Gruentzig's new abode, the playwright Eugene O'Neill built a twenty-two-room house in the early 1930s, in what he called "this island paradise."

For a divorced father wanting to shower gifts on his young daughter, the place indeed promised magic. The island was steeped in the broad sweep of the New World's history—ancient Indian settlements had yielded to the shifting occupations of the Spanish, French, and English, with the occasional pirate stopover thrown in. In the nineteenth century, the place hosted several sprawling slave-worked cotton plantations. Tranquillity was assured, because Sea Island was well removed from the mainland.

From Brunswick, Georgia, one first crosses a bridge to another island called St. Simons, whose private airstrip is also rich in history. Gruentzig, the pilot, would have quickly learned that the short runway there was Charles Lindbergh's last touch point before embarking on the first flight to Mexico. It was also often used by Eddie Rickenbacker, who gained World War I fame by shooting sixty-nine of Gruentzig's countrymen out of the sky and later became president of Eastern Airlines. From St. Simons's eastern shore, a long causeway proceeds over salt marshes and past a watchful police checkpoint to the smaller and far more exclusive fiefdom of Sea Island. The isle was transformed into a resort in 1928 by a Detroit industrialist named, as if by Eugene O'Neill, Howard Coffin. He commissioned one of America's top architects, Addison Mizner, to build the Spanish-style Cloister hotel with its red-tiled roof, shady inner courtyards, and high-ceilinged reception rooms. Then he sold off the surrounding lots at heady prices for the day—$550 and up. Although worth millions today, the value of such resort holdings plummeted throughout the U.S. during the Depression, and Mizner was mortally stricken by a heart attack in 1933. Coffin killed himself three years later.

His protégé, Bobby Jones, next took the reins, and Jones's offspring still control both The Cloister and the management of every inch of Sea Island, whose 500 homeowners all ante up heavy dues to the resort association. Most don't mind, for The Cloister is lacquered with the courtliness of a nearly vanished Southern gentility. Afternoon tea in the hotel's elegant Spanish Lounge was then—and still is—a stylish ritual that gave way to formal dinners and the twice-weekly ballroom dances where Andreas and Margaret Anne stood out like royalty themselves. The golf course, the seaside pools with their moonlit waltz sessions, the marina, the clay tennis courts, the skeet-shooting and stables—every facility that Sea Island offers is meant to exude a singular fineness.

Mary Jane Thornton relished the latest turn of her daughter's life, and presented the couple with a piano. Shortly afterward, Andreas phoned Heidelberg to tell his Mutti that his dream house had been outfitted with a musical instrument she would adore on her visits. But Charlotta Gruentzig was somehow dismayed. "She hung up the phone," Mary Jane Thornton recalled. "I don't know what she said, because he was embarrassed that we would know that she hung up on him. It hurt his feelings. . . . It was like someone slapped him."

There would nonetheless be plenty of visits by Andreas's Mutti, his brother Johannes with his wife and children, his aunt and uncle from Hanover—all evidently paid for by Andreas. In a likely fraternal conspiracy, Johannes sashayed onto Sea Island's beach in his own thong-like bathing suit—to the howling laughter of all present, save Mary Jane Thornton. But the Mutti sometimes waxed sullen. "I think she always felt that he should have married a German woman—that he needed a Hausfrau," conceded Mary Jane. "She made no bones about that," her husband added wistfully.

The romance flowered regardless. In November of 1982, Andreas and Margaret Anne visited Macon for a traditional American Thanksgiving feast. Afterward, they led the elder Thorntons out for a vigorous walk, whereupon Gruentzig began singing arias at the top of his lungs. "His voice was gorgeous . . . I wanted to tell people to open their doors and listen," Mary Jane recalled. "Then Margaret Anne made something up to sing with him and she would follow him and he'd sing, 'No more old women for me,' and she would sing, 'No more old men for me.' They were so cute together . . . Just being with them was delightful."

On one Sea Island weekend, he proclaimed to Margaret Anne's parents. "She looks so good, I'm going to have to kiss her." And not only did Gruentzig make an unabashed display of doing that, he also took his own photograph of the embrace by wrapping one arm around Margaret Anne's neck while extending a camera outward with the other. Gruentzig in fact performed that trick so many times that this compulsion—along with his love of being photographed by all comers—cried of narcissism. But there was more to it, insomuch as the man was clearly in love with his adored Margaret Anne and with celebrating life itself.

Some associates claim that he kept his real passion glowing in Zürich with his frequent visits there. "One could see that in his heart he was still in love with Michaela," said Maria Schlumpf. Margaret Anne reportedly cringed at his every phone call to his former wife, although this could be just hearsay.

Hundreds of photographs and countless anecdotes from that phase paint a fuller picture. They speak of a man smitten with the affection and zest for life that his beautiful young mistress bestowed upon him. A charge of eroticism runs through the albums, especially in the close-ups of a skimpily clad Margaret Anne sprawled out on their bed in the soft glow of morning light. Countless other photos from Gruentzig's own appreciative hand display her cavorting in the surf at one exotic destination after another, always decked out in her next-to-nothing bikinis.

Margaret Anne was certainly doing everything she could to please her prize catch. Several medical journal articles Gruentzig wrote in the early 1980s credited her as a coauthor, due to her contributions in polishing his English. At the prestigious Radiological Society of North America's meeting in late 1982, it was Margaret Anne, and not a close medical peer, who Gruentzig asked to deliver a sweeping overview of the state of his therapeutic breakthrough. When Sonja reappeared with Andreas's mother that Christmas, bearing a crèche as she did each year, Margaret Anne attempted to engage the wayfarers in their native German, having taken private lessons to become a true member of his family. At Sea Island, she took Sonja trail riding, then joined in the late-night serenades while Andreas and his mother played the piano. Meanwhile, she put her impending radiology training at Emory on hold, so as to have more time to look after her beau.

In 1983, the couple decided to marry and engaged the resident orchestra from The Cloister for their big day. That the bandleader Jack Hurd's regular work in Florida consisted of running a funeral parlor gave them no pause. In early April, they hosted an engagement party in Macon, with Margaret Anne decked out in tight-fitting white leather pants newly purchased in Rome. Her mother recalled that outfit with a smile. "She said, 'Mama, I'm going to have to show Macon that I am not just Macon anymore.' "

Like every mother of the bride, Mary Jane Thornton flew into a tizzy of arrangements. As the president of the local garden club, she had overseen the restoration of the grounds of the nineteenth-century cottage of the town's once-famous nineteenth-century poet, Sidney Lanier. So the wedding-eve rehearsal dinner was organized there. The finest example of antebellum architecture in Macon is called the Hay House, a grand residence with corner turrets and a sweeping marble staircase leading to high-ceilinged drawing rooms dressed out in ornate plaster and cornice scrollwork capped by ceiling roses, and with a grandiose cupola arising from the roof. The place exudes the pretensions to aristocracy for which the Old South is famous. A tunnel under the home once served as a secret storage chamber for Confederate gold. The Thorntons booked it for the magic day, and dispatched engraved invitations.

As the ceremony approached, Gruentzig wrestled with twin aspects of his nature—the newly indulgent one and the formerly impoverished side still fussing over every penny. Quietly, he drew up a prenuptial contract with his lawyers to guard his newfound wealth, just in case things should not work out. Though now rich, he had cut a meager child-support arrangement to help with Michaela's raising of Sonja back in Zürich. The prenuptial deal was cast in the same mendicant spirit, providing Margaret Anne with a fixed payment of $100,000, deliverable in monthly installments of $5,000 if the marriage lasted five years before dissolving. If a schism came any earlier, she would be entitled to a fifth as much. Another chilly formulation read: "All property hereafter acquired individually by Andreas Gruentzig by gift, device, bequest, or inheritance shall remain the separate property of Andreas." The document, filed just three days before the May 28, 1983 wedding, acknowledged that Gruentzig had made $400,000 in the previous year from Emory, with an additional $200,000 coming in from Schneider Medintag's direct sales and kickbacks from Bard/USCI. Considering that he was developing a taste for secret Swiss bank accounts and had drastically undervalued the worth of his Schneider stock, this accounting was perhaps incomplete.

Things nevertheless proceeded smoothly. The actual wedding ceremony was designed to be rather intimate. Gruentzig managed to garble the word for bridesmaids as he pronounced, "Well, I am so glad because I never liked to see all those mermaids coming down the aisle anyway." The Friday-night rehearsal dinner at the Sidney Lanier House was followed by a morning brunch for the out-of-town guests at Macon's River North Country Club. The afternoon wedding in the back garden of the Hay House saw Joe Craver, Spencer King, and Andreas's brother Johannes, serving as the groomsmen. Margaret Anne glowed in her elegant yet simple, white off-the-shoulder gown, purchased from the New York fashion temple of Bergdorf Goodman. Andreas shined in his tailcoat. Macon, Georgia, the birthplace of down-and-out rockers like Little Richard, Otis Redding, and the Allman Brothers, had not seen many ceremonies like it.

The celebrants traveled next to the reception at the Thorntons' country club, called The Idle Hour, where the crowd multiplied by several fold. At the top of the stairs awaited an expansive ballroom with parquet floors, floor-to-ceiling windows, and gilded chandeliers. The wedding feast played out there with champagne toasts and the cutting of the ceremonial cake, whose crowning layer was topped by chocolate icing drawn into the configurations of a whimsical catheter pointing toward two united hearts.

The sixteen-piece orchestra began to swing, and the guests drew into a circle in the middle of the ballroom, with their gifts laid before them on the floor in a kind of wreath, after the German style. To a crescendo of clapping, Margaret Anne and Andreas stepped forward and set to waltzing. That evening, the newlyweds departed in their chauffeured white limo—next stop not Paris or Rome, but their beloved new Sea Island cottage.

For all the laurels that had accrued to Andreas Gruentzig, his inner nature remained down-to-earth. His mother-in-law telephoned a couple of days later to check in and was told, "We went out crabbing today and caught one hundred and fifty of those big guys."

CHAPTER 15

BY 1983, THE DECADE'S ECONOMIC BOOM WAS IN FULL SWING, and venture capitalists and brokerage houses swarmed over proliferating investment opportunities in the medical-device field. Financial scouts swiftly realized that the Gruentzig quest to replace invasive surgery with delicate catheter work was being emulated by myriad medical-device manufacturers from coast to coast, and that this was fertile investment ground.

No one studied this scene more closely than Dan Lematrie, then a young point man for a private Boston banking firm called Cowen and Company. Interviewed high above that city's financial district, where he was recently working as a senior analyst for Merrill Lynch, Lematrie recalled the excitement that surged around the Gruentzig idea. The parlance of his trade labels each potential investment play a "story," so he used that word in saying, "I can tell you that this was *the* story. If you had to pick one area where you had to say 'I have to own this story, I have to understand what's going on in this market'—angioplasty was the story. It was a game-changer, it was having a dramatic impact on the markets."

Lematrie visited every important medical meeting that might divulge new trends. He frequently mingled with the likes of USCI's Prigmore, ACS's Simpson, John Abele, and Bill Cook, and the backers of a Minnesota company named Scimed that was itching to jump into the angioplasty game.

Meeting and greeting, Lematrie discovered that medical procedures from the brain to the toes were undergoing a radical rethinking, all thanks to the pioneering work of Andreas Gruentzig. One area of investigation involved replacing traditional reconstructive knee surgery with the much more benign and now routine procedure called arthroscopy. Invasive kidney operations were destined to undergo a similar transformation, as was the surgical reaming out of neck arteries in the hope of preventing stroke. It was plain to see that many lucrative markets waited in the wings. Seed capital flooded toward doctor-inventors at a volume that Gruentzig, in his time of endless rejection, could not have dreamed.

"There is no doubt that angioplasty became the model where people said, 'If you can figure out how to do things less invasively, patients are going to demand it,'" Dan Lematrie observed. Meanwhile, each new wrinkle in balloon angioplasty equipment sparked shifts in the purchasing of millions of dollars worth of equipment. Investors inevitably became fascinated by the horse race between USCI, ACS, and Schneider, each of which pumped up their catheters' prices until they ultimately neared $700 a go—*70 times* higher than what they had been a half dozen years earlier. Heavy royalty payments began rolling Gruentzig's way, what with the 20 percent arrangement still in place with Schneider, along with the indirect fees piling up from Bard/USCI. By 1983, he was pulling in roughly $1 million in royalties, even as his earnings at Emory grew beyond $500,000. At the same time, myriad companies with inventive ideas began offering hefty consultation fees. The boy who begged for eggs in Leipzig street markets could have caviar and champagne whenever he wished, and often consumed quantities of both as he jetted around the world.

In fact, Gruentzig earned so much that he was said to dispatch satchels of cash to untraceable Swiss bank accounts. Mary Jane Thornton acknowledged as much in relating, "Margaret Anne would deposit $250,000 every time she'd fly over there. In fact, she didn't deposit it herself. She would go over there and then Andreas would fly over and somehow they'd meet. I don't know how they did that. I know one time in particular I was very upset with her flying with that much money, and I said, 'Margaret Anne, I hope you're not doing anything illegal.' She said, 'No, because Andreas is not an American citizen.' I told Margaret Anne that she had to stop doing this, that it was too risky."

Heliana Canepa, by now virtually running the Schneider company, watched as Gruentzig bought his mother a Heidelberg apartment and regularly funneled money to her and his brother Johannes. "His brother always needed money for strange purposes, such as going to India for some kind of research. So Andreas would call me and say, 'When you make the invoice about my royalties, please deduct ten or twenty thousand Swiss francs and send it to my brother under the name Schneider and say that you are supporting his trip there, and don't ever tell him that I paid for it." Before long, she informed Gruentzig that Schneider Medintag—whose gross angioplasty earnings were approaching $10 million a year—would have to cut his royalty payments down to 10 percent. Yet this was not likely overly worrying, since USCI's earnings had catapulted past the $100 million mark, and it was evident that further riches were destined to come his way.

The first catalysts for the angioplasty revolution, Sones and Dotter, were not faring remotely as well. In fact, USCI's Prigmore worried about Mason Sones, by now battling lung cancer and struggling to make ends meet. "My relationship with Mason became strictly personal. I was here [at the Rhode Island beach home where he was interviewed] on vacation two different years when my secretary called and said, 'You've got to call Mason tonight.' The first time he said, 'I just want to let you know that I'm going to have lung surgery tomorrow. I don't have any reason to think I'm going to make it through, so thanks for everything.' I said, 'You're too mean a bastard to not make it through. We'll see you next week.' "

Prigmore sent his old friend an $80,000 draw against royalties to pay for a cruise around the world with his first wife, who had benevolently taken him back despite his long betrayal. A year or so after that voyage, Sones telephoned once more and Prigmore tried to make light of the situation's obvious gravity. "He said, 'I'm going for surgery again tomorrow and just want to say thanks.' I said, 'How many times are you going to do this?' He was a lonely, forgotten, broken guy by that point."

Meanwhile, René Favaloro, the man who had harnessed Sones's breakthrough to usher in the era of coronary bypass surgery, had founded his own medical institute in his native Argentina. That enterprise would eventually become embroiled in financial scandals. Soon the once-celebrated heart pioneer's life fell apart. The young woman he married while in his late seventies and certifiably infertile reportedly became pregnant by another man, and Favolaro shot himself with bitter irony, in the heart.

After a decade of defying mortality, Charles Dotter, now sixty-three, was struggling, too. He underwent extensive radiation therapy for Hodgkin's disease in the early 1970s, and capriciously gave up wearing his medical lead apron for the next five years, because he reasoned that he had been given only that long to live in the first place. Although he passed that milestone by a long shot, he later developed crushing angina and underwent coronary artery bypass surgery. "A week to the day after the operation, he walks back into the department with his shirt unbuttoned and this nice scar showing, strutting around," Fred Keller, a former colleague, recalled with a chuckle. "So months went by, and one day Enid Ruble [Dotter's secretary] called Dr. Rösch and me to come quick. Charles was lying on the floor in his office, in terrible pain and sweating. We thought he was having a myocardial infarction and we took him and put him on the angiographic table, then called a cardiologist. The cardiologist asked, 'Do you take any medicines?' And Charles said, 'Just a Dexedrine [a common amphetamine] every morning.' "

"The guy was up on the ceiling anyway. He was the last guy in the world who ever needed a Dexedrine, but it turned out he had a perforated ulcer." Dotter may have dodged the bullet that time, but as the years passed and as his bypassed coronary arteries began to congeal with fresh disease, his chest pain returned. By the end of 1983 his days of practicing medicine were nearly done.

Gruentzig was having no such problems. Like his rival from Kansas City, he had begun performing angioplasty more aggressively to treat evolving heart attacks and clear multiple coronary blockages at one sitting. Myriad new technologies began to capture his imagination, a number of them involving potential solutions to the riddle of "restenosis" (the reclosing of coronary arteries a month or so after the balloon procedure) a quest that by itself had begun to attract tens of millions of dollars of fresh investments. With his technique increasingly well-established, he set aside time to resume flying lessons and even started shopping for his own plane.

Meanwhile, Gruentzig's exotic travels continued. With Margaret Anne, he jetted to speaking engagements in Malaysia, Singapore, and Japan. Photographs from 1983 show a grinning Andreas decked out in a teal Malaysian tunic while drinking cocktails out of coconuts, fooling about with packs of monkeys, and handling poisonous vipers with a local snake charmer. In a Walter Mitty moment, he bruised his knee while leaping onto a rickshaw.

In Singapore, the accident-prone cardiologist collided with his hotel room's glass veranda door and banged up his nose; a few months before, he had fallen face-forward while attempting to leap a hedge. There were no grave mishaps in Tokyo, however, where Andreas and his young bride engaged a master to teach them the elaborately mannered protocols of the Japanese tea ceremony. Dressed in his-and-her blue kimonos, they stayed up until deep in the night in their hotel suite, kneeling before a low, black-lacquered table to get every nuance just right.

Back in Atlanta, Gruentzig broke his ankle after tripping over a hose while bounding toward his backyard pool. The master physician was developing a habit of getting in his own way. Mary Jane Thornton responded to the latest mishap by purchasing an inflatable raft so that her son-in-law could elevate his plaster cast above the water while floating beside Margaret Anne. It is difficult not to ponder what dynamic kept sparking these misadventures. There is no evidence that alcohol was involved, but a number of colleagues assert that Gruentzig was developing an appetite for heavy drinking.

Late in 1983, he was invited to give a presentation in Riyadh, Saudi Arabia. A formal dinner was thrown in Gruentzig's honor at the American embassy there. Meanwhile, he adopted a new guise that can perhaps best be labeled as "Andreas of Arabia." Gruentzig bought himself a glorious sheik's regalia of a white caftan with embroidered buttons up to the chin, and topped by a gold and black fez. The Omar Sharif look-alike sauntered into the course's crowning celebration—which took place in a desert tent—as though he had crowned himself some kind of prince. "Nobody else would have had the chutzpah to do that," observed Spencer King. "Humble was not a word in Andreas's dictionary—insecurity, maybe." But what prompted this display? Festering narcissism, or sheer disdain for small-minded conventions of every kind? The answer is elusive, since both of those characteristics were evident in the ever more chameleonlike behaviors of Andreas Gruentzig.

In Atlanta, trouble simmered. The tenfold increase in angioplasty procedures in the last three years piled both riches and pressures on the Emory cardiovascular staff, and with them came growing circles of jealousy and resentment. Tongues clicked over Gruentzig's most obvious self-indulgences, along with the special favors he kept extracting from the administration. Heliana Canepa remembers one particular dinner engagement gone sour when she attempted to arrange a meeting with a certain would-be inventor. "I had a young doctor once who said, 'Could you please introduce me to Andreas Gruentzig, because I have an interesting idea to discuss . . .' Andreas said, 'Well, come on, who is he? Now I have a Porsche and an airplane.' I kicked him under the table and said, 'Are you crazy? This is terrible.' He said, 'Well, I was treated myself the same way when I was young.' "

But others believe that this episode of churlishness was an aberration. "He would talk to anyone, and he was a fabulous conversationalist . . . He would find out what people's interests were, their dreams and concerns. He was an incredibly charming man," said Gary Roubin, an Australian cardiologist-in-training taken under Gruentzig's wing.

In a 1986 interview, Roubin spoke emphatically of the warmth and humanity that he saw in his mentor. "To stand by him when he was trying to solve a difficult problem for a patient was a wonderful experience. He was able to achieve things which you felt that no other angioplasty physician could have achieved at that time, perhaps even now . . . He did acrobatics regarding catheters. He was able to guide a guide wire with incredible precision. And when he put you up front, and allowed you to do the manipulations and to do the case, he would stand behind you so that you felt that you had to do what he wanted you to do and do it right, and do it quickly, and not make any mistakes. He was a wonderful teacher because he would whisper in your ear at just the right time to do something, and sometimes he would even move the catheter for you, and move the wire for you imperceptibly so that no one in the room but he and you knew that he had achieved the maneuver for you."

Speaking to a medical-video maker named Burt Cohen, Roubin continued, "He had an acute feeling for his patients. He was wonderfully soft for his patients. I was with him on the first occasion that he lost a patient, and he was shattered. He always felt that it didn't really matter that you didn't succeed; what mattered was that you didn't harm the patient. It's a principle that he was very, very keen to instill in all of us who worked with him.

"To go on ward rounds with Andreas in the afternoons, to see the patients for the following day, the patients he had just done, was a delight. He had incredible rapport with his patients and their families. The ward rounds were filled with humor. He had great empathy with the patients; they would ask him about all sorts of wonderful problems, for example, with their bladders. And he would sit down and talk to them about their bladders. For a man who was incredibly busy, and who was a wonderful technician, he was also a wonderful physician. There was never any question in our minds that his primary concern was for his patients and not for his own reputation."

But there is no question that Gruentzig's life, along with his personality, grew ever more complex. A few colleagues felt that Gruentzig's remuneration might be growing excessive. Charles Hatcher, the chief administrator of the Emory Clinic, was confronted by a collective venting of their discontent. "Several of the cardiologists came down and questioned what Andreas's income might be . . . I explained to them, 'What Andreas is making is only that amount of money that he produced that I feel is appropriate for him to retain. So you are not contributing to it, just get over that, not one doctor at Emory has ever given Andreas Gruentzig one penny. . . . You are all benefiting financially from the fact that he is here.' " Hatcher continued, "It's like the Atlanta Braves [the local major league baseball team], they all wear the same uniform and they have a captain of the team and they have a manager, but the highest paid guy out there is Henry Aaron, [the all-time home run leader] and why is he paid more? Because the crowd comes to see him play."

Gruentzig was well aware of this growing resentment. Junior staff let him know that the cardiovascular surgeons sometimes took their sweet time to perform emergency rescue operations when he wasn't around, signaling their displeasure at having their weekends interrupted to safeguard a technique that was eating into their own pockets. Gruentzig also seethed at the subtle disapproval directed at his young wife. Eventually, he set to exploring opportunities elsewhere, with his first goal being to become a full chief of cardiology in his native Germany. But the word circulated there that he had become too big for his britches. Heartbreak and triumph so often being interlinked, no German institution wanted him.

New friends circled around Andreas Gruentzig and, innocent of Emory's internecine politics, still found an inspiring visionary. One of these was a thirty-year-old radiologist named Rick McClees. Joining Emory in the summer of 1983, he discovered that he shared with Andreas a passion for both flying and exploring breathtaking new possibilities in imaging the heart.

The North Carolinian and his fellow radiologist wife Kristina Gedgaudas quickly became favorites. "We all seemed to hit it off," Kristina recalled. "Andreas, for all the international importance that he had, and he always carried himself in a very regal manner, was a very friendly, down-to-earth guy and took to Rick immediately."

The couple stood out for their kindliness toward Margaret Anne, who had only recently resumed her role as a resident-in-training in her own radiology department. To their thinking, she seemed to fear that she was being constantly judged and undermined by her colleagues. The Canadian-born Kristina Gedgaudas is a dark-haired and fine-featured woman who was similarly struggling to adapt to the then overwhelmingly male inner precincts of medicine. Margaret Anne's plight tugged at her. "She felt very isolated in her group of residents. They felt that she was privileged . . . There was a little bit of jealousy there . . . She was hurt. Her own profession and obligations took second stage . . . She was very much alone and I felt very badly for her. So I think that brought us even closer together."

Before long, the two couples started enjoying dinners at each others' homes and various restaurants. Rick McClees recalled, "Part of the reason that they liked being with us was because we weren't from the crowd of people who just haunted him." The newly-arriving radiologist found Gruentzig to be completely genuine, despite his wealth and fame. "I remember him recounting the first day he sat outside his curb [at his first Atlanta house] and he remarked that he looked at his trees and said to himself, '*These are my trees.*' He was just . . . very simple. It wasn't 'Look at that fancy chandelier,' it was the trees . . . One thing you have to remember when you talk to all these cardiologists, there was a huge amount of jealousy against him, because he was the golden boy."

McClees also warmed to Gruentzig's wife. "Margaret Anne was very outgoing and pleasant and always had a smile on her face. But I got a sense that she never totally relaxed. She was a lot younger, not only than Andreas but a lot of his friends, and there she always was with these Herr Professors. And she had a lot of growing up to do. It was a little bit like the Princess Diana situation. Only a lot of it was fairly new to Margaret Anne—I think she was maybe a little bit self-conscious. Where was there for her to be comfortable? If you knew her, you realized that she was fragile."

Kristina occasionally had to console Margaret Anne for her excruciating migraine headaches. "As a woman, I often wondered if there was an insecurity based on how long was this thing going to last? . . . that she always had to be interesting and beautiful and charming in order to please him . . . I am sure there were situations where women would have been very happy to have an affair with him. I think that was one of the reasons Margaret Anne chose to travel with him, because she was not stupid."

McClees lost little time in joining Gruentzig's flying club at Peachtree-Dekalb Airport, about ten miles north of Emory. There, Gruentzig's newly purchased Beechcraft Bonanza gleamed beside rows of corporate and personal planes that testified to Atlanta's booming new wealth. The Beech Bonanza model (along with its successor, branded the Baron) had been dubbed "the doctor killer," since it packed so much power for the wealthy few who could afford it—quite often doctors and lawyers—that weekend pilots often became seduced into thinking they were barons of the sky, and paid for that delusion with their lives.

By the end of 1983, Andreas Gruentzig was avidly pursuing certification as an instrument-rated pilot, in an arrangement that allowed him to forgo the boredom of recertifying the basic flying credentials he had already achieved in Switzerland. Gruentzig's primary instructor, Dan Emin, found his student to be capable and generally cautious. But Rick McClees believes that Emin and his co-instructor Jim Olin too often looked the other way when their trainee with the fancy Porsche showed lapses in judgment. "Gruentzig liked to fly in 'actual weather' conditions, he liked to get training and he liked to be proficient. So rain clouds would come, thunderstorms would be there, and he would go fly. A couple of times they might have told him, 'Look, flying in the rain is one thing, but when thunderstorms come around you should put your plane on the ground.' I just thought there were opportunities there for him and myself to be better educated. I don't want to say the instructors missed the call—but there were things in the flying system, that as a recreational pilot, as somebody who did not do this for a living, you just did not get the correct feedback to tell you when you were not using good judgment."

McClees believes that a root problem was a kind of cult of celebrity that infects many flying clubs, whereby the biggest fish, especially those who own planes that can be shared with other trainees, are considered such lures to gaining new members that instructors avoid criticizing them at all costs.

Despite such concerns, McClees thought Gruentzig generally exercised exemplary judgment, especially when the two doctors flew off to Tennessee with their wives and the visiting Sonja, now seven, to catch an end-of-year Michael Jackson concert. Fretting over a minor hiccup in his Bonanza's exhaust system, he fixed the problem himself with arch caution before flying back to Atlanta. Gruentzig's flying nonetheless worried a number of his colleagues, especially when it became apparent that he had neither taken out personal life insurance nor filed a will. He finally resolved the former issue, but strangely never bothered with the latter.

Gruentzig pursued life with so much passion that he saw nothing but fresh opportunity ahead. The surgeon Joe Craver put it this way: "He just didn't have time. That's exactly the gist of his life. He absolutely didn't think about dying tomorrow. That was the last thing on his mind . . . If you said, 'How would this guy die?' I thought he might be killed by a [jealous] husband or die in some dumb ass thing like a car race."

Gruentzig began shopping for a grander house. More wealth awaited him, since Schneider Medintag had become a hot target for acquisition, with Bard and a medical subsidiary of the Italian Agnelli conglomerate knocking on the door. Their bids were eclipsed by a $40 million offer from Pfizer, which spelled a $2 million windfall in the making for Gruentzig.

When it came to houses, he was therefore thinking *big*. He dithered for a while over a vast mansion in the most exclusive Peachtree Heights section of Atlanta's pricey Buckhead district, an area where every lane comprises a multimillionaire's row. This one on Cherokee Street—which Andreas pronounced "*CHAR-aw-kee*"—needed drastic renovation, however. So the Gruentzigs ultimately settled on a slightly smaller brick residence of roughly 7,000 square feet. As a symbol of status, it was still about as large as they come, lording as it did over a sculpted rise.

Purchased in late January of 1984, 8 West Andrews Drive came wrapped in sprawling lawns, ornamental trees, and a woodland with a meandering brook filtering past clay tennis courts. It also had a quirky pedigree, beginning with its construction in the 1920s by a former governor of Georgia, William Austell. The next owner was a bachelor named A. L. Weeks, who made his fortune from producing a syrupy confection called "Little Rebby Snappy Cakes." A. L. liked to begin his day by jogging around his four acres of paradise and then heading back up the long curving drive toward a waiting breakfast. One morning in 1983, Weeks finished his jog and, just after settling himself down at the table, keeled over and died from a heart attack.

Gruentzig evidently did not worry about any lurking omens. The handsome brick mansion with its white-columned front entrance manages to be regal without garishness. In fact, the house's true grandeur is hidden from those passing by. Its main approach lies beside a balustraded rear terrace. A Tuscan-style coach house sits back there beside an Oriental garden with fountains playing over a central fish pond. The rear entrance is sheltered by a shady back portico with fluted columns and a sculpture of the Greek goddess Daphne.

In *A Man in Full,* Tom Wolfe described a much similar and closely neighboring house: "Rolling lawns, absolutely perfectly cut, watered, landscaped, and ornamented by flowers and deep-green bushes, every leaf of which seemed waxed and polished by hand—rolling lawns swelled up either side of Valley Road, leading to stupendous piles of Georgian brick with real slate roofs or romantic but equally stupendous villas of Italianate stucco atop the crests. And even though it was nine o'clock in the morning, on a hot day in May . . . here in the real Buckhead all was serene and green and cool . . . The house numbers seemed to be mainly on mailboxes at the foot of the driveways. In a neighborhood like this, if you actually put the number on the house, no one would ever see it; it would be too far away from the road."

At 8 West Andrews, a pair of Palladian doors opened to an indeed very Italianate vestibule crowned by an ornately corniced ceiling painted in trompe l'oeil. The marbled floor here would soon be covered by a fine Persian carpet, one of many acquired from Andreas's visits to his Hamburg-based Uncle Konrad, a top German rug importer. A later appraisal put its value at $65,000.

Short steps led to beveled doors that opened to a banquet hall–like dining room, and the enormous living room boasted a stately marble staircase to the living quarters above. A drawing room gave way to a music room, where Gruentzig installed a grand piano. Off it lay a circular cocktail room with a mahogany-paneled bar. Accompanying her husband on a speaking engagement to Venice later in 1984, Margaret Anne visited the island of Murano, famed for its exquisite glasswork, to dress up her new refuge. There she purchased a set of deep-blue champagne glasses and accompanying ice bucket, all etched with gold and worth thousands of dollars.

The house's second floor was no less imposing. The master-bedroom suite boasted his-and-her dressing rooms that were larger than any office Gruentzig occupied in Zürich. Margaret Anne's had a marble fireplace, and her sumptuous bathroom was outfitted with a gold-plated sink. To complete her new domain, the young bride made frequent forays to antique dealers in New York, Paris, and New Orleans. She purchased an ornate Empire-style chair for her dressing room, low to the floor so as to help eighteenth-century ladies wriggle into their hooped skirts. Meanwhile,

Gruentzig took care of a few fantasies of his own, beginning with the purchase of a sleek new Jaguar to complement his Porsche. He also hired himself a full-time housekeeper, and cook. Lest this entourage lead to too much pampering, there was always the full exercise studio in the basement, with jet-streamed bath and sauna, to work off excess calories.

One vexing issue involved the need for a chauffeur. The one who had worked for the house's previous owner had a strike against him from the start, according to Mary Jane Thornton, in that Gruentzig had by now become viscerally wary of American blacks. "He didn't want a black in the house," his mother-in-law said. "He wanted burglar bars put up around the house, and . . . he was afraid of this man. I knew he wasn't going to get the job when he [the chauffeur] came up and put his hand on Andreas's shoulder and called him something familiar and said, 'We're going to do well together.' I thought, 'You just lost your job by putting your hand on Andreas's shoulder.' " So, despite his wife's protests, the ever loyal Sarge took care of that role.

For a time, Andreas and Margaret Anne were in seventh heaven, inviting friends and colleagues by to show off their new surroundings over champagne and Cuban cigars. Dressing up the residence became a passion, and Gruentzig pursued it with his typical zeal. He developed a consuming interest in acquiring art, especially after visiting Denton Cooley and discovering that his "Cool Acres" ranch boasted a masterpiece by Marc Chagall.

In February of 1984, Andreas and Margaret Anne flew to Manhattan with their personal interior decorator to visit the New York Antiques Fair. While strolling along a snowy Madison Avenue, they paused to admire a painting in the window of the Hilde Gerst Gallery, having no idea that this was run by a woman who had been feted in the White House and was a personal favorite of Prince Charles and Princess Diana. Hilde Gerst, an Austrian Jew who fled the Nazi takeover of that country in 1938, sold them the painting and utterly captivated Andreas Gruentzig. Before long, he and Margaret Anne would have her making regular visits to 8 West Andrews to provide in situ guidance on their art needs—the couple usually bought Impressionist paintings for tens of thousands of dollars.

For the newlyweds, these were heady days. Once a week, they whirled through lessons in ballroom dancing and especially their beloved tango. The couple had by now acquired a pair of Irish setters named Gin and Tonic, the name a nod to Margaret Anne's favorite drink.

When Emory's next angioplasty course came around in April of 1984, Gruentzig threw a formal dress evening at his new residence. Working with her caterers, florists, gardeners, and house staff, Margaret Anne set out to make this celebration outshine every other one her husband had ever hosted. Two hundred disciples were greeted by a violinist at the entrance, while butlers unburdened the gentlemen and ladies of their coats and wraps. Few would have minded the awaiting elegance, since a good number of these visitors had recently become inestimably richer themselves—thanks to the pioneering work of their host.

The tables were dressed in linen, fine china, and silver, and capped with showy floral arrangements. The guests were treated to their various courses in clusters arranged throughout the entire house, from the dining room to the music room and library, to the living room and private bar, and even Margaret Anne's personal, marble-hearthed dressing room. An orchestra enlivened the ambience. Aperitifs and wine appeared, even as some guests checked out a sporting event on the household's multiple TVs. Of course, these were days of industry-sponsored expense accounts such as earlier explorers of the heart, like sad Mason Sones, had never tasted.

On this spring night in 1984, no party in Atlanta could have been more electric with triumph. Cigars and cordials were proffered, and the orchestra stepped aside for the first swinging notes of a jazz combo. It was time for toasts. But where was Andreas Gruentzig?

Suddenly, the diners in the central rooms looked to the top landing of the sweeping marble stair. Here was their host, all right, decked out as a sheik with his fez and white caftan, tripping down the stairs to wild applause. The Prince of the Heart had come to claim his tribute, and to dance into the night.

CHAPTER 16

THE SEDUCTIONS OF FAME AND WEALTH brought fresh layers of intrigue into Gruentzig's life, pitting rivals against allies and even friends against friends. The frontal challenges were not all that daunting. Gruentzig faced down a growing clamor to use his balloon angioplasty technique as a magical replacement for coronary bypass surgery by more or less saying, "Fine, but we'll do it my way." His way was to call for a major clinical trial to test the long-term results of the procedure in a direct randomized comparison against those of the more established surgical approach. Nobody could argue with that, nor with the man's abiding love for rigorous scientific methodology. The problem lay in the larger frenzy of ambition that permeated his once-careful choreography.

By the beginning of 1984, Gruentzig was well aware that a whirlwind of ready cash was driving angioplasty and numerous spin-off procedures forward pell-mell, many of them hurtling forward with little of the restraint he so steadfastly advocated. He worried more than ever about whom he could trust. He saw—and sometimes perhaps imagined—signs of betrayal near and far. One of the keener observers of the situation was the Australian cardiology trainee, Gary Roubin, who joined Emory in March of that year. What he found upon arrival was a boss who seemed to be struggling with loneliness and suspicion while at the very peak of his celebrity. Freshly sep-

At the 1980 teaching course, Gruentzig hams it up with Charles Dotter to his right, Richard Myler, and Simon Stertzer. *(Courtesy Maria Schlumpf)*

David Williams stands beside Mason Sones at the 1980 Zürich teaching course. *(Courtesy Gary Roubin)*

At the wild rustic party at the final 1980 Zürich course, Gruentzig embraces Mason Sones before the latter falls into a ditch. *(Courtesy Gary Roubin)*

Andreas embraces his mother and Maria Schlumpf by the Zürichsee, circa 1980. *(Courtesy Maria Schlumpf)*

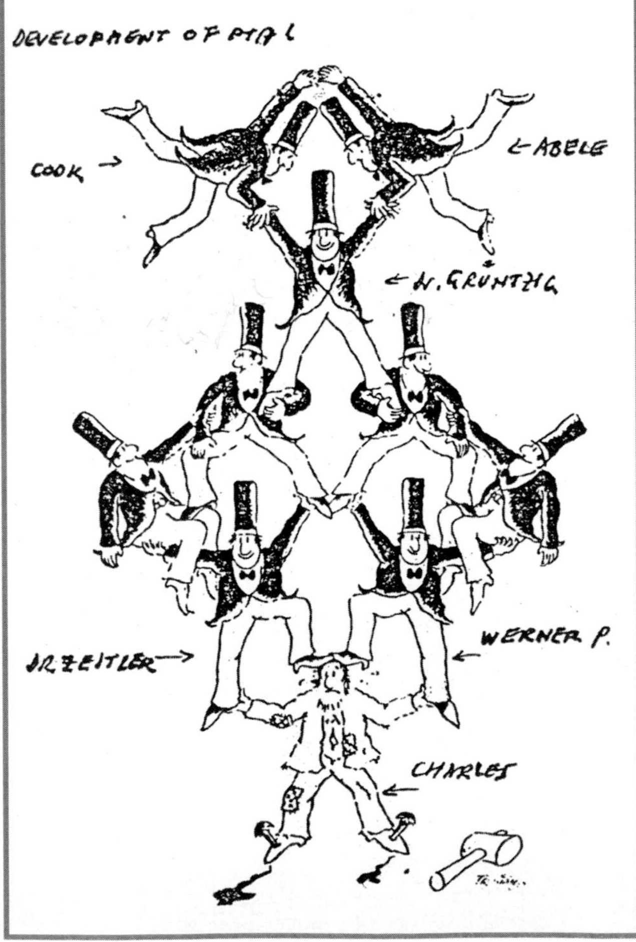

A Dotter doodle of himself in tramp mode impaled to terra firma as the balloon angioplasty procedure takes off, with Zeitler and Porstmann held up in his hands and Andreas Gruentzig topping them, and with the industry executives Abele and Cook cavorting over all. *(Courtesy Dotter Vascular Institute)*

Masters of the heart—Simon Stertzer *(left)*, David Clark, and Richard Myler, pose during the early 1980s. *(Courtesy the Author)*

Gruentzig with his surgical cap fashioned into a beret. *(Courtesy Gary Roubin)*

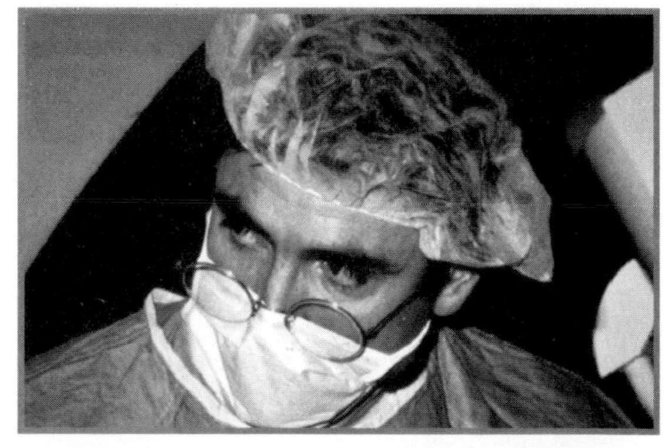

An angioplasty balloon is shown advanced through a stent. When the balloon is inflated, the stent will be embedded with the arterial wall, propping it open like permanent scaffolding. *(Courtesy Wilhelm Rutishauser)*

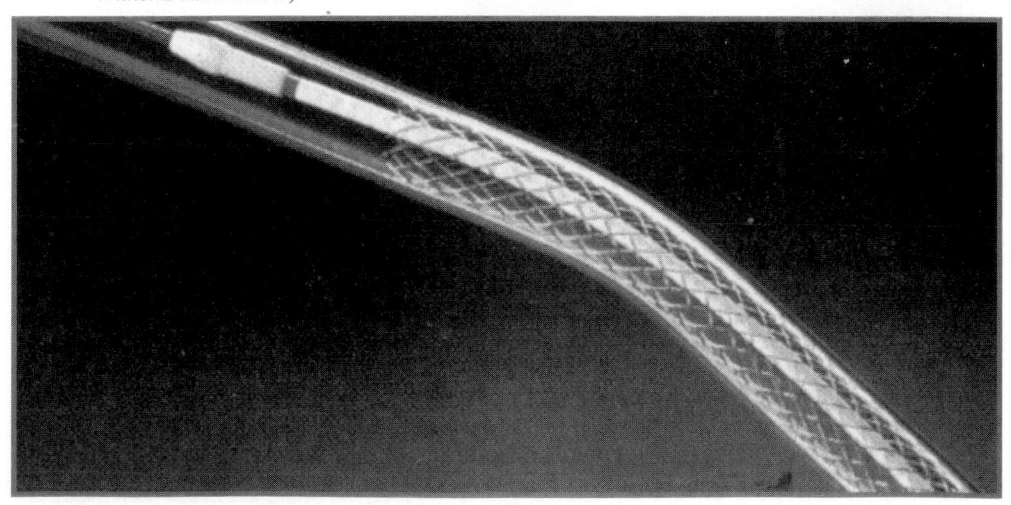

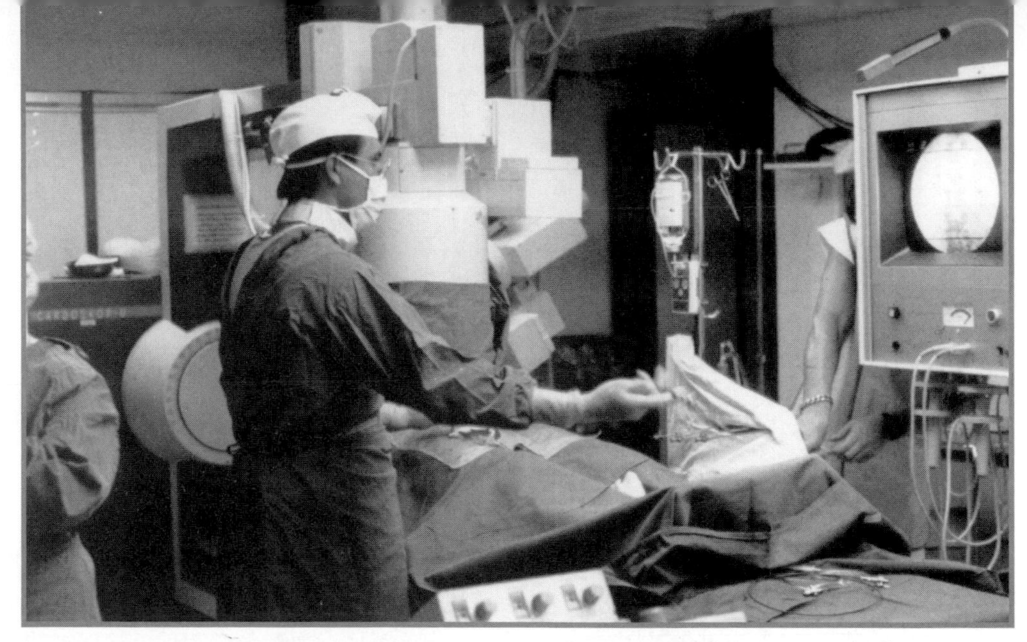

Gruentzig performs an angioplasty at Emory. The X-ray monitor at right is used to reveal the exact positioning of the balloon catheter within the heart. *(Courtesy Gary Roubin)*

The increasingly accident-prone Gruentzig nurses a cast-enwrapped broken ankle and a cocktail in an amphibious apparatus in his backyard Atlanta pool. *(Courtesy the Thorntons)*

With his-and-her bikinis, Andreas and Margaret Anne pose at poolside. *(Courtesy the Thorntons)*

Andreas Gruentzig in a soft moment with Margaret Anne around the time of their 1983 wedding. *(Courtesy the Thorntons)*

The bride and groom waltz at their 1983 wedding reception at Macon, Georgia's Idle Hour Country Club. *(Courtesy the Thorntons)*

Together again in Atlanta, Andreas hugs his visiting mother Charlotta and daughter Sonja. *(Courtesy the Thorntons)*

Andreas, wearing a tunic from his far-flung travels, plays the piano on a Georgia night. *(Courtesy the Thorntons)*

Margaret Anne, Sonja, and Andreas prepare to embark from Atlanta's Peachtree-Dekalb Airport on his single-prop Beech Bonanza, the first plane he purchased. *(Courtesy the Thorntons)*

The West Andrews Gruentzig home, seen on a December day long after its post-tornado restoration. *(Courtesy the Author)*

Andreas (in sheik mode) and Margaret Anne relax with the visiting Sonja at the wind-down of their evidently busy New Year's Eve 1984 party in their new West Andrews, Buckhead Heights, Georgia home. *(Courtesy the Thorntons)*

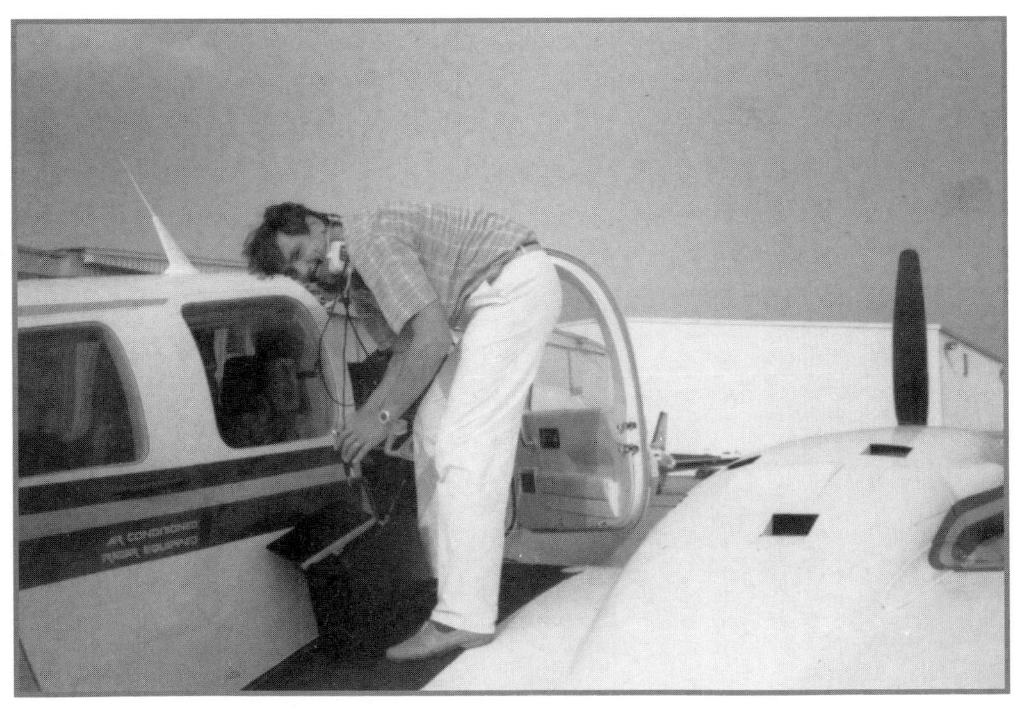

Andreas Gruentzig checking out his Beechcraft Baron and smiling at the visiting Maria Schlumpf on the runway at St. Simons Island one month before his fatal crash. *(Courtesy Maria Schlumpf)*

arated from his wife, the Australian had a rough-and-ready background and a maverick's nature himself, which immediately appealed to Gruentzig. As a teenager, Roubin had developed a love for working with horses and cattle and eventually won a scholarship to study veterinary medicine, thereby becoming the first member of his family to achieve higher education. After switching to cardiology, he was mesmerized by a talk Gruentzig gave in Melbourne in 1982 and eventually pursued the connection to land his junior position in Atlanta. With a disdain for convention and an outspoken personality, he was clearly a man apart from his other buttoned-down colleagues-in-training.

"I was broke. I arrived with a guitar, a pair of skis, a good education, and about a hundred bucks in my pocket. I had nowhere to stay. It was one of the loneliest and unhappiest times in my life," Roubin recalled. Struggling to get his feet on the ground, he weathered the typical Gruentzig humiliations as he fumbled to provide assistance on his first few angioplasty cases. However, during a gathering on the night of his thirty-sixth birthday, he received a gift of expensive Spanish brandy from his boss. He sipped it like a man receiving a benediction.

Before long, the new trainee was invited to move into his mentor's coach house while he got his affairs sorted out. There the two established such a natural rapport that some colleagues say they resembled brothers. They talked late into many nights about not only medical issues but also their own personal struggles and heartbreaks. "We used to joke about all the cultural things that we found so odd in the South . . . particularly in Atlanta in the early 1980s. Atlanta was incredibly 'good old boy.' At clubs like the Piedmont Driving Club, Jews and blacks were still not allowed to be members, and that was abhorrent to Gruentzig," said Roubin, who is Jewish himself.

Other associates say this assertion does not jibe with the racialist comments Gruentzig sometimes uttered. But the Australian clearly heard many of his boss's innermost laments. "There was the whole 'good old boy' behavior where they would be sweet as pie and invite you up to their country houses, but the sort of real friendship that you expect from people who behave like that in his culture and in my culture was not expressed. It was all superficial . . . I think he [had] learned that by then, that it doesn't matter what these guys say to you as they shake your hand and slap you on the back. It's a mannerism and has little meaning."

A medical pioneer in his own right, Roubin is now chairman of interventional cardiology at the Lenox Hill Heart & Vascular Institute in New York, where he owns a lavish apartment overlooking Central Park. Famous for extending angioplasty as a stroke-preventing technique to clear out clogged arteries to the brain, he too has ridden the Gruentzig technique to unexpected fortunes. Among his rewards is ownership of a sprawling equestrian center in the western resort of Jackson Hole, Wyoming, and a ranch down the road. So Roubin naturally retains a profound admiration for the German who changed his life. One evening at his refuge in the Rocky Mountains, he uncorked for his visitor the rare Spanish brandy he had treasured ever since Gruentzig presented it to him in 1984. "He was like a highwayman," he said, using a phrase referring to yesteryear's swashbuckling roadside robbers of the rich and stealers of ladies' hearts, as celebrated in stories and songs. "The young women on staff were all in love with him . . . His standards were incredibly high. He didn't like to be surrounded by inadequate people who would bullshit their way through life, and he would cut them down in front of anyone. He did this publicly and he could be tough and rough."

Roubin, who like Gruentzig can be challenging and exquisitely charming by turns, heard chapter and verse about the tensions playing through Emory, and by extension throughout nearly every redoubt of the angioplasty technique in the world. The Atlanta heart surgeons, he claims, were so full of resentment at the triumph of so-called "minimally invasive medicine" that they often procrastinated before performing rescue operations when trouble arose. "They gave him a hell of a time, and I know they did through many incidents where, every time a patient would go to emergency surgery, they would make a big deal about it. Not only that, they dragged their feet, particularly if it wasn't Andreas [performing the procedure]." Other senior staff members dispute this allegation, and assert that Roubin's viewpoint was tainted by a bitter falling out with them a couple of years later. But the Australian was privy to his mentor's deepest misgivings and perceived slights.

Something of a ladies' man back then, Roubin also wore revealing, skin-tight "Speedo" bathing suits at every pool party, to the merriment of his boss. "This was sort of a no-no in Georgia in those days. He didn't care and I didn't care—there were many parallels . . . He liked brandy and I liked brandy and we both liked good cigars. I was thirty-six years old, and he was only nine years my senior, so we hit it off—we clicked." Roubin in fact would roar with laughter at Gruentzig's forty-fifth birthday party in June, when the Emory nurses presented a fresh pair of Speedos, which the world-famous cardiologist held before his crotch as he pranced about the room.

So who was the real Andreas Gruentzig now? Was it the heroic if injured everyman Gary Roubin perceived, or some more self-centered being who cast aspersions upon the very people who had helped him the most? There is no easy answer, for Gruentzig's behavior was often enigmatic—robustly good-natured one moment, and imperious and even sullen the next. One peculiar photograph from that era showed the inventor sitting at the stern of a gondola in Venice during a medical meeting there, with his arms crossed magisterially before him and his jaw held high over a full-length white leather jacket. There was no warmth in his mien. His dark hair had recently been dyed jet black and permed into tender curls by a certain "Mr. Chris"—an indulgence that was often whispered about by his peers. The neatly clipped moustache had shed every trace of the dishevelment that once so disturbed his mother-in-law. Old friends wondered whether fame had warped him. His cheeks had grown fuller, his brows more furrowed, and there was a hint of dissipation about his lips. Indeed, Emory gossips had it that Andreas and Margaret Anne were now drinking to excess and that there were outright bacchanals at his new estate. Some even said he was becoming bitterly disenchanted with his marriage, that he was so manic on his upswings that he must be deeply depressive and lonely behind closed doors.

During a Miami medical meeting around that time, Spencer and Gail King called to Gruentzig's suite at the five-star Fontainebleau Hotel and had their heads spun by the newest incarnation of their former houseguest. "When he opened the door he had on a gorgeous silk smoking jacket. I thought, 'You know, he thinks he's Clark Gable.' The guy was just out in space and loving every minute of it," Gail King recalled. But then her own connections to the man had sadly changed, owing to increasing acrimony with Gruentzig's new wife.

Gary Roubin nurtured a different vision. "The people who were close to him were the staff," he asserted. His take is that Gruentzig had by now substantially withdrawn from his immediate colleagues and that his trust in most of them had worn thin. But why? Many associates glimpsed moments of unprovoked brittleness on Gruentzig's behalf, and surmised that he perceived betrayal in so many distant directions that he transferred his suspicion toward those closest at hand. Whatever his reasons, Gruentzig socialized less and less with his Emory peers, and instead bid the technicians, nurses, and cardiologists-in-training to gather at his Buckhead mansion for Friday-night release. "We would sing standing around his piano, Roubin recalled. "I used to take my guitar. I played Peter, Paul and Mary; Bob Dylan; seventies protest songs; and a lot of rugby ditties. He loved this stuff, drinking brandy with the whole group, singing . . . This guy treated us like family. He was the grand master and there he was amongst us. He was very humble and very capable of being part of his surroundings, very unpretentious."

Years later, Roubin rejected the split-personality theory of a man becoming unraveled by his own success, but he acknowledged that Gruentzig often seemed distraught back in 1984 and '85. "I don't think he lost any of his soul, at least to those of us who worked closely around him. To say that he lost touch with reality, with his past, with the principles he stood for, I think is totally incorrect. The fact that he enjoyed the Porsche and the private airplane and the Sea Island mansion and the Buckhead mansion and the clubs—was he changed? I really don't think so."

Many disagree. In any case, even his Australian protégé worried over Gruentzig's hunger for abandon, especially as the speeding tickets piled up from his breakneck journeys to and from work. At times, it appeared that this man, who devoted his every working hour toward extending the life of others, was trifling with his own. Even as he dutifully pursued his weekly formal dance classes with Margaret Anne, Gruentzig preferred boogying to "Rocky Top Tennessee." Sometimes, he behaved as though there were some hidden being inside him struggling to break free. His late summer of 1984 teaching course finished off with a rip-roaring grand finale at a resort in the north Georgia mountains called Big Canoe. As the night wore on, the drinking and dancing grew frenetic and culminated in a collective nude frolic in an adjacent waterfall. A complaint brought the police with screaming sirens to disperse the revelers.

Up in those mountains, Gruentzig also toyed with danger, according to Roubin. "He was a crazy guy. I mean he liked to take risks. I heard a story of him driving with one staff member in his Porsche after they had a lot to drink, and he hit eighty miles an hour on this dirt road in rural Georgia and turned the lights off. The guy said, 'Andreas, turn the damn lights back on. You're going to kill us both' . . . But I don't believe he would ever have done anything like this with his plane."

Perhaps Gruentzig's most quixotic aspect was that he remained a master of appearances, presenting himself in various guises as the situation required, and leaving those who thought they knew him best picking through the carefully refracted shards of his personality. It is hard to find any other explanation as to how Earl Thornton and his wife Mary Jane could insist that Gruentzig remained down-to-earth and was never remotely troubled—except, that is, for the day when he was shattered by the loss of his first patient around the beginning of 1985. The picture they paint is of a dutiful son-in-law who went out of his way to pay them Sunday visits in Macon. He typically arrived in a tracksuit after flying from Atlanta in his beloved Bonanza with Margaret Anne at his side. Upon arriving in their Macon house, he would get down on all fours to play on the kitchen floor with their poodle, Joy. On Mother's Day that year, the gallant doctor presented Mary Jane with a dozen red roses, just as he did for her daughter Margaret Anne at the end of every work week. In Macon, a favorite pastime was to hit the Ace Hardware store, where a cornucopia of gadgets unavailable in Switzerland inspired tinkering with household improvements. On these Sundays with the in-laws, there was no sheik of cardiology to be glimpsed.

As far as the in-laws could see, the Conqueror of the Heart was happy with his station. He could buy whatever he desired, travel anywhere he fancied, rid a half dozen patients of crippling pain in one session and soar off in his own plane just as free as a bird. Hilde Gerst alit on regular occasions with paintings by not-so-minor French impressionists bearing price tags in the $5,000 to $70,000 range. The trips to New York and New Orleans and Paris continued, yielding $10,000 walnut Louis XV commodes, $2,400 sterling salt-and-pepper shakers, a Regency cabinet for $12,500, a mantle clock with cherubs for $2,500, ottomans and silver soup spoons—before long the home-furnishings budget alone would hit a quarter of a million dollars.

Mostly, Andreas Gruenztig would leave the shopping to his wife. He would regularly throttle off to Macon to collect his in-laws before continuing on to the St. Simons Island airstrip, just a few minutes' drive across the causeway to Sea Island. From the cockpit, Mary Jane Thornton marveled at Andreas's proficiency at the controls. But a couple of incidents frightened her. One occurred when they were shuttling down to check on the rebuilding of the tidal wall in front of the Sea Island cottage. "We got halfway between Atlanta and St. Simons and all of a sudden it was pea soup, and I looked up and could not see anything outside the window of that plane. I thought, 'Well, we are going to crash into the ocean. I'm not a very good swimmer and they'll drown trying to save me.' We landed fine and we got over to the house and I said, 'Andreas, sweetheart, let me tell you something—anytime you know we're going to fly into pea soup like that again, you tell me about it ahead of time and I am not going.' "

Another scare occurred as they approached Charlie Brown Airport, a congested regional facility about half an hour south of Atlanta. "As we were descending, a little plane flew right under and in front of him and he just pulled his plane back up. He was calm and handled it beautifully, and I never saw him as a nervous pilot . . . But he said he would never fly into Charlie Brown Airport again."

Despite all of Andreas's seeming competence as an aviator, Mary Jane Thornton was not immune to a mother's anxieties, especially just after Margaret Anne had weathered a car crash. "The next morning, Andreas and Earl decided they'd fly down to Sea Island to see how the shrubbery was doing. They just wanted to go down and check on the house . . . But Margaret Anne didn't want him to go down there after having been in her accident. So when they got in the car to go to the airport Margaret Anne started crying in the kitchen. She said, 'That plane is just something to get killed in.' "

With the gift of hindsight, a number of colleagues say they, too, became increasingly worried about Andreas's flying, that the man seemed ever more Icarus-like in tempting fate. But the surgeon Joe Craver was not one of these. The former all-American football player flat-out admired Gruentzig's swashbuckling zest for action. Craver and his wife and Andreas and Margaret Anne enjoyed many afternoons together at the time, with the men riding horses and shooting skeet at Craver's 800-acre mountain getaway.

With Craver, Gruentzig began to elaborate on the dream he had once formulated during the night of torches in the Swiss forest, just before Mason Sones fell into a ditch to the ethereal tunes of the pipes of Pan. That was the evening when Gruentzig first spoke of creating a distant retreat where the greats of cardiovascular medicine could gather to relax and trade their shared visions. "Andreas had these grand plans. He was entranced with the beach and he was entranced with his own privacy, and he was entranced with his own ability to own a Caribbean or Bahamian island. He said, 'I am interested in buying and owning an island, and having it be private and first class.' He was into the first class after he hit his money . . . He was saying, 'We'll be able to do this regularly, we can get there in three hours on Thursday night after work and come back Sunday afternoon . . . He wanted to invite commercial people there to discuss business. . . . His idea was he was going to have an Andreas Gruentzig corporate jet—this was before leased jets. He was going to have one himself . . . That was right where he was headed. He needed an island big enough to land, he needed a mile strip, or a convenient place nearby, and that's where I think I would have gotten involved with him, and it could have killed me."

Craver has no truck with the theory that Gruentzig somehow lost his drive after hitting the pinnacle of triumph. "I never saw him when I thought he was depressed, and you could easily say he was manic every day of his life. I never saw him sad, but I saw him disappointed when there was a bad angioplasty result, or he was too tired, or frustrated with people [for] not providing the resources he needed."

A dreamer perhaps, but Gruentzig was anything but naive. It was by now becoming manifest that his procedure, born on a kitchen table, presaged the very future of medical therapy, so long as he could restrain its ardent young advocates from making overly chest-thumping claims. The cardiologists for their part were convinced that their professional adversaries, the cardiac surgeons who had driven them up the wall since the DeBakey, Cooley, and Barnard era, were doing the same thing themselves. Thomas Preston, M.D., a Washington state cardiologist, let it all out in an 1985 editorial in the possibly point of view-heavy *Journal of Holistic Medicine,* arguing:

> Nearly half of the bypass operations performed in the United States are unnecessary. A decade of scientific study has shown that except in well-defined situations, bypass surgery does not even save lives, or even prevent heart attacks; among patients who suffer from coronary-artery disease, those who are treated without surgery enjoy the same survival rates as those who undergo open-heart surgery . . . The concept of bypassing an obstructed artery to restore blood flow is simple enough for any eighth-grader to comprehend . . . So compelling [was] the hope of restoring diseased hearts, that physicians immediately embraced the idea of coronary-bypass surgery when the procedure was introduced in 1967 . . . A few physicians cautioned against an untested and unproven therapy, but most simply took it on faith that the operation would relieve angina, prevent heart attacks, and prolong life . . . The operation is the biggest revenue-producer in the health field. As a result the procedure has taken on a life of its own. It is not so much the public's health as the profession's wealth that now dictates its use.

My, my, Gruentzig might well have said before these booming (and perhaps excessive) assertions—there is a war afoot. With so much in play, his personal animus remained huge—and contradictory. He whirled into the Emory hospital wards drawing blushes and admiring oohs and aahs from junior staff, as Claire Rice, his ambitious young research nurse recalls. "You had to 'hup two' to keep up with him," she said, referring to the boot camp refrains of the American military. "I remember that lab coat flying in the breeze because he was long-legged and he would walk fast. And I walk fast and I had to really stretch it to keep up with him . . . He was flamboyant. If ever there was a time in my life that was magical, this was it."

She says their relationship always remained free of the fawning that some of the other nurses displayed. In an utterance that epitomized the situation, her boss at one point barked, "You are not the prettiest, but you are the smartest!" With soft green eyes and short auburn hair, Claire Rice is in fact quite pretty, but she also has a discerning intellect and often let Gruentzig know it as she observed his various conflicting traits. Sometimes, she pondered whether all his success had brought real happiness.

"He was always striving," she observed over coffee years later. "It was always the next thing, so I don't know that he stopped to think whether he was 'happy.' He was happy when his daughter came, I know that. That lit up his life, there was no question. Sonja was *it* for him, and I think he was happy when he made someone feel better. He was such a generous person—incredibly generous, and I think giving things made him happy. He knew how to live, he knew how to celebrate life. But he asked me once, because I was getting married, 'Well, are you happy?'

"I said, 'Yes!' I mean I was extremely happy. I had found the love of my life. . . . I said, 'Are you?' He said something like, 'Marriage is a convenience.' I think that's the way he looked at it."

In wondering about the swings of her boss's personality, she observed, "You know, he was a drinker—Jack Daniel's was his favorite—so he was often a little bit tipsy. He didn't eat much, and he was like a big figure. I think that was part of it—he was always up."

She found cardiology's new breed to be driven. "I think the more you have, the more you want. You know, they never figure out what true happiness is. They think that more is better, so you must conquer more and get more recognition. They are never really satisfied. . . . Andreas didn't trust anybody."

Gruentzig's old friend Richard Myler roundly agreed with that assertion. "He came to me one day—I was staying in his house actually—and he was crying . . . I said, 'What's the trouble?' He said 'You're my only friend.' He wasn't talking about angioplasty now. And I said, 'What are you talking about? Everybody loves you.' He said, 'No, they don't!' We chatted a little bit and he calmed down. I thought to myself, 'This is the craziest thing I have ever seen—the guy's got everything going for him.' "

Another time, Myler sat beside Gruentzig as they tooled around Atlanta in his Porsche and the visitor attempted to revisit the subject. "I put my hand on his knee. I was sitting in the passenger seat. I said, 'Are you okay?' just like that. Now I didn't make any comment, just 'Are you okay?' And he said to me, 'Why did you ask me that question?' " That, according to Myler, was as far as the conversation went.

With a number of close colleagues, Gruentzig could be all solicitude one day and chilly the next—letting people close was not his style. Toward Spencer King, the doctor who first threw out a welcoming carpet to his new American life, Gruentzig evidenced a steadily cooling affection. He called him "the politician" and seethed when King delivered a lecture at a medical course led by Geoffrey Hartzler. The courtly Georgian shakes his head wistfully about the anguish of that disaffection, saying only, "I never understood his definition of a friend." By this point, everyone at Emory should have been reveling in their shared triumph, what with King and his partner John Douglas performing approximately 450 balloon angioplasty cases each year, roughly a hundred more than Gruentzig himself. The major reason for that discrepancy was that the two primary partners constantly had to fill in for the star's ceaseless travel. And Gruentzig was now routinely leaving work at five P.M., which was simply not done by most medical professionals performing high-stakes procedures, then or now. Colleagues called these and a variety of other indulgences "Gruentzig's Rules." Tension reigned, however elusive in its cause.

As interventional cardiology transformed wide reaches of medical practice, its vanguard multiplied and shed its last vestiges of collegiality. Doctors searched for their own personal tickets to stardom and wealth, sensing that they could suddenly trump the heart surgeons in spades. In the midst of seemingly obscure medical practices in places like Ohio, Wisconsin, and Maryland, the newest protégés of the Gruentzig technique were pumping out 700 to 800 cases a year, the most aggressive among them earning more than $1 million in procedural fees alone. A few dons of the field joined Gruentzig in warning that his innovation was prompting a medical gold rush fraught with avarice and blind ambition.

By 1984, about 60,000 balloon angioplasties were being performed around the world and the arc kept shooting higher—toward 200,000 annual cases by 1987. Today, two million such procedures are completed in the heart every year, with at least a half million more in other circulatory vessels. Sensing the vast potential of the field, a new generation of disciples began patenting minute refinements of the associated technology. The former pill-pushers of cardiology, once unsung understudies to the glories of the DeBakey and Barnard era, also began to throw their weight around in the hospital corridors. With sometimes Machiavellian arm-twisting, they leaned on hospital gatekeepers to forbid radiologists from even illuminating coronary arteries. The cardiologists argued that they were the hottest revenue generators and that rivals should step aside. Heart surgeons who tried to slow the floodtide of patients to catheter procedures were often warned that referrals for their operations, which inevitably started with a recommendation from a cardiologist down the hall, could dry up. The deft catheter jockeys suddenly acted as though they were the ones with the big biceps now, and they no longer needed heart surgeons to stand in waiting in case something went wrong.

An aghast Gruentzig worried that things were moving much too fast, as he told an interviewer in early 1984. "Why am I so cautious?" he began rhetorically, and then answered his own question. "I am cautious simply because I know that damn balloon catheter; I invented it. It has problems. You can cause problems with the guide wire; with the catheter. You may disrupt the artery. You may close the artery. You may dissect the artery. You may dissect other arteries that were not diseased before. You can really mess things up."

The idea of dispensing with surgical support to stave off emergencies and, further, performing angioplasty procedures deep into the night troubled him to no end. "To do a dilatation at night, when the team is tired, the patient is tired, a surgeon is not available if something happens—why should I not intervene further in an improved situation and [not] risk disaster? That has been my policy all my life. There are times when you have to step back and let Mother Nature go to work," he told Sheila Stavish of *Cardio* magazine.

Finally, Gruentzig opened both barrels of his dismay. "I have seen too many complications in the past not to be cautious," he said. "I have gotten excited and jumped in to do dilatation. Then we put the catheter in, and boom, the artery closed immediately when I just touched it with the guide wire. The patient got a re-infarction [a second heart attack]. 'Oh, God,' I said, 'If only I had stepped back.'

"Now people are thinking they can do everything. But I remember one of my colleagues who started very early in coronary dilatation. After doing thirty-eight cases without any complication, he decided there was no more need for surgical standby. His next patient died. He's had standby ever since . . ."

He was equally scathing about euphoric notions for "magic bullet" success and prodigious cash-generating by half-tested laser contraptions that start-up companies, spearheaded by the notoriously aggressive Trimedyne Laser Systems, were pushing from the American West. "It's ridiculous the amount of publicity that the laser has gotten in the treatment of atherosclerotic disease. Not one patient has really been treated in the way they all claim. Atherosclerotic disease has been treated in dogs, in postmortems, in postmortem pieces put into dogs. But at the moment, no one would treat his own dog with the laser," Gruntzig told Sheila Stavish.

Gruentzig was not finished in remonstrating about what had become of the therapeutic field he had created with the help of daring forebears such as Sones and Dotter. "There is too much jumping into the limelight. The more one investigator jumps forward the more others feel pressed to do the same. Sometimes the push is not from the doctor, but from hospital or public relations offices. They push people, insisting that you have to attract patients, have to show that you're in the forefront. And of course, the press helps to get money."

Gone, gone were the days when Gruentzig could preside over a coterie of hand-selected acolytes, every one of them requiring his personal certification to proceed into the heart. Gone, too, were the days when USCI's Dave

Prigmore could scoff disingenuously at fellow medical-device industry executives that the procedure was a ruinous financial proposition. The genie was loose. In 1984, the pharmaceutical giant Pfizer powered through a $40 million takeover of the once miniature Schneider Medintag company—one they would flip fourteen years later for $2.1 billion to John Abele's emerging Boston Scientific conglomerate. Schneider Medintag had just ramped up its value by connecting with another German cardiologist-inventor named Bonzell who demonstrated that a little "monorail" system could make it a snap for one person to perform the procedure without an understudy whipping connections to the system behind his back.

The broadsides had barely started. The shot that truly resonated around the medical world was a sudden offer by Eli Lilly to purchase the fledgling ACS company for a whopping $500 million, as long as projected earnings were matched (as they indeed were) over the next five years. For John Simpson, the first rival to Gruentzig's dominance, this amounted to a personal windfall of $45 million. It was a nice take for the former "Keeper of the Cork," considering that Gruentzig himself, the inventor upon whose lifework Simpson piggybacked his refinements, had received an estimated $13,500 in the original deal with USCI only six years before.

Gruentzig was incredulous. For one thing, he was galled that Simpson's personal showcases of his new technological wrinkles often ended up with mediocre and even sorry results, thanks to inexpert maneuvers, according to Geoffrey Hartzler, by the equipment's inventor himself. Most unnerving was the fact that it was Hartzler—the so-called Flash from Kansas City—whose virtuoso displays of the Simpson system inspired the Eli Lilly windfall. Hartzler trumped Gruentzig again when his own innovations began generating a stream of what would eventually be millions of dollars in royalties. So exponentially would the profits grow that in 2006 John Abele's Boston Scientific company (the eventual outgrowth of his garage-born Medi-Tech) would gobble up the lion's share of the Lilly spin-off of ACS, called Guidant, for the tidy sum of *$27 billion.*

No one in 1984 could have foreseen what lay ahead. A venture-capital-bankrolled enterprise called Scimed materialized outside Minneapolis, displaying a knack for tweaking technologies just a shaving off its rivals' patents. Boston Scientific bought them, as well. The Miami-based Cordis Corporation, eventually to be amalgamated in Johnson & Johnson's drive to dominate the field, jumped into the fray.

Meanwhile, pharmaceutical firms from the U.K. to California poured millions into investigating catheter infusions of clot-dissolving compounds that might stop heart attacks, which was yet another outgrowth of the Forssmann-to-Sones, Dotter-to-Gruentzig axis. The sizzle surrounded a San Francisco start-up company called Genentech. This pioneering enterprise in genetic engineering was busy cloning a substance called tissue-plasminogen activator (TPA), an enzyme that disintegrates the threads that bind heart-attack-inducing blood clots. Such was Wall Street's excitement over the promise of this substance that Genentech's 1980 public stock offering saw its initial share price skyrocket from $35 to $88 in less than one hour. A major trial began in 1984 of the potential of reducing death from heart attack with TPA—which did prove to eliminate as many as one out of five fatalities (while itself inducing significant numbers of brain-damaging strokes in the process). Before long, Genentech's share price rose to *360 times* the fledgling company's earnings. When the drug was finally approved by the FDA three years later, Genentech lit up the San Francisco sky with a massive fireworks display. *Restraint*?

By the summer of 1984, the Bard/USCI group could feel the winds changing and realized that their hegemony was finished. With entrepreneurs egging them on, a new breed of young Turk cardiologists were keen to bypass Gruentzig's cautious injunctions and latch on to the stock-option-rich research schemes being whispered into their ears. Suddenly, it seemed that even trusted allies hungered for deals on the side, especially when two of the most celebrated angioplasty practitioners, Simon Stertzer and Gerald Dorros, entered into private negotiations with the medical-device giant Medtronic to start their own potentially blockbuster enterprise. An eventual deal to launch a start-up company called A.V. E. (for "Advanced Vascular Enterprises") earned Stertzer tens of millions of dollars. Meanwhile, the busy Stertzer and Richard Myler began trying to lure Gruentzig from Emory to join them in a California super-practice. All was in play. The medical-device industry was churning with visions: West Coast companies proffered laser catheters, ultrasound probes, and whirling arterial drills for eager Gruentzigs-in-the-making to test out on their patients. Dave Prigmore, the generally no-nonsense USCI leader, couldn't believe what he was hearing. "By this point people were getting really greedy. People were really upset that (John Simpson) had made many millions of dollars and they had only made . . . " here the droll retired executive chuckled, "millions of dollars."

These were heady days. The Cook Group from Indiana had begun experimenting with the potential of inserting little wire scaffolds—the newly christened "stents"—into diseased coronary arteries to hold them open for years after clearance by balloons. Bill Cook insists that his close friend Caesar Gianturco was the first to sketch this current bedrock treatment for the diseased circulation. Perhaps, but by then Ulrich Sigwart in Geneva was working on a similar concept, as was a French cardiologist named Jacques Puel, and as were the Texas duo of Richard Schatz and Julio Palmaz, who became involved with the eventual front-runner in introducing those devices, the Cordis division of Johnson & Johnson. Industry executives made constant pilgrimages to Emory seeking the imprimatur and insights of the man whose breakthrough launched this new world. "Everybody who had one of those ideas was trying to get to Andreas and every one of them was trying to offer stock options or whatever they could offer him," Prigmore noted.

The potential for a schism with USCI loomed, and Prigmore finally convinced Bard's parsimonious corporate leadership in New Jersey to offer Gruentzig proper royalties along with a hefty consultancy contract. Their inducements—which would grow to include an up-front lump sum payment of $3 million—were basically too little, too late. The former tinkerer with Krazy Glue had learned the game well enough to retain staunch legal representation. Prigmore was left in the awkward position of traveling back and forth to Atlanta with hat in hand in an agonizingly protracted attempt to close the deal. Bard's chief executive, Robert McCafferty, finally paid his first and only visit to the increasingly disaffected Gruentzig. It was as though this company, with its gift-wrapped proprietorship over the hottest medical development of the decade, had willed its own disaster. Through penny-pinching, Bard had missed the opportunity to buy out Schneider, preempt

Pfizer, and dominate the balloon angioplasty market for years to come. When presented with the opportunity to swallow the upstart John Simpson's advance, they flinched.

The industry's eyes began turning ravenously to the potential of using the balloon procedure to implant the little stent coils to prop open diseased coronary arteries more reliably than ever before. The idea was that inserting these tiny scaffolding devices would stop arteries from closing up again after angioplasty, and that this neat advance would draw thousands of new patients toward angioplasty and away from bypass surgery. Gruentzig himself became keenly interested in this experimental technique. After entertaining a number of discussions with the Cook Group, then developing its first prototypes, he appointed Gary Roubin as his point man in exploring the concept. Bard/USCI essentially ignored the whole idea.

Inevitably, the company imploded a couple of years later. When the going got intensely competitive, Bard introduced what its executives thought was the Ferrari of all angioplasty catheters in an attempt to regain market dominance, a product they called "The Probe." They at least twice soon hid its lethal malfunctions from the United States Food and Drug Administration, causing federal agents to invade their Billerica, Massachusetts manufacturing plant.

Sadly, Dave Prigmore and John Cvinar were eventually sentenced to eighteen months in a federal penetentiary (never to be served after an appeal), for defrauding the FDA, and shrouded in ignominy. Many colleagues felt this development was at least partially owed to populist grandstanding by the federal government, although the executives had clearly played loose with the regulations. Nonetheless, to this day, Bard/USCI remains a fallen entity, drummed out of the very field that it had pioneered. The heart had simply gotten too hot for its touch.

CHAPTER 17

WITH HIS BREAKTHROUGH ATTRACTING SUCH FEVERISH ATTENTION, Andreas Gruentzig ostensibly had little left to prove. His attention therefore occasionally shifted to the larger spectrum of reshaping the management of cardiovascular disease. Late in 1984, he became particularly interested in the idea that patients in their prime might be screened for signs of potential circulatory trouble, and that his technique might be better used at the beginning of the atherosclerotic syndrome, rather than as a last-ditch intervention for hearts already in irrevocable jeopardy. The best way to spare patients from the peril of last-ditch intercessions could be to identify and swiftly treat their disease at the earliest possible stage, Gruentzig reasoned. The first step was to prove that coronary angiography could be conducted in a screening session that was so benign that it would leave younger patients free to jump back into their daily routines hours later. Unfortunately, that cheery prospect proved to have a troubling denouement.

Gruentzig was in no danger of losing his sense of drama, however. At five o'clock on the afternoon of December 10, 1984, he abruptly burst into an Emory catheterization laboratory like a reborn Werner Forssmann and, without warning, commanded his assistants to perform coronary angiography on himself. When the staff protested that he was not ill, he told them to get on with it—he wanted to prove that the heart's depths could now be screened without fuss. Had not Mason Sones performed angiography on his own mother? What was good for the goose was good for the gander, Gruentzig insisted. Then he casually chatted away while a young understudy named Hal Whitworth advanced the diagnostic catheter through his coronary circulation, finding not one shadow of imminent trouble. The relieved Gruentzig lay back like any garden-variety patient afterward while pressure was held against the groin puncture site and a so-called sandbag was perched on the wound until sufficient coagulation was established. By this point, Gruentzig, still whirling with new ideas, had been drafting ideas for an ingenuous plug-like device to hasten this process; indeed, a man who later saw his rudimentary sketches would make millions from the concept. In any case, after resting for an hour, Gruentzig now hopped off the gurney to resume his affairs, exhibiting the carefree panache that embodied the point he sought to make: The human heart could be painlessly examined in the prime of life to check for incipient cardiovascular disease when it mattered most, thereby bringing coronary artery procedures to bear when they might do the most good—that was what he was arguing.

As far as Gary Roubin was concerned, Gruentzig exhibited no loss of vitality. "It seemed to me," he told Burt Cohen, "that he had . . . satisfied himself that angioplasty worked, and he was very busy [with new pursuits]. He had, I think, reached a point where he didn't have to defend the procedure in the community. But he was interested in moving on, and codifying the data he had accumulated. At the time I had arrived, he had done about two and a half thousand procedures, many of them multi-vessel disease patients . . . From almost the week I arrived, he asked me to start working on a randomized trial to compare angioplasty and bypass surgery in patients with multi-vessel disease . . . That was the main thrust of our research efforts at that time."

Roubin marveled at a man still enamored of countless fresh possibilities. "He wanted to develop technology that would allow us to direct a guide wire through totally occluded arteries, or direct laser beams through totally occluded arteries. Even when he had six or eight angioplasties to do on his list for the day, at our research meetings he would sit down and he would draw on the blackboard, and talk to us about these incredible devices that he dreamed of. I remember one morning, it was about 10:30, and we were about an hour and a half behind on the PTCA schedule. He was still drawing mountains on the blackboard, describing how, if the scientists could organize nuclear missiles that could travel sixty feet above the ground between the U.S. and Europe, then there was no reason that we couldn't have the technology to direct laser beams, or guide wires, or other devices through and within the coronary arteries. He was a dreamer."

The next stop after Gruentzig's adventure in self-catheterization was a department Christmas party. Andreas showed up in his French tuxedo, raised his champagne glass, and let his fellow revelers know what he had just done. By the end of the evening he amazed them all by tripping the light fantastic with Margaret Anne. "Are you believers now?" he literally asked the doubting Thomases. The next morning the couple flew to Sea Island, with their Irish setters in the back of their plane. There, Gin and Tonic broke loose, causing Gruentzig to chase them down the beach. Before long, he noticed that the previous day's puncture site was oozing prodigiously (the first telltale trickles began happening while he was dancing the night before), a condition that can result in serious complications unless promptly treated. His in-laws, who had joined him at the island cottage, teased that he had pushed himself much too hard. But the fact is that diagnostic angiography and even balloon angioplasty are now routinely performed in just the "in-and-out" manner that Gruentzig had demonstrated on himself, with patients often sent home within hours. Sixty years earlier, these same patients couldn't have had their hearts so much as touched without inducing death.

Christmas soon came around with its comforting family rhythms. Sonja arrived with Gruentzig's mother and elderly aunt from Leipzig, Alfreda Beier, who was known as "Tante (Aunt) Alf," along with his Uncle Konrad from Hanover and his wife Herta. Days walking the beach were followed by a festive party in the Buckhead mansion. There, mountains of presents for Sonja appeared around a towering tree. A comely nurse was assigned to hand them out. She had riotously rouged her cheeks and donned a short, frilly, green tunic over fishnet stockings in order to resemble a Nutcracker Suite vision of one of Santa's elves. Family, friends, and colleagues retired to the music room for carol singing beside Andreas at his piano. But a squabble broke out between Margaret Anne and one of the doctor's wives. The bubble of euphoria was pricked again.

A few days later, Tante Alf collapsed in cardiac arrest as little Sonja screamed the alarm. Andreas rushed up the stairs and administered CPR. He saved his beloved aunt's life but broke her ribs in the crushing frenzy of resuscitation. The mayhem that can beset a human heart was far from vanquished. The distraught nephew pondered what he should do next to further tame it.

The months turned, and Gruentzig kept hungering for a better position elsewhere, one where he might not only perform his signature procedures but preside over an entire academic cardiology department devoted to transforming the treatment of cardiovascular disease. The prospect of returning in triumph to his native land still excited him the most, but neither the administration at the University Hospital in Düsseldorf, where his brother Johannes lived, nor anywhere else got around to making an offer. That such a medical star would be brushed off so continuously seems utterly baffling today. But Bernhard Meier, Gruentzig's Swiss understudy in Atlanta who had recently returned to practice cardiology in Geneva, says the reasons for the rejections were as basic and old as human nature itself—jealousy and personal insecurity. "None of the European universities want a very bright figure in their midst. They are not money-oriented. You can't get a real big star [under recruitment] past your colleagues. Nobody wants to see a towering tree in his neighbor's garden that is casting a shadow over his house and find himself living in that shadow for the rest of his life . . . The well-being of the university is secondary here. Primary is your own well-being, and your well-being is better if when you look left and right, you don't see anybody better than you are. That's the way it is in nonprofit-oriented European universities."

With no prospects materializing across the sea, Gruentzig therefore continued his negotiations with his earliest American allies, Richard Myler and Simon Stertzer, who were each churning out more than 500 angioplasties a year and being showcased in their own medical presentations around the world. Unlike the Europeans, that duo knew that joining forces with the founder of the technique would create a powerhouse of cardiovascular medicine like none other. They were so stirred by this vision that they pulled political strings to have Gruentzig granted a nearly instant license to practice medicine within the state of California with no board examinations required. Once again, Gruentzig wanted to line up a suitable companion position for his wife, since Margaret Anne was determined to resume her medical career. After considerable lobbying, Myler thought he had found the ticket at the University of California at San Francisco, and she arrived for an interview. But things immediately went sour. Irritated at a failed arrangement to collect her at the airport, Margaret Anne fell into a snit. Her anger over this and other perceived slights effectively spelled an end to the proposed California cardiology triumvirate, according to Myler's wife Sharon. "She was hysterical about it. But the bottom line was she did not want to come to San Francisco. She wanted to stay near her folks."

Sharon Myler, who freely admits that she nursed a much deeper affection for Gruentzig's first wife, Michaela, hadn't liked what she had seen in recent Atlanta visits with Andreas and Margaret Anne either. "By this point it was a very unhappy house. She was drinking too much, and he was drinking too much. . . . They [her husband Richard and Gruentzig] were really close. Andreas trusted us. She tried to drive a wedge. She did anything she could to say, 'Don't listen to Richard, you don't want to move there.' She managed to put distance between anyone he had a relationship with."

Claire Rice agrees that ill omens abounded. She remembers Andreas coming to work one day and relating that he had just coaxed his ever-loyal but evidently suicidally distraught man Friday, Sarge, down from the roof. According to the Mylers, Sarge had been warning his employer that his Southern belle would bring him nothing but trouble, and Gruentzig had grown furious about this impertinence. For his part, Spencer King said that he sometimes thought his colleague was faltering. "I got the feeling he was tired of what he was doing—or tired of something. He didn't seem to have the same enthusiasm for pushing the envelope."

Mary Jane Thornton throws up her hands at most of these allegations. Her daughter's marriage remained vital all the way through, and Gruentzig was never despondent, she insists. "Was he depressed? I never saw him depressed one day in his life—except when he lost his first patient," she asserted.

The mother-in-law concedes that Gruentzig had hit a stretch where he yearned for some basic changes in his life. "I remember quite well that he wanted to pull back and not work as hard as he did, and he didn't want to travel as much." Peculiarly, Andreas and Margaret Anne took to writing messages to one another while cosseted in their first-class airplane seats. "On airplanes when they were flying, so they wouldn't have to talk so much, they would write to one another, sitting beside each other," Mary Jane explains of notes she later discovered. "The way he wrote one was: 'House running well. I am going to slow down. So will back you and you can do what you need to do.' "

Margaret Anne eventually returned to her medical residency in Emory's department of radiology. William Casarella, the chairman, found her to be a far more able and likeable young doctor than some others credited. "She had a lot of promise . . . She was a very capable resident. She was very attractive, very Southern. Your initial impression wouldn't be that she would be as strong and tough intellectually as she was. But she was really a strong resident and going to be a great doc, I think. She never really had a chance."

Early in 1985, Gruentzig abruptly traded in his flashy red Porsche for a sage green Mercedes SL with a beige interior, which he called his "old man's car." Meanwhile, he decided that he needed a new airplane, one that was far peppier than the Bonanza, and selected a twin-engine Beech Baron. His flying friend Rick McClees was aghast when he heard the news over dinner one night at West Andrews. "I remember my blood turned to ice when Margaret Anne said, 'I am so glad we got the twin. I am so glad we got that second engine.' She said, 'We are going to be so much safer now with twin engines!' I was like, 'God!,' because the data doesn't prove that; and in fact, for low-time twin pilots, that's a dangerous situation."

McClees's wife Kristina added, "We talked about it on our way home. Rick said, 'I know he's got the money; but just because he can afford it, it doesn't mean he can fly it.' And he could afford it. He was already talking about how he was going to eventually become a jet pilot and get a Lear. He could write it off and make it part of a corporation and so on and so forth. And the Baron was just one step in that direction. I remember that things were going mighty fast."

There was no turning Andreas Gruentzig back once he was enthralled by a new passion. Before long, Andreas and Margaret Anne were heading out to Beechcraft's corporate headquarters in Wichita, Kansas, to preview their specifications for the plane's interior. To complement the leather seats, they ordered a rich interplay of deep reds and pale blue. The registration number they selected was 583AM, the nomenclature evidently designed to signify "Andreas and Margaret, married on the fifth month of 1983."

Andreas Gruentzig had most everything going his way as the spring of 1985 approached: a new plane under contract, a new job for his wife, and a $3 million payout at last due to materialize from Bard/USCI. But one April evening the Georgia skies were seized up with a fury like that which had obliterated his native Dresden. This time, it came at the hand of God. He was sitting at the dinner table as Margaret Anne finished up some preparations in the kitchen, when the sound of lashing wind and pounding rain gave way to a rumble that soon roared as though a mammoth train engine was headed their way. Recognizing an approaching tornado from childhood experiences in Macon, Margaret Anne screamed, "Andreas, grab the dogs and get in the basement!" No sooner had they slammed the door than the cyclone began cracking trees into bits across every square yard of their expensively manicured grounds. Mighty trunks catapulted into the air and crashed through the mansion's tile roof. An entire corner of the house crumbled into rubble and Margaret Anne's gilded dressing room was blown to smithereens, with a delicate French painting ripped in half. The wrought-iron fencing that surrounded the palace fell twisted under the weight of fallen tree limbs, and the dream swimming pool under construction resembled a bomb crater. Within minutes, the ultra-affluent West Andrews neighborhood had been reduced to chaos, with utility poles crumpled, live wires singeing the asphalt, and mansions shattered, like some vision from Stephen King.

The next day happened to be Easter Sunday. Margaret Anne and Andreas decided to go ahead with a dinner with her parents and hired a portable generator so that they could cook and carry on with their lives. But there were gaping holes in the roof. Equanimity was scarcely restored. Shortly, the couple were faced with the agonizing process of hiring workers to salvage some order. In the anxious days that followed, one of them haphazardly unplugged the generator that also oxygenated their fish pond, according to Mary Jane Thornton. "The poor fish were lying on their sides belly up, and Andreas plugged it back in and said, 'Don't ever do that again, you killed our fish.' " He reportedly massaged a number back to life, even making a show of breathing into the mouths of a few in a theatrical display of his rejuvenating powers.

It could have been worse. Although distraught, the couple had the means and the insurance proceeds to move into temporary digs at the posh Ritz-Carlton Hotel, followed by a months-long stay in an upscale townhouse off the Buckhead district's fashionable Lenox Square, while their ruined house underwent repairs.

The next month, Gary Roubin threw himself a thirty-seventh birthday party in his new apartment, inviting numerous senior and junior staff. The wine flowed, the guitars came out, and Andreas and his host began to sing. But at the bar, an old acrimony suddenly hit the boiling point between Margaret Anne Thornton and Gail King. The ladies went after each other with a vengeance, with Gruentzig's wife railing that she had never ever been treated with the respect that was her due. "I went up to her and I said I am sorry, I didn't realize how badly your house had been destroyed, and she looked at me and called me 'a lying so and so,' for saying I didn't know," Gail King said. The next thing everyone knew, the women were showering each other with drinks.

Yet another chilling experience occurred when Gruentzig brought his young wife back to the East Germany of his childhood. Following a medical meeting hosted by Eberhard Zeitler in Nürnberg, he decided to visit his beloved Tante Alf, who was now languishing in a hospital in Leipzig. The East German police state nursed a dossier-rich memory about every soul who had fled the Iron Curtain, particularly when their talents blossomed into fame in the West. Rather than being greeted ceremoniously, Andreas and Margaret Anne were held up for eight menacing hours at the Checkpoint Charlie border crossing in East Berlin. The implicit threat was that the only way he could go home again was in shackles as an enemy of the state. Eventually, Gruentzig was cleared to proceed. But there was little nostalgia to be mined from revisiting the innocent days of his youth. Andreas met his failing aunt all right, just days before her death. When that played out, his last connection to his roots was gone, seeing as his mother had long since departed for Heidelberg.

A succession of other fatalities soon haunted Gruentzig as well, beginning with that of Charles Dotter. The original "heart plumber" underwent a second coronary bypass surgery in early February, but never woke up from the anesthesia. He ultimately succumbed more or less to congestive heart failure, thanks to a suicidal syndrome compounded by amphetamine withdrawal.

Bill Cook was there just before the end on a pilgrimage to Dotter's bedside with a couple of the marquee names of cardiovascular medicine at the time, including Caesar Gianturco and Kurt Amplatz. "He killed himself—it was that simple. He just pulled the plugs. He stopped the breathing apparatus and died. By the time they got into the room he was gone. . . .

"He was on amphetamines, and when he went to Milwaukee to have the surgery he didn't tell anybody, so he went into shock after the surgery. They didn't know what was wrong with him until Josef Rösch came out and told them that he was on amphetamines. They fed him on amphetamines and a week later he was recovering, but he had lost about two-thirds of his breathing capacity. His lungs just disintegrated."

A few months after Dotter's demise, Mason Sones's lung cancer finally played itself out. Then Melvin Judkins, the Seventh-day Adventist who had ingeniously simplified the catheterization of the heart, succumbed to his own myocardial infarction. One by one, the greats of cardiology were vacating the clamorous new medical era. With Forssmann now deceased for six years, Gruentzig was the last great catheter pioneer left alive.

Despite all these ill tidings, the summer of 1985 began with bright prospects for Andreas Gruentzig. The rebuilding of his ruined mansion was by now well underway, with Margaret Anne fastidiously researching the house's architectural pedigree in order to bring it back to an even higher standard. In June, she and Andreas hosted a lavish weekend at their Sea Island cottage for a collection of Emory cardiology fellows, nurses, and technicians. Bikinis and Speedos were donned and drinks were poured in the afternoon on the Gruentzigs' terrace above the beach. Gary Roubin, wearing an Australian Akubra bush hat, strummed his guitar. Gruentzig eventually stretched out for a siesta on a chaise lounge for all to see. Roubin took this to be a sign of unaffected humanity. "It was three or four in the afternoon and he was in an unbecoming bottle-and-a-half-of-wine pose. We were all just chatting and he fell asleep and was snoring. He didn't care, he was the master."

Nobody bounced back better than Andreas Gruentzig. At night, he hosted the entire crowd at The Cloister for a formal dinner followed by ballroom dancing. The next morning, an Emory staff photographer brought down for the occasion snapped a portrait of Andreas and Margaret Anne staring adoringly into each other's eyes in the house's airy new sunroom. For all his rumored philandering, it would be difficult to find a more emblematic portrait of starry-eyed love. Indeed, Andreas and Margaret Anne were said to be finally seriously discussing the idea of having their first child together. Optimism reigned, and he told friends that he was sure to eventually win the Nobel Prize. Meanwhile, he also decided to make peace with Geoffrey Hartzler, telling his former nemesis that he had come to recognize that he was a legitimate pioneer and had advanced interventional cardiology in directions that needed to be conquered.

Sonja, nearing nine, arrived for another halcyon summer on the Sea Island beach, swimming in The Cloister's pools, riding horses, and playing tennis with Margaret Anne and Andreas. The weeks passed happily and on July 11, Gruentzig gained his coveted clearance to fly multiengine aircraft, a necessary step for handling his new Beech Baron. Seven days later Gruentzig took delivery of his custom-appointed aircraft in Wichita, Kansas.There he waved aside the free flying lessons that came with the package, saying he would have his own people look after that. Gruentzig did attend the flight simulations they provided, but the staff later claimed that he was cavalier about completing the necessary homework. One instructor, no doubt fearing an eventual lawsuit if things went wrong, lamented that he "appeared more infatuated with his radar and avionics than with acquiring knowledge of his aircraft."

Life was nonetheless on the upswing, and Gruentzig enjoyed his first supervised flights in his fast new plane in late July and early August, soaring off to his Sea Island weekends with a copilot sitting close. When the weather was good, he made solo runs back and forth from Atlanta. According to Mary Jane Thornton, he became so concerned for the Baron's security that he tried to buy a private hangar at St. Simons's little Malcolm McKinnon Airport. When told the hangar could only be leased and not purchased outright, he gave up the idea. There were plenty of ways to spend his money, however, and he bought Margaret Anne a spectacular ring with a giant rectangular emerald set in gold. Her mother now wears it every waking moment like some talisman of all that she has lost.

Another profound circle of affection promised to close around Gruentzig's life at this time. Ascending a shopping mall escalator with her stepgrandfather, Sonja suddenly blurted, "Did you know that I am going to live with Poppy and Margaret Anne?" Her husband Earl, Mary Jane Thornton says, replied, "Well, Sonja, that would be wonderful, you can help me wash the cars every Saturday." And she laughed and giggled because she called Earl " 'the world's greatest grandfather.' " The Thorntons were nonetheless stunned when Andreas confirmed that the prospect was being actively negotiated with Michaela.

With summer drawing to a close, the Thorntons purchased an expensive sapphire ring in honor of Sonja's forthcoming ninth birthday in September. Margaret Anne was having her stepdaughter's portrait painted at the time, and the item was brushed in as a finishing touch. "Do you see how much we loved her? She was our only grandchild," Mary Jane said with a distraught look crumpling her face. Sonja nonetheless departed, true to the usual pattern, for the beginning of the next school year back in Zürich.

As autumn approached, Gruentzig's days were full. Even as his procedures piled up and his intensive consulting resumed, he found time to climb into the cockpit of his beloved Baron and log a few more hours as a solo pilot. The plane boasted not only extraordinary moxie but extremely sophisticated avionics, including an autopilot system for night and bad-weather flying. Many observers say that it is just the kind of package that lures weekend pilots into overconfidence. But his instructor, Dan Emin, head of Quality Aviation at Peachtree-Dekalb Airport, waves this judgment off. "For a 500-hour pilot flying the equipment he was flying, and coming out of a Bonanza into a Baron, which is a very nice, comfortable transition, I would rate him at least an 8 or a 9 [on a proficiency scale of 1 to 10]. Nobody's perfect, but he was very confident for his skill level. And he really enjoyed it." Supporting this assessment is the fact that Gruentzig easily passed a September 5 examination of his proficiency, landing the Baron under instrument controls.

Bill Walker, a seasoned flight instructor who owns Golden Isles Aviation on St. Simons, nonetheless found this characterization a bit nonchalant. He himself has logged 16,000 hours in the air, starting as a teenage barnstormer taking off from Sea Island's beach. "A low-time pilot who goes into a sophisticated airplane and probably learned on a pretty nice airplane, they're almost entirely oriented toward these sophisticated instruments. So if they lose those things, they find themselves in a real jam, because they never really learned on basic instruments, whereas older guys like myself, we learned on basic instruments. That's all we had—dead reckoning."

Walker never personally met Gruentzig, but he knew his aircraft inside out and was a denizen of Malcolm McKinnon's runways at the same time. "The Baron's a quick airplane. They're fast, and if anything goes wrong, you've got your hands full. Things happen fast on a Baron, and plus it's got differential problems. If one engine quits, it wants to curve in the other direction because the engine is pulling from one side. They're whizzing along at a fast speed—220 miles an hour. A Baron is not a plane for a low-time pilot. A low-time pilot gets into a Baron, the airplane will get ahead of him. He won't have the pilot skills developed yet to handle the airplane in an emergency."

Walker himself nearly perished once when a copilot became disoriented when the airspeed indicators went to a false reading of zero due to an ice storm. His Beechcraft Queen Air plummeted 6,000 feet due to the copilot's overreaction and Walker had to wrest command to employ the aerobatic instincts that he had learned in his barnstorming youth. He therefore remains very wary of junior pilots who are entranced by the avionics of their flashy new planes. "Buying and selling airplanes for years, I've flown with lots of pilots, transitioning them into another airplane from what they've been flying. Even if it's a bigger, faster, nicer airplane, I am always amazed how far behind the airplane many of them are in terms of basic skills. You just about have to retrain the guy completely. Then you worry him and point him to an instructor to make sure that he gets trained enough that he doesn't go out and kill himself flying the airplane."

Gruentzig's flying friend Rick McClees felt the same trepidation. "I personally think somebody should have pulled him aside and said, 'Gruentzig, a twin Baron is a little bit of hot airplane.' " Dave Prigmore had been forced to demand a so-called key man insurance policy on Gruentzig's life before handing over Bard's recent $3 million payment. When he told seasoned corporate pilots about Gruentzig's new purchase, he received this unequivocal response, "That man has no business flying that plane."

However, proceeding timidly was not Gruentzig's style. His teaching course in the second week of September proved to be the most lavish of them all, culminating in a gala formal dress party in downtown Atlanta's newly opened and determinedly opulent High Museum of Art. Tom Wolfe managed to paint this scene vividly as well: "The atrium was an immense space, almost fifty feet high and pure white, like the building's exterior. Up a great curved window wall with white industrial muntins rose a series of curving ramps, one above the other, with white pipe railings and white wire grills instead of balusters. Spotlights and floodlights beamed down from all over the place."

On to a beaming Andreas Gruentzig, surrounded by acolytes from around the globe, he might have added. Among them was his devoted Zürich muse, Maria Schlumpf, still assiduously following the long-term health of the first patients to receive balloon angioplasty. When the last fluted champagne glass was quaffed and the course was done, the two managed to enjoy some mirth at Sea Island. A photograph shows a tipsy-looking Gruentzig cavorting about in Maria's nightgown there and sporting a hilarious grin. One late afternoon he took her up for a spin in the Baron. While dipping in and out of the clouds, he suddenly became alarmed by some quirk in the plane's instruments. But there was no great fretting, and the two dispatched happy postcards to several mutual Swiss friends.

After Maria departed, Andreas flew on September 24 to a medical meeting in Chicago with a copilot closely watching over him. All went well, and the very next day his plane received a clean bill of health after a rigorous safety inspection, which was mandated after fifty hours of air time. A couple of days later, he and Margaret Anne boarded an ocean liner called the *Island Princess* for a cruise from Miami to Acapulco on Mexico's Pacific coast. This was none other than the "Love Boat" from the fabled and eponymous 1977–1986 TV series featuring romance and hanky-panky on the high seas with cameo appearances by the likes of Gene Kelly and Cab Calloway, Tom Hanks and Jamie Lee Curtis. Andreas gamboled on board in his Panama hat and crisp summer whites, with his young wife looking gorgeous in a short, clingy green dress. They returned, bronzed and renewed, to Atlanta at the end of the first week in October.

Gruentzig next flew to a meeting in Washington, D.C., with an instructor at his side. Not a problem was encountered. A few days later he felt sufficiently confident to embark on a nearly thousand-mile solo flight for another meeting in Syracuse, New York, which he also completed safely. The return journey was another fair-weather, daytime affair—nonetheless, some evident irregularity in the avionics unnerved the plane's new owner. Setting back to work at Emory, he told Spencer King that he suspected that the navigational indicators went errant at one particular setting and he had lodged a complaint with Beechcraft. After a careful inspection on October 18, the company replaced the so-called radio magnetic indicator, whose directional needle points a pilot toward the path indicated by radio signals from control towers, with a loaned instrument. They also cross-checked every aspect of the plane's systems that could be examined on the ground, which necessarily left out the autopilot. Gruentzig seems not to have been overly worried, since he flew the plane to Sea Island that very night.

The next week was as busy as ever, culminating with Gruentzig performing a string of seven coronary angioplasty procedures on Friday, October 25. One of these patients happened to be a vascular surgeon and therefore a representative of one of the rival fields that had criticized balloon angioplasty from the start. But all seemed to proceed without a hitch, and at the end of the afternoon Andreas scrambled off to the airport with Margaret Anne to depart for yet another getaway to Sea Island. Knowing that a hurricane was gathering strength far south in the Gulf of Mexico that very night, he wanted to arrive before dark.

Chapter 18

On Saturday, October 26, Hurricane Juan began roiling the Gulf of Mexico. Within a few hours, the entire Southeast, including Sea Island, was drenched by torrential rains—a not uncommon event for this time of year. More disturbing to Gruentzig was the news he had received when telephoning Emory. The vascular surgeon treated the afternoon before had just suffered an abrupt closure of his newly opened coronary artery, sparking the onset of waves of erratic heart rhythms and piercing angina. Spencer King and John Douglas tried to salvage the situation with a repeat angioplasty. When that failed, the man had to be rushed off for emergency bypass surgery.

Gary Roubin heard his mentor's voice thicken on the other end of the line. "To not be able to open a fouled angioplasty was to Andreas the greatest measure of the failure of his technique, a personal failure . . . And when he had an angioplasty arranged for a vascular surgeon, it had to friggin' work. He asked me how the guy was doing and I said, 'He's okay. He's still alive and well.' Andreas said, 'Shet, this is terrible! I feel bad.' He used to say, 'This looks shetty!' whenever there was a problem doing cases. He said, 'Well, I will come back and see him Sunday night.' "

The next morning, a fretful Gruentzig confronted further tumult. Before his cottage, the surf groaned and moaned, thanks to the frontal assault of Hurricane Juan several hundred miles due west. The normally enchanting view across the strait to Jekyll Island was blackened by driving rain. Ever the devoted doctor, he nonetheless wanted to rush back to Emory to attend to his ailing patient and began checking the weather forecasts, which by now confirmed that the eye of the storm was nearing landfall outside New Orleans. Hurricane Juan would kill sixty-three people and cause $1.5 billion in property destruction.

An agonizing decision awaited Andreas Gruentzig. He had so far accumulated almost no solo bad-weather experience in his peppy Beech Baron, although he had plenty in the simpler and slower Bonanza. So what to do? He made a series of phone calls to weigh his responsibilities in Atlanta against the potential risks en route. By early afternoon, the meteorologists said that Juan's fierce peripheral squalls were already dissipating over Georgia itself. Back at Emory, he knew a patient had gone through a weekend-long hell, triggered by his own hands, and that he had five patients waiting for his attention the next morning. The clock ticked. At 1:23 P.M., the cardiologist called the control tower at St. Simons's Malcolm McKinnon Airport. The route northwest, he was told, was not especially violent, although visibility levels were dim. But his Beech Baron's elaborate instruments were meant to navigate through all that—in fact, this had been the point of acquiring the machine.

Totting up the odds, the renowned doctor, so careful when treating patients but also possessed of some sense of personal invincibility, made his decision, and decided to take off around three P.M. The die was cast. He and Margaret Anne began their preparations, packing suitcases and pouring a Benadryl solution on the dry food of their Irish setters to prevent them from vomiting should things get choppy on the way home.

The couple were nearly out the door when the phone rang an hour later. This was Mary Jane Thornton, imploring her daughter to tell Andreas not to fly, saying she had heard a very different forecast—that deluges were still flooding half the state. Margaret Anne brushed her mother's fretting aside. "Everything will be fine, Mama, don't you worry. Andreas has checked it out and we have to run," she replied with her sweet Southern lilt, before adding, "Bye-bye now."

The couple sped over the causeway above the gloomy marshes to St. Simons's airstrip, all of five minutes away. Hurrying through the boxy, one-story terminal, they ordered fifty gallons of fuel, climbed into their new Baron and signaled an imminent takeoff. As was her custom, Margaret Anne opted to sit in the back beside Gin and Tonic and then stretched out a protective net to prevent them from roving into the cockpit. Gruentzig was told to take off toward the ocean with the wind at his rear and then circle back toward the mainland—which was standard in rough conditions. Air-traffic controllers from Jacksonville, Florida, kept watch over the flight's swift ascent to 11,000 feet and saw no sign of incipient trouble as the Baron headed off toward Atlanta.

The first leg of their route was normally not all that daunting, with about ten initial minutes over flat terrain and through uncluttered skies leading toward a lock-in with the control towers just south of Macon. At that point, approaching pilots are meant to receive precise directions forward through the congested skies that surround Atlanta, home to 1.5 million landings a year. When things flow smoothly, the funnel onward from Macon lasts but another twenty to thirty minutes, depending on the thickness of aviation traffic.

Andreas and Margaret Anne knew the routine well. Official reports from October 27, 1985, show no indication of severe turbulence. Whatever weather he found two miles high, Gruentzig voiced no distress as he flew toward Macon—nor did any other pilot in the vicinity. However, given the uncertain weather, few private pilots were aloft. The official meterological records indicate that the conditions over mid-Georgia were less than atrocious—at most, belts of heavy rain alternating with dense fog. The National Oceanic and Atmospheric Administration (NOAA) says the average wind speed was only seven knots, with cloud cover beginning at 600 feet and visibility in most places extending to three miles. Not bad at all for an experienced pilot.

But average conditions are often deceptive and never more so than when a hurricane is wreaking havoc not all that far away. Ordinary individuals could readily see that Georgia was still being buffeted by a wildly unstable storm front. Spencer and Gail King, for example, were driving toward

Macon that very same afternoon and found the weather so appalling that they turned back. "There was blinding, horrible rain and dark clouds. There was a squall line that went down through the middle of the state and there were reports of tornadoes that were spinning off," remembers Gail. Taking off from south Georgia at approximately the same time as Gruentzig, an Argentinean doctor named Duval Ferrari put down at the nearest available airport and took a public bus back to Atlanta, which was something this experienced pilot almost never did. People on the ground in the Macon area said the fog was so thick that they could barely see 200 feet. Meanwhile caskets buried above ground in the Louisiana bayous began floating toward the Gulf.

Nonetheless, the progress of the Baron, 583AM, seemed steady enough as it neared Macon Approach Control. At about 3:14, a controller there named Jerry Pace instructed Gruentzig to descend to 5,000 feet and enter into a 360-degree rightward loop in order to queue for his final descent north of Atlanta. Pace's trade is frenetic, so he issued his guidance in the clipped jargon known to every pilot: "Baron eight three Alpha Mike, I'm gonna have to make a right three sixty with you for spacing into Atlanta Approach, is gettin' busy. If you'll give me about, uh, five miles on the downwind leg, I'll be able to turn you back in. Descend and maintain five thousand, make a right three sixty."

Gruentzig responded with the expected brevity: "Okay, down to five thousand and, uh, three sixty to the right." No sooner had 583 Alpha Mike commenced its designated banking descent than the Macon controllers radioed that Gruentzig should switch to a heading 128 degrees northwest, dubbed "radial Husky," toward Peachtree-Dekalb Airport. In other words, he had a ticket to land. But something was already going very wrong. Gruentzig in fact turned 128 degrees in the exact opposite direction, to the southeast. Radar monitors revealed a squirrelly path. At 3:19, Jerry Pace warned a colleague beside him, "We (sic) just now finding out, I think he's tracking the wrong way."

Five seconds later, Pace asked Gruentzig to radio in his heading. The response was professionally terse, but his voice was strained. "Heading is 128. I have a malfunction in my autopilot system." The worried air traffic controller quickly ordered him to complete a 300-degree turn to the north, and was relieved to see that this course was indeed pursued for the next two and a half minutes. Then things grew weird. Gruentzig, now about twenty miles north of Macon, abruptly headed west and two minutes later swiveled inexplicably back to the southeast. A shift of personnel was underway in the Macon control towers. The distraction may explain why several minutes passed before Pace next radioed his concern, "And, uh, eight three Alpha Mike, you back on the autopilot again? You're headed south," Jerry Pace warned.

Scanning the array of instruments before him, Gruentzig responded at 3:24:47: "No, uh, negative I have to fly with my, uh, backup. The autopilot and the whole first system is gone." A frightening syndrome was now in evidence. Air-traffic controllers know well how fiendishly small pockets of airspace can change their demeanor at a moment's notice, and that whatever 583 Alpha Mike was confronting was certainly not blue sky. It soon became clear that before their eyes dangled the fate of a pilot who was altogether losing his way.

A worried Jerry Pace called for a 340-degree turn northward, which Gruentzig tersely repeated on cue. Yet his tone was quaking now. Whether Gin and Tonic were yelping in the mounting confusion, or Margaret Anne was screaming, no one will ever know. But the radar records paint a ghastly picture. Gruentzig's next movements became wildly erratic, his plane turning 390 degrees while plummeting down from 5,500 to 4,900 feet, followed by a bouncing up to the same elevation all over again. Pace muttered to a colleague, "Watch eight three Alpha Mike." Thomas Chesley stuttered, "I, I'm watching him."

The German cardiologist Ulrich Sigwart has flown in all kinds of weather in myriad planes since he was fifteen and offered one possible explanation for the alarming confusion. "He had never done bad-weather flying; he was a fair-weather flyer. I am sure that Andreas always flew by the seat of his pants, in angioplasty and everything. He went by his feeling and you can't do that in bad weather; you have to be absolutely stupid. Bad weather makes you stupid and you have to function like a computer; you have to process information mechanically even when it's exactly against your common sense. The earlier you start the better you do this, because it's like learning a language. You have to be like a trained gorilla, you have to follow commands by the letter and do what they say. You have to make a click and get over it. It's a matter of personality whether or not you can accept certain things that you do not really believe in. Blind flying by instruments is totally inhuman." The fact is that by now Gruentzig had logged a grand total of only 448 hours aloft. Of these, only 58 had been in his sophisticated Beech Baron.

Bill Walker had himself just flown back to St. Simons from Atlanta that afternoon, but took an easterly route and landed back on his home turf without event. Having trained innumerable less-experienced pilots, he recognized a sadly familiar pattern. After examining radar records and the on-and-off voice transmissions from Gruentzig's calamitous flight, he observed, "He's probably not even talking on the radio. He's sweating bullets trying to control the airplane and doesn't want to take the time to handle the microphone and talk to anybody. The airplane was probably getting ahead of him . . . and he lost it—the same thing that happened to JFK, Jr. Eventually, it can get to the point where you just simply can't resurrect it, because you don't have any idea how to resurrect it. Whatever you do might exaggerate the problem in the movement of the controls."

The fact is that 583AM dipped below radar level at 3:27:42. Walker ventured one guess as to why: "Think of it: He's in these clouds. He doesn't know whether he's right-side up or upside down. He's looking at his instruments and they're doing this," here he made a wobbly motion with his hands. "He's not sure which way he's headed, and he thinks, 'If I can just get below the cloud layer I will be able to see and I can level the airplane out and pick my way to a highway or a little airport somewhere and land.' That's a very common tendency."

Panic in the skies can have some very technical ramifications, Walker said. "If you start messing with the trim, running the trim up and down to try to help you fly the airplane; or if you lost an engine and that caused an out-of-synchronization situation with your controls; or if you made violent alterations in the controls of the airplane to resurrect a situation where you suddenly found yourself in a nose-down condition and you pulled way back on the yoke to stop it—then you'd find yourself in a nose-up condition. Then you'd stall, and then you'd find yourself in a nose-down position. If you found yourself in a steep turn, which can easily happen, then you might overcorrect, which would then stall the inside wing, the lower wing. And the upper wing would still be flying, and that's when an airplane gets into a spin and will spiral right into the ground."

The above is mere conjecture, but something disastrous was unfolding, and Mary Jane Thornton says she could feel it on the ground. Just before half past three that afternoon, she was showing a house just a few miles north of her own to a wealthy prospective purchaser. The afternoon was unnaturally mild, what with the vast tropical air mass having been pushed ashore by the landfall of Hurricane Juan. But Margaret Anne's mother remembers, to this day, a chill going down her spine at this time.

The air-traffic controllers in Macon and Atlanta were simultaneously exchanging numerous alerts, and calling for urgent satellite tracking. At 3:30, an Atlanta controller asked a nearby colleague if he had obtained a reading. "Nope, where is he?" was the response. The next minute was filled with more frantic interchanges, and finally a Cessna pilot heading north past Macon was asked to attempt contact. Dead silence was all Gruentzig's frequency yielded. For the following five minutes, the Macon and Atlanta air-traffic controllers fired messages back and forth regarding the seeming disappearance of 583 Alpha Mike.

Finally, just before 3:38, Atlanta Approach asked Macon, "Y'all ever get in contact with this five eight three AM yet?"

Jerry Pace responded, "Naw, we hadn't got him yet."

James Hathcock, Jr., of Atlanta Approach demanded, "Y'all hadn't, you don't know where he is?"

Pace rejoined, "We sure don't."

Hathcock responded at 3:38:18, "No, we don't see a thing in there. We've been looking and we don't see a cotton pickin' thing."

In a pine forest in rural Bollingbroke, Georgia, about twelve miles north of where a worried Mary Jane Thornton was showing her house, a group of deer hunters were perched in a tree stand just off a remote logging road. They were having no success, since they could barely see 200 feet through the ghoulish murk. Suddenly, they heard the screeching of an airplane descending like a dive-bombing Stuka and then they felt the earth shake with a tremendous "wumph." Nobody had to tell them what happened. They knew and shortly thereafter informed the police, who had already been alerted to the shrieking descent by a resident on the nearby Pea Ridge Road.

The local sheriffs slogged into the woods and found a scene of devastation, with wreckage and clothing dangling from broken trees and a thirty-eight-foot crater strewn with the core of a demolished airplane. Almost nothing that resembled its original form survived the crash, save a few pages of a manuscript Gruentzig was writing about how to best repair human hearts, some useless dollars, and a sapphire ring set in gold that he had so recently presented to his young Southern belle, Margaret Anne. It is the same ring her mother wears to this day.

"Bye-bye now," were the last words that Mary Jane Thornton ever heard from her daughter's lips. Not one more would the world hear from the lips of Andreas Gruentzig, the man who changed the course of medicine.

Epilogue

The tragedy of 583 Alpha Mike was not communicated to the outside world until the next day. When Gruentzig failed to show up at work Monday morning, Emory staff called his house and learned that he had never returned from Sea Island. Spencer King next telephoned Malcolm McKinnon Airport on St. Simons and gleaned that Gruentzig had departed in such haste that he had not even filed a flight plan. A representative of the Federal Aviation Administration suggested that King phone the Monroe County Sheriff's Department, since there had indeed been a crash in that vicinity. Dialing the local authorities, the cardiologist learned that the wreckage was so complete that the police had been unable to identify the victims before darkness fell and that they would therefore appreciate some help. The distraught cardiologist agreed to drive to the scene with Gary Roubin. By the time they arrived, 583AM's tail numbers had been isolated and the proof of who had perished was positive. The vista that Spencer King discovered was ghastly. He took on the difficult duty of informing the Thorntons, already troubled by a night of unreturned phone calls to their daughter.

Earl Thornton received the awful news while at work at his textile finishing plant north of Macon. The bereaved father lurched out of his office and struggled home. In the driveway, he met his wide-eyed wife, trying to make sense of her husband's appearance in the middle of the day. She demanded, " 'Earl, have you lost your job?' He said, 'Janey; It's worse than that.' Then he told me and I didn't know that any human being could scream like I did."

With Spencer King at the Thorntons' sides, the wider telephoning began, engendering reactions of horror and disbelief around the world. Across Germany and Switzerland, Gruentzig's shattered family and friends boarded airplanes for the funeral, soon to be organized by the band leader from Andreas and Margaret Anne's wedding, whose day job was to serve as the local undertaker. None could have been more distraught than little Sonja, who arrived with her mother and Maria and Walter Schlumpf. Since Andreas's grief-stricken brother Johannes was adamant, the Thorntons reluctantly agreed to visit the crash site with him and his mother, Charlotta. The trip, under the escort of a sheriff, proved to be a ghoulish exercise. "The plane was totally obliterated. You couldn't tell it was an airplane," said Earl Thornton.

"Andreas's body was so badly damaged that they couldn't tell whether the remaining bones belonged to a man or a woman; it was so horrible," his wife added. At the sight of her daughter's lingerie dangling from a nearby tree, Mary Jane Thornton became so unglued that she asked the sheriff to shoot them down.

Doctors and industry executives descended upon Atlanta from all over the globe, with many gathering to mourn at the home of Spencer and Gail King. Outside sat the little 1970 Volkswagen Beetle in which Andreas Gruentzig had first darted around Atlanta, a symbol of carefree days now vanished forever. Another cell of anguish coalesced at Gary Roubin's apartment, where the junior cardiologists, nurses, and technicians struggled to come to terms with the sudden annihilation of their medical Camelot. That night they found their way to a particular anthem with which to celebrate their fallen hero. This was a recently released ballad called "The Highwayman," sung in turn by the greats of American country and western music Gruentzig had so admired: Willie Nelson and Waylon Jennings, Johnny Cash and Kris Kristofferson. The mourners played the lyrics until deep in the night, while flipping through photographs that epitomized the swashbuckling nature of Andreas Gruentzig.

"It's about people who die seeking new horizons, and it's haunting," Roubin said, as he played the Jimmy Webb song one Wyoming night while showing photographs of Gruentzig in all his guises: grinning with his cocked surgical cap, dancing in his tuxedo, climbing aboard his plane . . . Looking always rakishly larger than life itself.

In the meadow outside, a herd of elk silently materialized in the darkness as Gary Roubin offered a taste from the same treasured bottle of Spanish brandy, Conde de Osborne, that Andreas Gruentzig had given him long ago. His eyes grew moist as he tried to explain how even the song's line about "the bastards hung me" had deeply resonated on the tear-stricken night of October 28, 1985, so incredulous was the gathering in his apartment about Gruentzig's death. "The sense that somebody 'got him' was in the air, because nothing about the story seemed right, even today it doesn't seem right . . . that it was calm weather but with no visibility," Roubin's gaze wandered as though he were looking to the elk for reason. "People who are in grief and distress can come up with far-fetched explanations for why somebody that you honored, who was the most technically expert, the most careful, the most brilliant figure, who could never make a mistake, yet could make this fatal mistake. It was beyond belief."

As it happened, Johannes Gruentzig would later take to proclaiming that his brother might have been somehow deliberately sabotaged. But the eventual Federal Aviation Administration and National Transportation Safety Board reports found not an iota of evidence in that regard, attributing the crash instead to pilot error.

The funeral at Macon's Ingleside Baptist Church was packed with more than 300 mourners as twin cadres of Andreas's and Margaret Anne's families and friends carried in the eerily light caskets. Eulogies hailed the pioneering spirit of the greatest medical figure anyone gathered that day had ever known, along with the glowing ebullience of his young wife. Spencer King summoned up the recent reappearance of Halley's Comet, last seen on the day of Mark Twain's death on April 21, 1910, in this eloquent allusion: "In this year of the comet, we are reminded that he came into our lives like a comet and burned briefly but brightly. Oh, how this light has ignited the imagination of others throughout the world. All future therapies for the treatment of obstructive coronary and peripheral arterial disease will be built on the foundation he laid."

Gruentzig's old Swiss friend Felix Mahler later summarized the catastrophe even more succinctly. "It was a pure tragedy. The higher you fly, the bigger is the tragedy. That is his life story—a Shakespearean tragedy."

The remains of Andreas Roland Gruentzig, forty-six, and Margaret Anne Thornton, twenty-seven, were buried that afternoon of October 30 at a cemetery a couple of miles down the road in Macon, Georgia. Hard by the ceaseless roaring of Interstate 75 and at the end of a stretch of flashing neon fast-food outlets, it seems an ignoble resting place. The question had in fact been raised whether Gruentzig should more properly be lain to rest in his native Germany. But he had no meaningful home there, so his family consented to the Macon burial and his mother merely brought a vial of German soil to toss over the casket. Andreas's brother said at the funeral, "They lived together, they loved together, and they died together. And they should be buried together." Michaela, magnanimously, did not object.

Despite that grace note, bitter acrimony lay ahead, with lawsuits against Beechcraft and between Michaela (on behalf of Michaela's daughter Sonja) and the Thorntons dragging on for the next seven years. At issue was an estate valued at $16 million, its proper disposition confused by Gruentzig's cavalier failure to complete a will, or at least one that anyone could ever locate. Other complications materialized with the appearance of his illegitimate daughter, Katrin Hoffman, as an additional plaintiff. Despite the NTSB's verdict regarding pilot error, Beechcraft settled, and the mess, which enriched nearly every major Atlanta law firm, was finally resolved in October of 1992. Sonja Gruentzig received the lion's share of the liquid assets, Katrin Hoffman a third, and the Thorntons half the proceeds from the real estate. By that point, Charlotta Gruentzig's own ashes had been dropped out of a plane over the North Sea, as was the final wish of this woman who had seen so much grief and dislocation in her life—all beginning in the sky.

Gruentzig's legacy would be celebrated many times in the following months and years, with a huge ceremony at Emory followed by another tearful one at the next annual meeting of the American Heart Association in November of 1985. At Emory, he is remembered by plaques and the renaming of the cardiovascular center in his honor. Alas, his legacy is all but forgotten at the University Hospital in Zürich, save by a few devoted friends and colleagues from the kitchen-table days. Michaela Gruentzig still sees troubled patients and dines in the same apartment where her husband's contraptions were first stitched together. Sonja has grown into a young beauty performing as a classical actress to acclaim in Berlin, while Katrin Hoffman, since renamed Gruentzig, traverses between a number of pursuits.

Meanwhile, Andreas and Margaret Anne's star-crossed Sea Island cottage was purchased by the chief of pediatric medicine at Emory University, George Brumley. In 2003, he and three generations of his family all succumbed in the crash of a private plane transporting them across Kenya to a safari. Then again, a daring young aviator named Paul Redfern had once flown off their very beachfront and Andreas Gruentzig's to attempt the first North American flight to Brazil, only to crash to another tragic death.

Sadly, the world of medicine is beginning to lose its collective memory of the great "highwaymen" of the human heart—from Werner Forssmann to Mason Sones, and Charles Dotter to Andreas Gruentzig—who defied convention and risked all for the sake of saving lives. The quiet celebrations of their legacy at the end of 1985 soon gave way to a rush forward into countless variations on the original Gruentzig idea, using lasers, ultrasound, vibrating energy, "Roto-Rooters," and especially wire coils called stents. Heart surgeons saw their incomes drop by half as patients flocked to the kinder alternatives Gruentzig introduced, and today, despite performing countless still-vital procedures, they cannot even find enough candidates to replenish their ranks without wholesale recruitment overseas. Meanwhile, Bard/USCI fell by the wayside once the FDA pounced on the company for pushing its medical devices forward without due reporting of cases gone wrong. Yet competitors, now numbering in the hundreds, jumped into the stream, and billions of dollars were ultimately made off refinements of an idea born on a kitchen table in Zürich. For all the field's excesses, millions of lives have also been extended or spared wrenching pain as a result of all this innovation and enterprise. One thing for sure is that any patient with access to modern medicine would never be left to languish as President Dwight D. Eisenhower once did.

Yet for those who were close to the inspired arc of Andreas Gruentzig's path, all of these facts remain somehow dry. They mourn to this day the loss of the last of the great explorers of the human heart, the man who has left doctors around the world turning their surgical caps into berets without even knowing why.

END NOTES

INTRODUCTION:

2 **Mason Sones, "Doing research is like kicking down a door:"** *The Miracle Finders: The Stories Behind the Most Important Breakthroughs of Modern Medicine,* Donald Robinson, Chapter 6, "The Way to a Human's Heart," David McKay Company, Inc. New York, 1976, p. 178.

4 **$105 million gift to Emory:** The amount of this gift in Coca-Cola stock has been validated by many sources, including Emory University's Web site's discussion of the history of the Woodruff Health Sciences Center.

4 **"Omar Sharif, Clark Gable":** Author interview with Emory University research nurse Claire Rice, December 19, 2003.

4 **$300 billion industry:** U.S. Department of Commerce data indicated that the global market for medical devices in 2000 was $169 billion, a figure reported in such venues as the University of California, Irvine, Biomedical Conference on November 12, 2004, and the Web site of *Medical Device & Diagnostic Imaging* magazine. The medical device industry's sales are widely believed to have grown by approximately 10 percent a year since 2000. A Reuters overview of the medical equipment and supplies industry put 2004 sales for this sector at $240 billion, with 10 percent of this total accounted for by cardiology and diagnostics equipment. The latter market is certain to remain particularly lucrative, since 200 million people have clinically diagnosed cardiovascular disease today, with 16.7 million of them dying from it each year, according to the World Health Organization's *2004 Cardiovascular Global Atlas.*

CHAPTER ONE:

5 **Charles II background:** *Devils, Drugs and Doctors: The Story of the Science of Healing from Medicine-Man to Doctor*, Howard W. Haggard, M.D., Halcyon House, New York, 1929, pp. 334–336. *Explorers of the Body,* Steven Lehrer, M.D., Doubleday, New York, 1979. (This fascinating account of some of the more bizarre and significant developments in the history of medicine is still available on the Internet.) *The Story of Medicine*, Albert & Charles Boni, New York, 1931, Chapter 9, "Medicine in the Seventeenth Century," by Victor Robinson, M.D., pp. 28–29. *Magic, Myth, and Medicine,* D.T. Atkinson, M.D., Fawcett A World Library, 1958. Also, "The Mistresses of Charles II," Brenda Ralph Leis, Brittania.com, LLC, 1999: www.britiannia.com/history/charmist.html and www.bbc.co.uk/history/historic_figures/charles_ii_king.sht ml.

6 **Pope Innocent and fatal blood transfusion:** *Transplantation,* Chapter 19, "Historical Aspects of Transplantation," by R. Randal Bollinger, M.D., and Delford L. Stickel, M.D., pp. 370–379. *The Rebuilt Man: The Story of Spare Parts Surgery*, by Fred Warshofsky, Thomas Y. Crowell Co., New York, 1965.

7 **Christopher Wren:** *Exploring The Heart: Discoveries in Heart Disease and High Blood Pressure*, by Julius H. Comroe, Jr., M.D., W.W. Norton & Company, New York, 1983.

7 **Lower's booze hound:** *Explorers of the Body,* Steven Lehrer, M.D.

7 **Jean-Baptiste Denis:** *Magic, Myth, and Medicine*, "Transfusion of Murder: Jean-Baptiste Denis," pp. 232–242. Also discussed in *The Rebuilt Man* by Fred Warshofsky.

7 **Eisenhower's heart attack:** *Eisenhower's Heart Attack: How Ike Beat Heart Disease and Held on to the Presidency,* Clarence G. Lasby, University Press of Kansas, 1997. Also, "Eisenhower's Billion-Dollar Heart Attack—50 Years Later," Franz H. Messerli, M.D., Adrian W. Messerli, M.D., and Thomas F. Lüscher, M.D., *New England Journal of Medicine*, Vol. 353, No. 12, September 22, 2005, pp. 1205–1207; and "Amazing Changes in Heart Care: If 'Ike' Had a Heart Attack Today," University of Colorado Health Sciences Center press release, December 21, 2003.

Also see *Heart Failure,* Thomas J. Moore, Random House, New York, 1989, Chapter 9, "A Mirror for Medicine," pp. 150–151; and see article, "Shorter Stay for Myocardial Infarction," Gregory D. Curfman, M.D., in *New England Journal of Medicine,* Vol. 318, April 28, 1988, pp. 1123–25.

Many sources elucidate the fact that the American president was long wracked with advanced cardiovascular disease, among other health problems, which he somehow managed to prevail against for years. In November of 1965, Eisenhower suffered a second major heart attack, leaving him bedridden for five weeks, and endured subsequent gallbladder surgery and chronic ileitis, followed by another two heart attacks in the next three years. Medical historians believe Ike may have weathered a total of seven heart attacks, before his final one killed him on March 28, 1969, at age seventy-nine, at about the very time modern medicine was discovering methods for repairing obliterated coronary arteries.

9 **Early Alexandrians**: Cited in *Explorers of the Body,* Steven Lehrer, M.D., Doubleday, New York, 1979, Chapter One, "Muscle and Blood."

9 **Galen, Vesalius and Servetus:** *Mavericks, Miracles, and Medicine: The Pioneers Who Risked Their Lives to Bring Medicine into the*

Modern Age, Julie M. Fenster, Carroll & Graf, New York, 2003, "The Art of Medicine: Andreas Vesalius," pp. 3–17. *The Scalpel and the Heart,* by Robert G. Richardson, Charles Scriber's Sons, New York, 1970, Chapter 2, "The Circulation," pp. 11–19. *Heart Failure,* Thomas J. Moore, Random House, New York, 1989, Chapter 2, "Hearts Then and Now," pp. 14–18. *Doctors: The Biography of Medicine,* Sherwin B. Nuland, Vintage Books, New York, 1988, Chapter 3, "The Reawakening: Andreas Vesalius and the Renaissance of Medicine," pp. 61–93.

10 **William Harvey:** *Bypass,* Jonathan L. Halperin, M.D., and Richard Levine, Times Books, New York, 1985. *Magic, Myth, and Medicine,* "Master of the System: William Harvey," pp. 244–256. *A Man After His Own Heart,* Charles Siebert, Crown Publishers, New York, 2004, Chapter 16, pp. 210–236. See *The Scalpel and the Heart,* and *Exploring the Heart: Discoveries in Heart Disease and High Blood Pressure,* pp. 280–292; and *Doctors: The Biography of Medicine,* Sherwin B. Nuland, Chapter 5, "Nature Herself Must Be Our Advisor," pp. 120–144.

11 **Harvey's underground "thinking chamber":** *Explorers of the Body,* Stephen Lehrer, Chapter One.

11 **"I found the task so truly arduous:"** *The Story of Medicine,* Victor Robinson, 1931.

11 **Henry Power (1623–68) quotation:** Cited in *Doctors: The Biography of Medicine,* Sherwin B. Nuland, Vintage Books, New York, 1988, p. 136.

12 **Stephen Hales's experiments:** *Magic, Myth and Medicine,* 1958.

12 **Priestly and Lavoiser:** *Exploring the Heart.*

13 **Screaming operating theaters:** *Explorers of the Body.*

13 **Insanity of anesthesia pioneers:** *To Mend The Heart,* pp. 114–117. Also. *Mavericks, Miracles, and Medicine,* Chapter 15, "Ether Frolic:

Horace Wells, William T.G. Morton, and Charles Jackson," pp. 199–215. See particularly pages 214–215:

> It is odd to think that in delivering millions of people from suffering, they brought so much of it on themselves: All three died in a state of madness. Disappointed by the failure of nitrous oxide to rival ether as an anesthetic . . . Wells turned to chloroform, testing it on himself—not with the self-control of a scientist, but with the abandon of an addict. During a bender in New York in 1848, he was arrested for throwing acid on a prostitute. While in jail, he came to the sad and self-perpetuating realization that he was no longer sane. He wrote lucidly of his condition to his wife and then took his life by slitting a major artery in his leg, taking care to anesthetize himself first.
>
> Morton also died in disarray in New York City . . . He waited until the doctors left his home and then set off in a frenzy that ended when he jumped out of a moving buggy to plunge his head into a lake in Central Park. Dragged out of the water, he fell unconscious and died at the hospital the same day.
>
> Jackson went more quietly . . . He was committed to an insane asylum near Boston, where he died in 1880, after seven years' confinement.

As quoted in Steven Lehrer's *Explorers of the Body*, Doubleday, 1976, an early investigator of nitrous oxide, the English chemist Humphrey Davy, had a giddy time of it himself, turning his friends Samuel Taylor Coleridge and Peter Mark Roger, compiler of the Thesaurus, on to nitrous oxide. He also penned this mid-1795 ode to his self-experiments:

Not in the ideal dreams of wild desire
Have I beheld a rapture-wakening form:
My bosom burns with no unhallow'd fire,
Yet is my cheek with rosy blushes warm;
Yet are my eyes with sparkling lustre fill'd;

Yet is my mouth replete with murmuring sound;
Yet are my limbs with inward transport fill'd;
And clad with new-born mightiness around.

13 **The history of heart surgery:** For overviews of the long struggle to begin to understand and repair the human heart, see *To Mend the Heart,* Lael Wertenbaker, Viking, 1968, Chapter 4, "A Long, Long Road, pp 47–68, (including **Stephen Paget** quotation, p. 50). Also *The Story of Medicine,* Chapter Nine, *The Scalpel and the Heart,* by Robert G. Richardson, Charles Scribner's Sons, New York, 1970, is revealing in Chapter 3, "Wounds of the Heart," pp. 20–33, especially with reference to dismal nineteenth century success rates and despair of that era's surgeons. See further, *Great Ideas in the History of Surgery,* by Leo M. Zimmerman, M.D., and Ilza Veith, Ph.D., Dover Publications, Inc., New York, 1961.

13 **Block suicide:** *The Scalpel and the Heart,* page 26.

13 **Billroth quotation:** *The Scalpel and the Heart,* page 28.

CHAPTER TWO:

14 **Dates of early twentieth-century developments:** *The New York Public Library Book of Chronologies,* Bruce Wetterau, Simon and Schuster, New York, 1990.

14 **Wilhelm Roentgen:** *Mavericks, Miracles, and Medicine*, Chapter 2, "A Peculiar Light; Wilhelm Roentgen," pp. 19–33.

15 **Alexis Carrel:** *The Culture of Organs,* by Alexis Carrel and Charles A. Lindbergh Paul B. Hoeber, Inc., Medical Book Department of Harper & Brothers, New York, 1938. See also, *Transplantation,* Chapter 19, "Historical Aspects of Transplantation," by R. Randal Bollinger, M.D., and Delford L. Stickel, M.D.

15 **Forssmann's early (and later) years.** *Experiments on Myself: Memoirs of a Surgeon in Germany.* Werner Forssmann, translated by Hi-

lary Davies, preface by André Cournand, Saint Martin's Press, New York, 1974. Particular references from this text include: "I let a few days go by," p. 84; "With the speed of light," p. 85; "I had a mirror placed," p. 85; and "there are hunters and shooters," p. 99. Original German edition 1972, Droste Verlag GmbH, Düsseldorf. Also see, article by Forssmann's daughter Renate Forssmann-Falck, M.D., of Richmond, Virginia, about her father, "Werner Forssmann: A Pioneer of Cardiology," in the *American Journal of Cardiology,* Vol. 79, Issue 5, March 1, 1997, pp. 651–660.

15 **Conditions in Germany during and immediately after World War I:** *A History of Modern Germany: 1840–1945,* Hajo Holborn, Alfred A. Knopf, New York, 1969; *The First World War,* John Keegan, Knopf, New York, 1999; *Einstein in Berlin,* Thomas Levenson, Bantam Books, New York, 2003.

The latter narrative observes that the economy in Germany grew so disastrously inflationary that by 1922–23, children were makings towers of blocks of paper money, while conditions in the Berlin slums reached appalling levels. The author notes that the licentious decadence of the era meanwhile reached such extremes in Berlin that Josephine Baker became the darling of the town as she strutted about all-but-nude in the notorious El Dorado nightclub; while the mostly naked "Haller Girls" arrayed themselves on stage as a debauched pantomime of the iconic statuary above the Brandenburg Gate. Transvestites flooded the streets at night, and nudist retreats flourished. The English poet Stephen Spender later remarked that the excesses of that Berlin era produced "dragon's teeth . . . poised at the heart of Europe."

16 **Norwegian 1901 heart experiments, followed by the work of Friedrich Trendlenburg in Leipzig in 1908:** *The Scalpel and the Heart,* pp. 55–56, 64–67, 80–83, 91–92, 97–105.

17 **Incredible pumping power of the heart:** For elaborations on these statistics, see *Bypass,* by Jonathan L. Halperin, M.D., and Richard

Levine, p. 28, in which the authors note that the heart beats 115,200 times a day, 42,048,000 times a year, 2,943,360,000 times in seven decades, and that this is "enough to fill the Yale Bowl or lift the battleship *New Jersey* fifteen feet out of the sea." On page 30, the authors offer a wonderfully succinct portrait of the organ's workings:

> The spent, used up blood then travels back to the heart, arriving at the right atrium, the body's drain. As that chamber contracts, the oxygen-poor, indigo blood is transported through the one-way gates of the tricuspid valve into the right ventricle. The ventricular contraction forces the blood through the pulmonic valve into the lungs, where it flows through 600 million capillaries around 300 million air sacs. Here the haemoglobin shifts into reverse and trades its carbon dioxide—the lungs will exhale it—for a new supply of life-sustaining oxygen. The now vermillion blood flows into the left atrium, through the mitral valve and into the left ventricle, the heart's main pumping chamber. When its powerful walls contract, the blood is propelled through the heart's main exit, the aortic valve, and out into the body. All this opening and closing is executed "so harmoniously and rhythmically," William Harvey observed, that "only one movement can be seen."

Also see *The Miracle Finders,* Donald Robinson, David McKay Company, Inc., New York, 1976, Chapter 6, "The Way to a Human's Heart," pp. 161–162.

17 **Horse catheter experiments by Chaveau:** The work that captivated Forssmann is mentioned in his autobiography, with further amplification on the experiments of Claude Chaveau and Etienne Marey in *The Scalpel and the Heart,* pp. 108–109.

19 **Henry Souttar:** *To Mend a Heart,* Chapter 5, "Milestones," pp. 69–77, gives substantial background about the first attempts to physically manipulate heart halves by the likes of Elliot Cutler and Souttar, making it clear how such new daring set the table for the bravado of Forssmann.

23 **Ferdinand Sauerbruch:** *Master Surgeon,* Ferdinand Sauerbruch, translated by Fernand G. Renier and Anne Cliff, Thomas Y. Crowell Company, New York, 1954. Also see, *The Scalpel and the Heart,* pp 88–91; and *To Mend a Heart*: Chapter 8, "Ten Minutes to Look," pp. 113–130. On Page 116, the latter narrative describes the late nineteenth century explorations by Paul Bert in Paris of primitive, room-sized, pressurized operating chambers:

> During the late 1800s Paul Bert in Paris conceived the idea of a primitive room-sized pressurized operating chamber. Two men pumped lustily to keep the atmospheric pressure in the chamber high, increasing the nitrous oxide effect and allowing adequate oxygen while patients were unconscious inside it. Bert was certified insane by the French authorities, labelled a criminal maniac, and forbidden his experiments. He put his chamber on wheels and moved it ahead of the police around the environs of Paris. Plenty of patients were glad to evade the law and have fractures set, amputations performed or teeth pulled without pain, and they followed him on his peripatetic rounds.

26 **Cournand and Richards:** *The Scalpel and the Heart,* pp. 111–113; and "Cournand & Richards and the Bellevue Hospital Cardiopulmonary Laboratory," by Yale Enson and Mary Dickinson Chamberlin, *Columbia Magazine,* Fall 2001.

27 **Forssmann and the Nazi party:** See letters, *American Journal of Cardiology,* December 15, 1997, Vol. 80, Issue 12, pp. 1643–1647; and *Magic, Myth and Medicine,* "Picture of Youth: Werner Forssmann," pp. 35–48. In *Experiments on Myself,* Forssmann describes his escalating disenchantment with the Nazis as the 1930s wore on, culminating in his witnessing, and trying futilely to intervene against, SS men indiscriminately shooting 600 Russian peasants on the Eastern Front on Whitsunday, 1942.

27 **Forssmann's Nobel Prize:** The official Nobel Prize Web site still carries his acceptance speech from December 11, 1956.

CHAPTER THREE:

28 **Gibbon, Lillehei, Carrel, Lindbergh and the story of the heart-lung machine:** *Magic, Myth, and Medicine,* "Long Way to Bypass: John H. Gibbon, Jr.," pp. 257–268; and. *Exploring the Heart,* Chapter 24, "Heart-Lung Machines," pp. 221–234. Also see *The Scalpel and the Heart,* and *The Artificial Heart,* Melvin Berger, Franklin Watts, New York, 1965. Further background in *The Risk Takers,* Hugh McLeave, Holt, Rinehart and Winston, New York, 1963; and *The Miracle Finders,* Donald Robinson, David McKay Company, New York, 1976.

29 **Lindbergh and Carrel:** Web site, www.charleslindbergh.com/heart/index2.asp.

30 **Mason Sones's childhood and early years:** Key information came from an interview by the author at the Cleveland Clinic with Emeritus professor William Proudfit, October 21, 2003, as well as from a variety of materials in the Cleveland Clinic Foundation's archives.

30 **Cardiology "a nothing specialty":** *The Miracle Finders,* Donald Robinson, 1976, page 177.

31 **Sones's typical appearance:** Proudfit, Shirey, Clayton interviews.

31 **Sones with pediatric patients and uproarious behavior with staff:** Author interview with his long-time secretary, Elaine Clayton, October 21, 2003. Also, William Proudfit, same day.

31 **Sones publicly erupting at major medical meetings:** Proudfit interview on October 21, 2003. Also, *The Superdoctors,* bibliographic info found on p. 301, p. 104, and p. 100.

At another conference, Sones heard a doctor from another institution acknowledge that his mortality rate in diagnostic angiography was over 1 percent and erupted: "Any son of a bitch who has that kind of mortality shouldn't have his hand on a catheter for the rest of his life."

32 **On conflicts with Don Effler and Willem Kolff:** Proudfit interview, October 21, 2003.

33 **Sones's antics while smoking:** Cleveland Clinic interviews with Proudfit, Clayton, and Early Shirey, October 21, 2003. See also, *The Superdoctors*, Roger Rapoport, Playboy Press, Chicago, 1975, Chapter Six, "Cough," page 81.

34 **Sones's drinking:** Proudfit, Shirey, and Clayton interviews, October 21, 2003, and numerous other sources, including Werner Niederhauser interview by the author, Bülach, Switzerland, September 16, 1994; and Dave Prigmore interview by the author and David Williams, July 21, 2004.

34 **"That son of a bitch Kolff":** Proudfit interview above.

35 **"Shirey, I want to talk with you":** Earl Shirey interview with the author, October 21, 2003.

35 **Sones's collaborations with industry:** "Cinematic Medicine: Method to Take Movies of Coronary Arteries Brings Diagnostic Gains," *Wall Street Journal*, Jerry Bishop, December 22, 1964, page 10.

36 **Sones's early imaging of heart:** Sones letter to E. P. Roy of the John A. Hartford Foundation, Inc., April 17, 1963; and various other sources from the Cleveland Clinic Foundation archives.

36 **Sones's historic breakthrough:** Interviews by the author with Proudfit, Shirey, Clayton, October 21, 2003. *The Superdoctors*, Chapter 6, "Cough," pp. 90–106. *Saving the heart: The Battle to Conquer Coronary Disease*, Stephen Klaidman, Oxford University Press, New York, 2000, pp. 51–62.

37 **"We just revolutionized cardiology":** Elaine Clayton interview, October 21, 2003.

38 **Sones being shortchanged by USCI company:** Joint author/David Williams interview with former Bard/USCI group president Dave

Prigmore at his Rhode Island home, July 21, 2003; author interview with John Cvinar, former USCI president, in Cambridge, Massachusetts, May 22, 2004, and joint author/David Williams interview with Marcia Schallehn, former USCI clinical education director, in Providence, Rhode Island, April 16, 2003.

39 **Sones's cautious documentation of his procedure:** Countless academic articles and biographical sketches in medical journals attest to his bedrock scientific integrity, including: "Selective Cine Arteriography: Correlation with Clinical Findings in 1,000 Patients," William L. Proudfit, M.D., Earl K. Shirey, M.D., and F. Mason Sones, M.D., *Circulation* , Volume XXXIII: June 1966, pp. 901– 910; "Clinical Course of Patients with Normal, and Slightly or Moderately Abnormal Coronary Arteriograms: A Follow-up Study on 500 Patients," Albert V. G. Bruschke, M.D., William L. Proudfit, M.D., and F. Mason Sones, Jr., M.D., pp. 936–945, *Circulation,* Vol. XLVII, May 1973; "Progress Study of 590 Consecutive Nonsurgical Cases of Coronary Disease Followed 5–9 years," *Circulation* XLVII June 1973, Albert V.G. Bruschke, M.D., William L. Proudfit, M.D., and F. Mason Sones, Jr., M.D., pp. 1154–1163. Also illustrative was "F. Mason Sones, Jr., M.D. (1918–1985): The Man and His Work," William Proudfit, M.D., *Cleveland Clinic Quarterly*, 121–124, Summer 1986.

39 **Dyslexic Sones:** Proudfit interview, October 21, 2003.

40 **Sones's neglect of his family:** Proudfit and Clayton interviews.

40 **"If you have to go home in middle of day":** Clayton interview, October 21, 2003.

40 **Issue of no Institutional Review Board to restrain experiments:** *Saving the Heart: The Battle to Conquer Coronary Disease,* Stephen Klaidman, Oxford University Press, New York, 2000, various pages.

41 **"Quint! Get up off your lazy ass!":** Proudfit interview, October 21, 2003. Also, *The Superdoctors,* p. 98.

41 **Kaltenbach studies under Sones:** Author interview with Martin Kaltenbach, Frankfurt, Germany, September 4, 2003.

41 **"I am being paid 12 cents an hour":** Author interview with Arnoldo Feidotin in his suburban Atlanta, Georgia, home, February 4, 2004.

42 **Melvin Judkins background:** Author interview with Josef Rösch, M.D., at the Dotter Vascular Institute at the Oregon Health Sciences Center University, March 19, 2004.

42 **On the birth of coronary bypass surgery and its cruder precursors by Beck and Vineberg:** *To Mend the Heart,* pp. 189–191; *The Scalpel and the Heart,* Chapter 12, "New Blood Supply for the Heart," pp. 126–137; *Saving the Heart: The Battle to Conquer Coronary Disease.* pp. 48–50.

43 **René Favaloro's introduction to Cleveland Clinic:** *The Challenging Dream of Heart Surgery,* René G. Favaloro, The Cleveland Clinic Foundation, pp. 1–50. Also William Proudfit interview, October 21, 2003.

43 **Favaloro's early personal struggles and apprenticeship in Cleveland:** *The Challenging Dream of Heart Surgery;* also *Saving the Heart,* Chapter 5, "Accidents and Innovations," pp. 63–73.

44 **Early one-time-only performers of bypass surgery:** Proudfit interview, October 21, 2003.

44 **First Cleveland Clinic attempt at bypass operation:** *The Challenging Dream of Heart Surgery.*

45 **Jubilation after the procedure:** Interviews with Proudfit, Shirey, and Clayton, October 21, 2003. "New Operations Hold Promise of Normal Life For Numerous Patients," *Wall Street Journal,* May 17, 1966.

46 **Sones travel to Brazil and Berlin:** Numerous pieces of correspondence and actual invoices in the Cleveland Clinic's archives document Sones's far-flung travels, often with either Effler or Kolff, or both, alongside.

46 **Staff calling their road show a vaudeville act:** Elaine Clayton interview.

46 **Sones invitation to Worshipful Society of Apothecaries of London:** Proudfit interview, and July 11, 2006, e-mail from A. M. Walling-Smith, clerk, Worshipful Society of Apothecaries of London. Cleveland Clinic records show that he was then so sick with lung cancer that the Galen Medal had to be presented to him at his home. In November of 2003, Sones was also awarded with the prestigious Albert Lasker Award. A transcript of his acceptance speech, following an introduction by the famous cardiac surgeon Michael DeBakey, reads:

> I'm glad they carefully defined for you the fact that this is a clinical award. God knows I'm not a scientist. I'm a doctor. I thoroughly relate to humans in trouble, particularly humans with coronary artery disease. All I've managed to do was follow my own hobbies. I thoroughly enjoyed techniques for pushing little tubes around in the human vasculature, especially since it really didn't hurt much. And I've always been a camera nut. And faced with the requirement of using an X-ray tube instead of sunlight or an incandescent bulb. What a fascinating time it was to develop high-speed X-ray-motion picture photography—60 X-ray pictures a second! My God, you can slow down what goes on inside the pump and even an idiot can sit there and look it over and over again and finally figure out what the hell's going on. So I'm most grateful for this experience. It's awe-inspiring. I might even go home and go back go work. Thank you very, very much.

See also "Laskers to Giants of CAD Portrayal, Vaccines, and Learning," Nathan Horwitz, *Medical Tribune*, November 23, 1983, p. 3; "Cleveland Clinic for World's Powerful," Richard D. Lyons, *New York Times,* May 8, 1984, Section C; page 2.

47 **Stouffer Medal tragic-comedy:** Author interview with cardiologist Ralph Lach, M.D., in his Upper Arlington, Ohio, home, October 22, 2003, followed by e-mail correspondence April 14, 2006. While dying of lung cancer in 1985, a gaunt and wasted Mason Sones was asked by a fellow patient why he was in acute care, what with his being such a prominent doctor. According to a biographical sketch in the journal *Clinical Cardiology* 17:405–407 (1994), "F. Mason Sones, Jr.—Stormy Petrel of Cardiology," by his colleague William Sheldon, M.D., Sones silenced the interlocutor by roaring: "I have AIDS!"

CHAPTER FOUR:

49 **Gruentzig's essential biographical facts:** Interviews by the author with widow Michaela (Seebruner) Gruentzig in her Zürich apartment, September 20, 1994 and November 15, 2003; also multiple interviews with his former research assistant Maria Schlumpf, the final two conducted with David Williams in Zürich, November 18–19, 2003; the former in September of 1994, and September 8, 2003.

Also, a German radiologist named Andreas Beck recently self-produced a substantially-researched monograph entitled *Andreas Grüntzig: Eine Idee Verändert Die Medezin,* ("An Idea That Transformed Medicine"), Clio-Verlag, Konstanz, Germany, undated, apparently 2001. Chapter 11, "Grüentzig's frühe Jahre," (Gruentzig's early years), pp. 123–139, is most pertinent biographically.

50 **Gruentzig's father's role in the Luftwaffe:** Numerous figures who became close to Gruentzig heard occasional, brief stories about his father's role in the Luftwaffe, but there were conflicting accounts as to whether Wilmar Gruentzig was in fact a pilot or otherwise employed in weather reconnaissance. Sources on this issue include Michaela Gruentzig, Maria Schlumpf, Katrin Hoffman, Spencer King, and Richard Myler.

50 **Dresden at the dawn of World War II, Berlin at its end, and looters rampant in German chaos:** See *Berlin—the Downfall 1945,* Antony Beevor, Penguin, London, 2004.

50 **Gruentzig family flight to Rochlitz:** "Angioplasty From Bench to Bedside to Bench," Spencer B. King III, MD, *Circulation* 1996; 93: 1621–1629. In this generally biographical sketch, based on years of close familiarity with Gruentzig, King describes the family's sequestering in a house that would later become the headquarters of the American commander. Research suggests this was the Villa Carolla, images of which can be found on the Internet.

50 **"Plate Rack Force:"** "Was Dresden an atrocity?" book review by Art Shay of Frederick Taylor's *Dresden: Tuesday, February 13,* 1945 in the *Chicago Sun-Times,* February 8, 2004. Also see, "Firebombing: A new look at the destruction of Dresden," by Gabriel Schoenfeld, Sunday *New York Times* book review section, p. 20, May 2, 2004.

50 **Pincer of Russian and U.S. armies to Rochlitz:** Soldier William J. Given account (with reference to *New York Times* publishing the core description on April 23, 1945) and other background facts from "The Brief American Occupation of Parts of East Germany, 1945," and "Meeting with the Russians, Rochlitz in 1945," www.koch-athene.de/6th/rochlitz/russians.htm.

51 The Hotel Goldener Löwe in the center of Rochlitz was definitely the site of a historic meeting between commanders of the Russian and American armies, whose agreement on that day changed the fate of the entire region, and of the Gruentzig family, for decades.

51 **Andreas's mother's post-war struggles, and her waiting for trains of returning ghost soldiers:** Interviews with Michaela Gruentzig and Maria Schlumpf.

51 **Three generations of Gruentzig's ancestors killed in defense of their homeland:** Author interview with Michaela Gruentzig, November 15, 2003. On this day, Michaela was met by the author in the Tiefenau Hotel and the interview continued with a stroll to her apartment, followed by lunch in a corner restaurant where she and Andreas frequently dined.

51 **Forssmann's travails at war's end:** In *Experiments on Myself,* Forssmann recollects that he was first incarcerated at a place called Herford, where 50,000 Wehrmacht soldiers were held in a single soccer stadium. On May 3, 1945, the future Nobelist was transferred to a barb-wired field in Büderich to join 100,000 other captives, where GIs dispensed food rations into their prisoners' hats, often consisting of powdered eggs, dehydrated rutabaga, and a spoonful of dried milk, supplemented by a dollop of fat. At night, the exhausted former conquerors of the Wehrmacht leaned against each other in human pyramids, so as to keep upright above a sea of excrement—which, according to Forssmann's calculations, might be replenished by as much as fifteen to twenty tons of new feces every day. Denton Cooley at this time was about to embark on occupation duty in the region, where he recalls in his biography, *Cooley: The Career of a Great Heart Surgeon,* Harry Minetree, Harper & Row, New York, 1973, being feted with much champagne.

52 **Charlotta Gruentzig careering about with her sons:** Interview with Katrin Hoffman Gruentzig in Zürich, September 8, 2003; Michaela Gruentzig, November 15, 2003.

52 **Traumatic early experiences of Andreas Gruentzig:** The story about gathering mushrooms was told by Gruentzig to Spencer King; the boyhood trauma of being teased with live bullets and sold ruined stockings at the market was told by him to Richard Myler.

52 **Flight to Argentina:** This information was outlined to the author by Michaela Gruentzig on November 15, 2003, and corroborated by Maria Schlumpf, and is also iterated in the Andreas Beck book.

52 **Return to Leipzig's Küstner Strasse and attendance at Thomas Schüle:** Andreas Gruentzig's brother Johannes reviewed this period in an unscheduled phone discussion in August of 2003, at which time he expressed strong reservations about being interviewed in person unless the subject of this book was changed into an investigation of the circumstances of his brother's ultimate death. Johannes continued on this same theme in subsequent letters. The parents of Andreas Gruentzig's second wife, Earl and Mary Jane Thornton, corroborated and expanded upon certain details, as did Michaela Gruentzig and Maria Schlumpf.

53 **Travails of student years in East Germany:** Eberhard Zeitler, a fellow refugee from a nearby region in East Germany, provided an overview of what it was like to be a student in the DDR in the same era, recorded during a two-day interview with the author in Zeitler's apartment and other locations in Nürnberg, Germany, on September 5–6, 2003.

53 **Early school excellence and flight to the West:** Personal interviews with Michaela Gruentzig and Maria Schlumpf.

54 **Plight of the Ostis in the West:** From the Zeitler interviews, as well as a interviews with Bernhard Meier, director of cardiology at the Canton Hospital in Bern, Switzerland, conducted alternately by David Williams and the author in early and mid-November of 2003, along with an interview of Meier by the author on September 16, 1994.

55 **The life of the Gruentzig brothers in Heidelberg:** Interviews with Michaela Gruentzig, Maria Schlumpf, Bernhard Meier, and Eberhard Zeitler, as cited.

55 **Their escapade to Spain:** Michaela Gruentzig interview, November 15, 2003; also a viewing of her personal photographs.

56 **Gruentzig's contested fatherhood:** Copy of *Sozial und Jugendamt* form from Heidelberg, 31 July, 1968, acknowledging paternity, on file in Fulton County, Georgia, Courthouse: Vol. 820, pp. 214–221. Katrin's mother was Ulrike Sieglinde Friedemann, then a student residing at Hauptstrasse 174, Heidelberg. The daughter is sometimes called Katrina. The mother married and became a teacher, changing her name to Hoffman and taking residence in Sandhausen, according to Fulton County records, Vol. 820, pp. 206–213.

56 **Gruentzig's first meetings with Michaela:** Michaela Gruentzig interviews on September 20, 1994, and November 15, 2003.

57 **Michael DeBakey background:** *Hearts,* Thomas Thompson, Fawcett Publications, Greenwich, Connecticut, 1971. Also, *To Mend the Heart,* Chapter 12, "The Texan Titans"; and *The Superdoctors,* Chapter Seven, "You Can't Handicap Cardiac Surgery Like You Do Golf," 108–122; plus, *The Living Heart,* Michael DeBakey, M.D., and Dr. Antonio Giotto, M.D., David McKay Company, Inc., New York, 1977.

Among many awards, DeBakey received the Congressional Gold Medal, bestowed with exceptional exclusivity, in 2005, when he was then ninety-seven. In 1996, he performed quintuple bypass surgery on Boris Yeltsin, President of the Russian Federation, who called him "a magician of the heart."

57 **DeBakey's medical empire:** *Hearts,* pp. 12–20. See also *New York Times* five-page profile by Allen R. Myerson, February 13, 1994, of DeBakey and Cooley as dueling medical superstars, "It's a Business. No, It's a Religion: Rival surgeons personify the debate within medicine over how to deal with a cost-obsessed world."

57 **"Anybody else out there want an operation?":** *Hearts,* p. 293.

57 **"Early Bird" worldwide TV extravaganza:** *Hearts,* pp. 159–160.

58 **"You practice this procedure by circumcising gnats:"** *Hearts,* p. 111. Also, *To Mend the Heart,* pp. 194–204.

58 **Cooley's fame:** *Cooley,* p. 20.

58 **Denton Cooley's charisma:** *Cooley* Also, Thomas Thompson's *Hearts.*

58 **Cooley's house is named "Cool Acres:"** *Cooley,* p. 249.

58 **"Suture inside a matchbox":** *Cooley,* p. 124.

58 **"With either hand":** *Cooley,* p.124.

59 **Cooley "often obliged to experiment" in his patients:** *Cooley,* p. 124.

59 **"If that's an aneurysm, I'll eat it":** *Cooley,* pp. 141–142. Thomas Thompson, on p. 94 of *Hearts,* notes that Cooley's office boasted a wall plaque quoting Theodore Roosevelt:

> The credit belongs to the man who is actually in the arena, whose face is marred by dust and sweat and blood . . . who knows the great enthusiasms, the great devotions and who spends himself in a worthy cause. . . . who at the end at best knows the triumph of high achievement and at worse fails while daring greatly, so that his place will never be with those cold and timid souls who know neither victory nor defeat.

58 **"Would you like to see Michael DeBakey's face in the mirror?":** *Hearts,* p. 70.

59 **Christiaan Barnard background:** *One Life,* Christiaan Barnard and Curtis Bill Pepper, The Macmillan Company, Toronto, 1969. *The Scalpel and the Heart,* Chapter 30, pp. 292–302. The author also interviewed a then-arthritic and no-longer-operating Christian Barnard and a South African understudy, Otto Tahning, in a Paris hotel on September 29, 1994, at which time Barnard asserted, "I was kidnapped by NBC and went to do *Face the Nation.*" Other medical industry personnel, including Lee Hibbs of PLC Medical Systems, alleged at this meeting that Barnard always expected female companions to be provided for him on his paid-for consultancy trips. Barnard's wife's autobiography was rife with similar allegations.

60 **Barnard and two-headed dog:** *One Life,* pp. 233 and 249.

60 **Barnard on the first heart transplant:** "With each step the weight of my doubt grew": *One Life,* p. 290.

61 **Washkansky wanting "no Mickey Finns":** *One Life,* p. 303.

61 **"Washkansky's heart came into view":** *One Life,* p. 301.

61 **"We could see it with our eyes . . . a heart for history":** *One Life,* p. 309.

61–62 **"The collapsing heart":** *One life,* p. 313.

62 **"I had never seen a chest without a heart or such a hole":** *One Life,* pp. 313–314.

63 **The global race to transplantation:** *To Mend the Heart,* Chapter 13, "Second-Hand Hearts," pp. 205–233; *Cookey,* pp. 153–169.

64 **The dismal results of heart transplantation:** *Hearts,* Section 2, "The Transplant Era," pp. 158–243, with line-by-line summaries of every transplant center's results, pp. 292–315. *To Mend the Heart,* pp. 223–235, also provides details on the sorry statistics. *Cooley* on p. 155 cites Jacob Zimmerman, M.D., of New Jersey's St. Barnabas Hospital saying during a conference on the ethics of heart transplantation, even as Washkansky was dying: "It is medically and morally wrong for us as doctors to stand by a dying patient's bedside, hoping he'll get it over with quickly so we can grab his heart." See also, *The Superdoctors,* Chapter Eight, "I Left My Heart in Palo Alto," pp. 124–136, with a quotation on page 131 from Norman Shumway, M.D., on the surgical nightmare he had unwittingly unleashed: "It really turned into a kind of geopolitical contest. It seemed as if national prestige was on the line. Quite clearly, transplant operations were taking place in many areas that hadn't conducted the proper research [or] developed the right capabilities or supportive services."

65 **Bitter fall out from the Haskell Karp procedure:** See *Cooley,* pp. 223–243; *To Mend the Heart,* pp. 201–203, and *The Superdoctors,* p. 112. In *Hearts,* pp. 226–235, Thomas Thompson discusses in detail how Cooley himself believed he had honorably employed a very different mechanical heart than the one De Bakey had sought to develop with Domingo Liotta, since theirs was meant to be a very brief holding device until a human donor heart could be located. He told a committee of physicians investigating the Haskell Karp episode on behalf of the National Heart, Lung, and Blood Institute that he was merely trying to buy a dying patient a perhaps life-saving extra day. According to Thompson (p. 232), Cooley told a reporter, "I have done more heart surgery than anyone else in the world. . . . Based on this experience, I believe I am qualified to judge what is right and proper for my patients. The permission I receive to do what I do, I receive from my patients. It is not received from a government agency or one of my seniors."

66 **Gruentzig's first skiing trip to the Bavarian Alps:** Author interviews with Michaela Gruentzig, September 20, 1994, and November 15, 2003.

67 **Gruentzig in Darmstadt:** Ernst Schneider interview by David Williams and the author, November 19, 2003, at Zürich University Hospital conference room, followed by dinner. Eberhard Zeitler interview, September 5–6, 2003.

CHAPTER FIVE:

68 **Gruentzig's arrival in Zürich:** Author interview with Alfred Bollinger at his Fasche, Switzerland, apartment, on September 9, 2003, followed by meetings on November 17 and November 19, 2003. See also, "Typisches Veränderungen in Achillessehnenreflex bei Claudicatio intermittens," A. Grüntzig und A. Bollinger, *Schweizerische Medizinische Wochenschrift* 100: Nr. 41, 1970; pp. 1730–1733; "Andreas Grüntzig's balloon catheter for angioplasty of peripheral arteries (PTA) is 25 years old," A. Bollinger and M. Schlumpf, *VASA*, 1999, Vol. 28: pp. 58–64; "The Reliability of True Half-Relaxation Time (TRT) and Maximal Contraction Force (Tmax) of the Calf Muscles in Intermittent Claudication," Andreas Gruentzig, M.D., Maria Schlumpf, and Alfred Bollinger, M.D. *An-*

giology, Vol. 23: No. 7, July-August 1972; pp. 377–391; "Prä-und postoperative Messung der systolichen Knöchlarteriendrucke mit Doppler-Ultraschall," A. Grüntzig, U. Brunner, W. Meier, A Bollinger, *Aktuelle Probleme in der Angiologie*, Band 13, 1971, pp. 177–187.

71 **Felix Mahler remembering Gruentzig's early Swiss days:** Interview by David Williams in Bern, Switzerland, November 23, 2003.

73 **Zeitler background and Gruentzig's pilgrimage to Aggertal Clinic:** Author interview with Zeitler in Nürnberg, Germany, September 5–6, 2003. Corroborated by joint author and David Williams and author interviews with Maria Schlumpf, November 2003. See also, unpublished historical book chapter drafted by Andreas Gruentzig and Maria Schlumpf, September 1985.

73 **Charles Dotter background:** "They All Laughed," Martin Mungia, *Oregon Health*, Volume 2, Autumn 2000, pp. 33–37. "Charles T. Dotter: A Pioneering Interventional Radiologist," Thomas B. Kinney, M.D., *Radiographics*, May 1996, pp. 697–707. Also, archival material from Charles Dotter Interventional Institute, University of Oregon Health Sciences Center. In "Inventive interventionalist: Dr. Dotter and his catheter," by Peter Ogle, *Diagnostic Imaging*, November 1983, pp. 106–109, Dotter described his work as "wedging, just as you would put a nail through a piece of cheese." Interviews with associates, delineated below.

74 **Seldinger background:** *The Catheter Introducers*, Leslie and Lanelle Geddes, Mobium Press, Chicago, 1993, various pages.

74 **Dotter's early experimentation, including work with body-plumbing guitar strings:** Interview with technician in Dotter's pioneering years, Manny Robinson, March 19, 2004. Robinson, wearing a tracksuit and gold medallion on a chain around his neck, clasped arms with Rösch when joining in the double interview, as if remembering years of esprit de corps. A transcript of an interview by

Burt Cohen, corporate documentary film maker, with Robinson on August 26, 1986, was also made available to the author.

75 **Bill Cook initial meeting and early association with Dotter:** Author interview with Bill Cook in his corporate Cook Group office, Bloomington, Indiana, on October 23, 2003. Aged seventy-two at the time, he was still boyishly crew-cut and dressed in an unassuming fashion, in a gray cardigan sweater with no tie and tan slacks. Yet, to reach this founder of a multibillion-dollar empire, one had to enter a vast cement block and glass palace, about a fifth of a mile long, with the entrance wreathed in Ionic columns, and a statue-enshrined reflecting pool.

Cook's expansive office was a curiosity shop. Among the lifelong pilot's memorabilia on display there are models of at least a half dozen planes he has flown, including a replica of Skymaster gunship and a Galaxy Astra. He used the latter to transport as personal pilot both the American President Ronald Reagan and the actor Charles Bronson, when the latter's wife, Jill Ireland, required emergency care while dying of lung cancer. Cook's shelves include a framed program from the first performance of his smash-success *Blast!* revue, and an original invoice for $3 for selling his first guide wires to Sven Seldinger in August of 1963.

The Cook Group's headquarters, which arose from a random rest stop in a blizzard, are set in an unusual terrain. Up the road from the charming university town of Bloomington, Indiana (where Cook is a noted benefactor), lies Beanblossom Creek, and just beyond that sits a Pentecostal church proclaiming: LORD, MAKE ME A TABERNACLE FOR THEE. Across the way a sign beckons for BRAD'S DISCOUNT GUNS. Further up the road a sign reads: AVOID HELL, REPENT, TRUST JESUS TODAY, followed by PAID FOR BY JACK REYNOLDS.

77 **Dotter's first "reaming" procedures:** "Transluminal Treatment of Arteriosclerotic Obstruction: Description of a New Technique and a Preliminary Report of its Application," by Charles Dotter, M.D., and Melvin P. Judkins, M.D., *Circulation*, Vol. 30:11, 1964, pp.

654–670. Also see "Charles Dotter: Interventional Radiologist," Steven G. Friedman, M.D., *Radiology*, Vol. 172, Number 3, part 2, pp. 921–924, which discusses not only his first cases, but also his fascination with saving endangered condors, and his various illnesses. See also "Portraits in Radiologys: Charles T. Dotter, M.D.," *Applied Radiology*, January-February 1981, pp. 115–116.

77 **Dotter kicking dog turds:** Josef Rösch interview by the author, March 19, 2004, in his emeritus office in the former Portland fire station, which has been transformed into a top medical research center called The Dotter Interventional Institute, thanks to a multimillion dollar gift from Bill Cook.

78 **Dotter becoming a laughingstock:** *Life*, "Clearing an Artery, Plumbing-style 'snake' restores blocked circulation," August 14, 1964. Elaboration on the outcome of this apparent fiasco was provided by the transcript of an interview by Burt Cohen with Dotter's longterm secretary, Enid Ruble, also conducted on August 26, 1986. Brian Hayes, vice president of the Cook Group, provided amplification during the interview by the author with Bill Cook on October 19, 2003.

78 **Dotter's wild antics and drug abuse:** Fred Keller, M.D., director of interventional radiology at Oregon Health Sciences University, was interviewed by the author on March 18, 2004, in his office at the crest of Portland, Oregon's "Pill Hill." He told one story concerning Charles Dotter's penchant for racing his cars against all comers. The incident saw the esteemed physician, with a young medical resident named Jan Anderson riding "shotgun," making taunting faces and gunning his engine in a "race me if you dare" fashion to a younger male gunning his own hot rod alongside. The latter became so annoyed that he stepped out of his car at a traffic light and punched Dotter in the face. According to Keller, Dotter promptly sped through the red light and said to Jan Anderson, "Ahem, we won't mention that any of this happened!"

In another aside, Keller, who spent nine years working under Dotter's wing, noted that his boss had a strange reaction to being diagnosed with lymphoma and hearing that his life expectancy was at most five years. Reasoning he had nothing to lose, Dotter ceased wearing backbreaking lead aprons to shield himself from exposure to potentially carcinogenic X-rays for the next half decade. In this same period, Dotter underwent coronary bypass surgery, and Dr. Keller offered the following anecdote regarding the situation then:

> A few months after his bypass operation, one day Enid Ruble called Dr. Rösch and me to come quick because Charles was lying on the floor in his office, sort of in terrible pain and sweating. We thought he was having a myocardial infarction and we took him and put him on the angiographic table and started an IV and called a cardiologist who came to see him. The cardiologist asked, 'Do you take any medicines?' And Charles said, 'Just a Dexedrine every morning.' The guy was up on the ceiling anyway. He was the last guy in the world who ever needed a Dexedrine, but it turned out he had a perforated ulcer, he didn't have an MI. . . . [Later] he went to have a second operation which he probably didn't need and he shopped around for a surgeon because nobody would operate on him here. He got operated on in Milwaukee and he never woke up after surgery. They called here and Josef told them that he had been on Dexedrine, so they gave him some Dexedrine and he woke up . . . He was like wired and like flying around the ceiling and very excitable. Maybe that was because he was on Dexedrine . . .

Speaking of the *Life* article's impact on Dotter's reputation, Keller added, "Would you let that guy touch you? With his red and white checkered drapes on the patient—it looked like a tablecloth from an Italian restaurant. He looked like a wild man."

Dotter's drug-taking was sometimes conducted for what he imagined to be the advancement of medical science. As he told Peter L. Ogle in the November 1988 edition of *Diagnostic Imaging,* p. 108:

> I have taken half a million units of streptokinase myself, just for kicks, right on that couch while watching *Nova.* I wanted to see what would happen. I have great confidence in what it can do. It should take about a year's work in Portland to prove that we can cut the mortality of acute coronary occlusions by 25 percent. That's what I think it will do, if we get it in soon enough and in enough volume . . . You've got to get this sort of agent into the ambulance. You can't afford to wait till they haul a patient to the hospital. The therapeutic window on this is very early because myocardial tissue begins to die twenty minutes after you've deprived it of oxygen.

It bears reflecting that, partly due to the advocacy of Charles Dotter, Portland, Oregon, has become one of the foremost rapid-ambulance-response municipalities in the world.

81 **Josef Rösch and Manny Robinson:** Author interview with Dr. Rösch, with Manny Robinson, March 19, 2004.

83 **Zeitler visits Dotter:** Author interview with Zeitler in Nürnberg, Germany, September 5–6, 2003.

84 **Gruentzig trip to Aggertal Clinic:** Author interview with Maria Schlumpf, September 8, 2003, followed by joint author and David Williams interview with Maria, November 19, 2003. Author interview with Michaela Gruentzig, November 15, 2003. Author interview with Eberhard Zeitler, September 5–6, 2003.

84 **Other Dotter background:** "Interventional Radiology: New Paradigms for the New Millennium," Frederick S. Keller, MD, *Journal of Vascular and Interventional radiology,* 2000:11, pp. 677–681; *The Catheter Introducers,* Leslie A. Geddes and LaNelle E. Geddes, Mobium Press, Chicago 1993; "Radiology History Exhibit, Charles T. Dotter: A Pioneering Interventional Radiologist," Thomas B. Kinney, MD, *Radiographics,* May 1996, pp. 697–707. "Portraits in Radiology: Charles T. Dotter, MD," *Applied Radiology,* January-February 1981, pp. 28–29.

CHAPTER SIX:

85 **Zeitler performs Dotter method with Gruentzig in Zürich:** Author interview with Zeitler, September 5–6, 2003; and Alfred Bollinger, September 9, 2003. Also, draft historical book chapter compiled by Maria Schlumpf and Andreas Gruentzig on Sea Island, Georgia, September 1985.

86 **Gruentzig monograph quotation:** Cited in "Angioplasty From Bench to Bedside to Bench," Spencer B. King III, *Circulation,* Vol. 93, 1996, p. 1622.

86 **Gruentzig track record with Dottering procedure:** Schlumpf/Gruentzig draft chapter, 1985.

87 **Krayenbühl erupting at Gruentzig:** Bollinger interview, September 9, 2003.

87 **"My colleagues laughed at Dotter":** "PTCA Trials in Progress," by Sheila Stavish, *Cardio* magazine, 1984, pp. 50–55. In the same issue of *Cardio,* p. 80: "It was my cardiology colleagues who were so strongly opposed. They knew dilatation couldn't work. They knew it couldn't work because it wasn't their idea. But of course that's true for all developments. And I think it's healthy."

88 **Gruentzig and the rabbi:** Alfred Bollinger interview, September 9, 2003.

89 **Katrin Hoffman:** Interviewed by the author in Zürich, September 8, 2003. Katrin, who has similarly swarthy features as her father, told of her own poignant quest, and revealed that the last time she saw her father was when she was six months old. She related over dinner that Andreas sent her a letter when she was five or six, but that her mother never allowed her to see it. After her father died when she was seventeen, she began wondering what she had lost, and eventually made her way to visit Martin Kaltenbach in Frankfurt. It was only then that Katrin discovered that her grandmother lived in Heidelberg, whereupon Katrin wrote Charlotta Gruentzig requesting a meeting. After a considerable delay, the illegitimate daughter, who has since changed her last name to Gruentzig, finally received an invitation to visit.

Katrin, who was then thirty-four and contemplating a pilgrimage to India to study yoga, was wistful in talking about her deceased father, and convinced that she shared many similarly adventurous traits, among them being a daredevil skier herself.

90 **Maria Schlumpf's early balloon-catheter making:** Personal interviews with Maria Schlumpf as cited. Also see, *Andreas Gruentzig: Eine Idee Das Verändert Medizin,* and draft historical chapter completed with Andreas Gruentzig, September 1985.

91 **Marko Turina early collaboration with Gruentzig:** Turina was interviewed by the author on September 6, 2003, and then again conjointly with David Williams on November 19, 2003. Prior to the first interview, Turina (who appeared far younger than his sixty-five years), had just weathered an overnight flight from Washington, D.C., followed by an entire day of emergency heart operations with no break for lunch, and was scheduled the next morning to fly to his native Croatia to perform multiple operations there at no charge.

91 **Gruentzig's arterial drill:** "Angioplasty from Bench to Bedside to Bench," Spencer B. King III, M.D., *Circulation* 1996; Vol. 93, pp. 1621–1629.

92 **Gruentzig hosting his brother in Zürich:** Interview with Michaela Gruentzig, November 15, 2003.

92 **Further details on early balloon catheter development:** Maria Schlumpf interview, November 19, 2003. Also, Maria Schlumpf draft historical chapter created with Andreas Gruentzig.

94 **Bernhard Meier quotations:** Interviews by the author in Bern, Switzerland, September 16, 1994, and November 14, 2003, with an additional interview by David Williams in Miami, Florida, in early

November of 2003. See also, "Zum 10 Todestag von Andreas R, Grüntzig 1939–1985," Bernhard Meier, *Schweizerische Ärztzeitung*, Band 76, Heft 48, November 29, 1995, pp. 1989–92.

94 **Wilhelm Rutishauser quotations:** The former departmental head for Andreas Gruentzig was first interviewed by the author in Dallas, Texas, in November of 1994, and a follow-up interview was conducted by the author and David Williams in Geneva, Switzerland, on November 22, 2003. The anecdotes regarding an early departmental party date to the 1994 interview. Other background: "A Swiss perspective of percutaneous coronary interventions: historic aspects," Wilhelm Rutishauser, *Kardiovaskuläre Medizin,* 2001, Vol. 4, No 3, pp. 293–306; and a transcript of Rutishauser's "Memorial Tribute to Andreas R. Gruntzig," given at the American Heart Association annual scientific meeting in Washington, D.C., November 12, 1985.

96 **Gruentzig's first balloon angioplasty procedures in the legs:** "Die perkutane Dilatation chronischer Koronarstenosen—Experiment and Morphologie," by A. Grüntzig, H. J. Schneider, *Schweizerische Medizin Wochenshrift*, Vol. 107, No. 44, 1977. See also, "Technique of Percutaneous Transluminal Angioplasty with the Grüntzig Catheter," Andreas Grüntzig and David A. Kumpe, *American Journal of Radiology*, 132: pp. 547–552, April 1979. Background on these procedures and developments to come was also provided by "Angioplasty From Bench to Bedside to Bench," Spencer B. King III, M.D., *Circulation* 1996; Vol. 93, pp. 1621–1629; and "The Development of Interventional Cardiology," Spencer B. King III, M.D., *Journal of the American College of Cardiology,* Vol. 31, No. 4, Supplement B, March 15, 1988, pp. 64B–88B. Also see, "Coronary and Peripheral Angioplasty: Historical Perspective," Richard K. Myler and Simon H. Stertzer, *Textbook of Interventional Cardiology,* edited by Eric Topol, W. B. Saunders Co., 1989, pp. 187–198.

CHAPTER SEVEN:

100 **Gruentzig's early flying:** Alfred Bollinger interview, September 9, 2003, Michaela Gruentzig interview, November 15, 2003.

102 **David Kumpe quotations:** The author visited the Kumpes' house in the foothills of the Colorado Rocky Mountains in March of 2004 and was treated to dinner by Rose Marie Kumpe, but Dr. Kumpe became unavailable, due to having to perform an emergency brain procedure on a patient with a cerebral hemorrhage. A phone interview was completed with Dr. Kumpe a year later in March of 2005.

103 **Hugo Schneider quotations:** The medical device manufacturer was interviewed by the author at the Claridge Tiefenau Hotel in Zürich on November 15, 2003.

103 **Werner Niederhauser quotations:** The balloon-catheter maker was interviewed by the author during a meeting of the European Congress of Cardiology in Berlin in mid-September, 1994, with a more detailed discussion following on September 21, 1994, in his corner office in Schneider Medintag's gleaming Bülach, Switzerland plant. Niederhauser, rising to become vice president of research and development, grew exhausted with the demands of his work, quit his job, and moved to Australia.

105 **History of Gruentzig's association with the Cook Group:** A variety of archival correspondence was made available by the Cook Group chairwoman of the board, Phyllis McDonough, including a May 12, 1975 letter from Andreas Gruentzig to Mr. C. Simonsgaard, director William Cook Europe A/S, Solberg, Denmark, outlining Gruentzig's understanding of their agreement to proceed with the collaboration; a June 3, 1975 letter from Simonsgaard to Gruentzig reviewing their understanding; and an October 7, 1975 Andreas Gruentzig letter to Simonsgaard with Gruentzig sketches of desired catheter improvements. Other records included: a November 3, 1975 letter in which Gruentzig billed Simonsgaard for his travel to a German Society of Cardiology meeting, plus reference to a previous paid trip to Florence; a December 10, 1975 letter from Bill Cook to Simonsgaard describing possible technical refinements for the Gruentzig catheter system; and a December 31, 1975 letter from Bill Cook to Simonsgaard offering further design iterations. A May 12, 1976, letter from Andreas Gruentzig to Simonsgaard lamented that the "W. Schlumpf Company, Zürich," meaning Maria's husband, Walter, could no longer keep up with the demands of making catheters by hand, but added that the Schneider company had offered to step in. An April 29, 1977, Andreas Gruentzig letter to Simonsgaard complained that he had learned from a third party that 2,000 orders had already been placed for his leg catheters, and requested "the payments we agreed upon." A May 5, 1977 response from Simonsgaard to Gruentzig announced the transfer of Sfr. 25,000 to his account, and asserted that only 200 orders had been received for Gruentzig catheters to that point. Simonsgaard added that the Cook Group did expect to finish producing 2,000 catheters by the end of 1977, and would pay the expected remaining fee of Sfr. 25,000 at that time.

Meanwhile, a February 2, 1978 letter from Simonsgaard to Gruentzig indicated that there were problems in transferring the second Sfr. 25,000 payment and that Cook Europe was still struggling to complete production of its promised 2,000 leg catheters. An April 19, 1979 letter from Gruentzig to Bill Cook referred to his recent letter of agreement with USCI to allow that company to handle distribution of his new heart catheters in the United States, Canada, and Japan. A May 5, 1979 memo from Simonsgaard to Gruentzig referred to the fact that Peter Rentrop of the University of Göttingen in Germany was already busy developing catheters to deliver drugs to the heart that might stop myocardial infarctions.

106 **Background on Itzhak Bentov:** *The Ship in the Balloon: The Story of Boston Scientific and the Development of Less-Invasive Medicine*, Jeffrey L. Rodengen, Write Stuff Enterprises, 2001, pp. 30–32.

Eleven-page single-spaced Bentov letter, with drawings, from Boston Scientific archives, November 1, 1974, beginning "Dear Maharishi," and closing with "Both Mirtala and I are sending you our love, Jai Guru Dev, Ben." Also Boston Scientific archive Bentov memos on aura-measuring instruments and investigation of resonant behavior of the body in higher states of consciousness. "Itzhak Bentov: John Abele's Recollection of Medi-Tech's founder," *Tech-Gazette* (in-house Medi-Tech publication) October 1984, Vol. 1, No. 2. *Middlesex News,* "He's dead, but his genius lives on," by Ron Davis, Sunday, June 22, 1980, pp. 1A and 12–13D.

107 **John Abele's childhood illness:** *The Ship in the Balloon,* page 33.

107 **Abele's early interest in catheters that might arrest heart attacks:** "Interview with Richard K. Myler" by Laurie Gustafson, *The Journal of Invasive Cardiology,* Vol. 13, No. 1, January 2001, pp. 63–74. A November 19, 1970 letter from Abele to Richard Myler, at the time a cardiologist at St. Vincent's Hospital in Worchester, Massachusetts. A responding letter from Myler to Abele on November 24, 1970, discussed the prospects further. A May 8, 1972 Abele letter to Myler, by then a cardiologist at St. Mary's Hospital in Daly City, California, explained that he had recently talked to Charles Dotter, who expressed concern about probing the heart with any catheter that had not first been tested in the leg circulation. An Abele memo to his Medi-Tech colleagues on November 15, 1973. A May 30, 1974 letter from Abele to Myler mentioning fresh contacts he had made about possible new heart catheters, among them James Dow in London and Martin Kaltenbach in Germany. The subject of this and later correspondence concerned prototypes for an open-mesh coil device, dubbed the "Myler arterial wall expander."

107 **Abele's first impressions of Andreas Gruentzig:** *The Ship in the Balloon,* pp. 57–60. Author interview with Abele at Boston Scientific corporate headquarters in Natick, Massachusetts, September 30, 2003.

108 **Gruentzig in Eggli:** Interviews with Michaela Gruentzig and Maria Schlumpf. A visit to the weekend cottage hosted by Alfred Bollinger, November 19, 2003.

109 **Abele collaboration with Gruentzig:** A John Abele letter to Andreas Gruentzig dated January 28, 1976, discussing an upcoming February 17, 1976 visit to Switzerland by a Medi-Tech sales representative. An Abele letter to Gruentzig on February 27, 1976 apologizing for his company's first crude attempts at manufacturing angioplasty balloons. An April 11, 1976 letter to Abele from Gruentzig from a Venice hotel with the letterhead referring to the Lido Spa on Belle Island. Various 1976 correspondence regarding an upcoming meeting at the June 1976 convention of the European Congress of Cardiology in Copenhagen, with a May 17, 1976 letter from Gruentzig critiquing Abele's prototype catheters and requesting a visit to Medi-Tech to explore better production methods. A July 2 letter from Gruentzig informs Abele that he will arrive in Boston the following Thursday. On September 9 and 23, 1976, Abele wrote Gruentzig to explain various catheter refinements in progress. A memo from John Abele to his staff, dated November 23, 1976, reviews his meetings with Gruentzig on November 15–16 during the American Heart Association convention in Miami and mentions Gruentzig's growing dissatisfaction with the Cook Group. On December 2, 1977, Abele wrote Gruentzig urging that he plan for a careful clinical trial and institute guidelines to protect against misuse of the new heart procedure. On December 23, 1977, Abele again wrote Gruentzig to discuss his company's latest balloon catheter refinements.

110 **Krayenbühl remonstrating to Zeitler:** Author interview with Eberhard Zeitler in Nürnberg, Germany, September 5–6, 2003.

111 **Gruentzig quote on legs being only his testing ground:** "Andreas Gruentzig: Cautious Innovator," *Cardio*, February 1984.

111 **Gruentzig presentation at the American Heart Association:** Author interview with John Abele, September 30, 2003. Also, Richard and Sharon Myler were interviewed by the author and David Williams in their seaside Carmel, California, home on March 22, 2004.

CHAPTER EIGHT:

114 **Veterans Administration coronary bypass surgery results:** *Heart Failure,* Thomas Moore, Random House, New York, 1989, pp. 112–125. Also see "NHLBI Disputes Certainty That Bypass Valuable for All," *Medical Tribune,* April 23, 1980, p. 1; "Wider Heart Bypass Indications, Pro and Con," *Medical Tribune,* October 15, 1980, pp. 1 and 14.

114 **Braunwald on a new bypass surgery industry:** *Saving the Heart,* pp. 109–110.

115 **Gruentzig return to Zürich:** Bernhard Meier interviews, September 16, 1994, November 2003; also Marko Turina interview, September 2003.

115 **"This crazy Gruentzig" quotation:** Alfred Bollinger interview, September 9, 2003.

116 **Kaltenbach meets Gruentzig:** Interview with the author in Kaltenbach's stately Dreiech Bülach, Frankfurt, Germany home, September 4, 2003, which is surrounded by flowering hedges, fruit trees, and a fountain, and is filled with subtle artwork inside. After the interview in an upstairs office, a brisk post-lunch walk was embarked on through a nearby woodland, for therapeutic purposes, since Kaltenbach suffers chronic lower-back pain due to all his years of performing catheter procedures under the weight of lead aprons. Afterward, Kaltenbach's wife, who is a psychoanalyst, served tea with home-cooked German desserts and offered these observations about Andreas Gruentzig: "His eyes were strömlin (brilliantly glowing); his presence filled the room; he was captivating. He remarried so quickly it astonished us. Everything he did was in a rush. He was involved everywhere, and he tried to jump too high . . . When he skied, he competed—everything for him was a competition." Asked whether Gruentzig's story was ultimately about the pursuit of fame, wealth, ego satisfaction, or the urge to help mankind, she responded: "It was all of that."

117 **Dotter in Nürnberg:** Eberhard Zeitler interview, September 5–6, 2003; David Kumpe phone interview, March 2005; Richard and Sharon Myler interview, March 22, 2004.

118 **Myler goes to Zürich:** Above Myler interview.

119 **First failed Gruentzig case in Zürich:** Marko Turina interview, September 2003.

120 **Myler collaboration begins:** Myler interview March 22, 2003. Myler also wrote Gruentzig on November 23, 1976 (Boston Scientific archives), expressing his enthusiasm: "As I am sure you already know from John Abele, I have been interested in this problem for the last eight years and have been in conversation with John—in fact trying to use certain prototypes, but without much success. I feel extremely encouraged by your work and feel that you have perhaps solved one of our major problems."

123 **Simpson saying that "the concept could be either awesome or just bizarre:"** *Saving the Heart,* p. 157.

123 **Dolf Bachmann's first name:** Dolf is a shortened form of Adolf and the patient was born in 1939, at the peaking fame of the German Adolf in question and before his notorious legacy had played out. Bachmann's physician, Bernhard Meier, related in a November 15, 2003 interview his understanding that his patient's parents chose the name due to its enormous popularization in German-speaking countries thanks to Hitler's ascendancy, but that Bachmann himself later dropped the "A" with all its associated taint.

123 **Dolf Bachmann background:** *Blick Zürich,* "Ich sah den Ballon in meinem Herzen!" (I saw the balloon in my heart), February 8, 1978, p. 4. Also, interviews with Bernhard Meier, who treats Bachmann to this day and marvels that his first-ever dilated artery remains wide open.

123 **Dolf Bachmann quotations:** See cover story "Ist der Infarkt-Tod besiegt? Der Kampf ums Herz" ("Has heart-attack death been conquered? The War Over the Heart), *Schweizer Illustrierte,* 13 February 1978. Bachmann's quotations were translated by the author.

125 **Observations on first coronary angioplasty:** Niederhauser interview, September 1994. Maria Schlumpf, Marko Turina, and Richard Myler interviews. See also, "The First Coronary Angioplasty as Described by Andreas Gruentzig," J. Willis Hurst, M.D., *The American Journal of Cardiology*, January 1, 1986, Vol. 57, pp. 185–186; and *Coronary Arteriography and Angioplasty,* "History of Cardiac Catheterization," Chapter 1, King, S.B., Douglas, J.S. editors, McGraw-Hill Book Company: New York, 1985.

128 **Gruentzig "I began to realize that my dream had become true":** See "Angioplasty From Bench to Bedside to Bench" Spencer B. King III, No. 9, *Circulation*, Vol. 93, 1996, p. 1624.

129 **Martin Kaltenbach's first procedures with Gruentzig:** "*Die erste PTCA in Deutschland,*" speech to German Cardiology Society, 1987; and "Evolution of Interventional Cardiology in Germany," Martin Kaltenbach, M.D., *Journal of Interventional Cardiology*, Vol. 15, No. 1, February 2002, pp. 33–39.

130 **Sones at Gruentzig's AHA presentation of his results:** Myler interview, March 22, 2003. Also, Richard Myler interview with Burt Cohen on August 27, 1986, and telephone interview with the author from Daly City, California, July 1987; and Burt Cohen interview with Cleveland Clinic chief of cardiology, William Sheldon, M.D., October 28, 1996.

CHAPTER NINE:

130 ***The Red Balloon:*** Director Albert Lamorisse, starring his son Pascal, 1956.

130 **Myler quotations:** Interview March 22, 2003. Also see "Interview with Richard K. Myler, M.D.," by Laurie Gustafson, *Journal of Invasive Cardiology*, Vol. 13, No. 1, January 2001, pp. 63–74.

132 **John Abele warning Gruentzig about emerging rivals:** December 2, 1978 letter from Boston Scientific archives.

133 **Simon Stertzer arrives in Switzerland:** Stertzer interview with Burt Cohen, August 27, 1986, in which Stertzer asserted that the first case he witnessed was an ultimate success after considerable struggle.

134 **First International Dilatation Society meeting:** Interview by the author with Lamberto Bentivoglio at his Philadelphia home, on a blustery January 30, 2004. Interviews with Richard Myler and Maria Schlumpf, and *Saving the Heart,* pp. 136, 139, 158.

134 **Lancet letter:** Feb 4, 1978, p. 263.

135 **Swiss popular press jumps on the story:** All publications were viewed in the archives of Boston Scientific, Natick, Massachusetts, on October 1, 2003. See, *Blick Zürich*, "Zürcher Arzt gelang Medizin-Sensation," February 8, 1978, cover story; *Schweizer Illustrierte,* "Der Kampf ums Herz," February 13, 1978, cover story; *Tages-Anzeiger*, "Medizin-Sensation: Ballonsonde gegen Herzinfarkt," cover story. A more soberly scientific account appeared on March 22, 1978, in the *Neue Zürcher Zeitung.*

135 **Krayenbühl saying, "Patients will die":** Maria Schlumpf interview.

135 **Swiss Society of Cardiology resistance to Gruentzig talk:** Wilhelm Rutishauser interview.

136 **Myler's difficulties with staff opposition and subsequent negotiations with various authorities to proceed:** Memo from Albert Fri-

etzche, M.D., February 16, 1978, to Richard Myler, regarding executive meeting within the hospital on February 13, and a phone call Frietzche made the next day to Mr. Lawrence Pilot, associate director for compliance at the Bureau of Medical Devices of the United States Food and Drug Administration (from Boston Scientific archives). Evidently buying time, Pilot had noted that draft regulations of medical devices had only been "published" in August of 1976 for public response, and such comments were still being compiled, with more expected. "However, this will be probably a year or two from now . . . At the present time," the Frietzche memo continued, "Mr. Pilot stated that there are *no* specific FDA regulations or need for review or approval of the devices or procedures planned for coronary PTA."

Myler next convinced Sister Anthony Marie, a chief administrator of his hospital, to submit a letter of inquiry to St. Mary's Hospital's legal counsel in San Francisco to make certain that the performance of balloon angioplasty in their hospital would be in compliance with federal medical-device law, while iterating that the device in question was not "commercially available" in the United States. The attorney's reply on February 27, 1978, went straight to the legal niceties saying the use of balloon angioplasty catheters was "not inconsistent with federal regulatory authority, despite changes in the federal Medical Device Amendments of 1976, since a) exemptions for devices manufactured or altered for own physician use, or b) devices solely for research, teaching or analysis . . . Under both subsections (d) and (f) Doctors Grüntzig and Myler are not required to register as manufacturers of medical devices. Similarly the specially prepared catheter is exempt from pre-market notification procedures." The attorney noted that Myler had made clear that balloon catheters constituted a "custom device . . . and [is] thus totally exempt" from requirements for rigorous "pre-market approval" authorization by the FDA.

136 **Myler and Stertzer performing first U.S. angioplasties on same day:** Burt Cohen, long-term chronicler of the angioplasty field, has reported on his Web site (www.PTCA.org) that both participants denied to him that any prearrangement occurred in their tandem case scheduling. Stephen Klaidman in the highly researched *Saving the Heart,* Oxford University Press, New York, 2000, p. 141, stated that "Myler vaguely remembers some coordination, possibly through Gruentzig, but Stertzer has no such recollection." Given the constant communication among the inner cadre of Gruentzig's disciples in 1978, this author feels that the timing of these cases suggests some form of coordination. In any case, even though the procedures did occur on the same day, Stertzer has ever since noted that he performed the first coronary angioplasty in the United States.

136 **Assertion that Stertzer's procedure was overly aggressive by Gruentzig's lights:** Author interview with Lamberto Bentivoglio. Maria Schlumpf concurred that Gruentzig was upset about the conduct of this procedure, but was no longer sure why at the time of her July 2006 e-mail regarding this issue.

136 **Continuing controversy over the legality of early procedures in the U.S.:** Discussions within St. Mary's Hospital and with the federal government continued for months, even after Myler and Stertzer had begun performing their first coronary angioplasties. An August 16, 1978 letter from Mr. Glenn A. Rahmoeller, director of the division of cardiovascular devices of the Bureau of Medical Devices of the FDA, to Richard Myler (newly incorporated as a member of the San Francisco Cardiovascular Medical Group) refers to a telephone conversation of July 31, 1978. Rahmoeller notes that his understanding is that angioplasty catheters have been used to date in only five or six centers in the United States [evidently including St. Mary's Hospital and Lenox Hill], and concludes:

> I would also like to know if this device is used in the peripheral vessels. If so, is its use in the peripheral vessels also considered experimental and is its distribution limited to clinical investigations? Given my present understanding of the investigational use of this device, I

> do not believe that it is in violation of the Federal Food, Drug, and Cosmetic Act. However, I would appreciate a response to this letter.

The letter evidences an earlier period of such light regulatory control of the medical device industry that its author, charged with federal oversight of the industry, is asking a doctor with vested interests in a new procedure to inform the regulatory official what is and is not "experimental."

An August 22, 1978 response from Myler to Rahmoeller's inquiry is significant for its specific references as to whom had done what regarding angioplasty to this point—a subject which has been contested over the years. Myler noted that, in the U.S., only Peter Block of Massachusetts General Hospital; Lamberto Bentivoglio of Philadelphia; and Steven Meister of the Medical College of Pennsylvania had started performing the procedure in the coronary arteries (i.e., besides himself and Simon Stertzer). But he went on to refer to his rapidly developing collaboration with John Simpson of Stanford, and William Parmley of the University of California at San Francisco. Myler continued, "I do have some feeling for the fact U.S.C.I., Edwards Laboratory, Cook, and Cordis are all interested in developing such a system."

Myler then emphasized that Gruentzig had already performed his angioplasty procedure in 300 leg patients in Switzerland and that these peripheral artery efforts were being advanced in the U.S., by Dr. Waltman of Mass General, David Kumpe, William Casarella at Columbia, and Thomas Sos at Cornell University, New York.

137 **Cavalcade of early visitors to Gruentzig's door:** William Casarella interview in Atlanta, Georgia, December 19, 2003; Niederhauser and Maria Schlumpf interviews. Also the Boston Scientific archives hold a cheerful letter from John Simpson to Andreas Gruentzig, dated May 25, 1978, referring to his earlier visit to Gruentzig and explaining work on his guide catheter, about which he wished to show Gruentzig his newest design.

137 **Sigwart meets Gruentzig, begins angioplasty:** Ulrich Sigwart was interviewed by the author in his Geneva office on September 10, 2003, and jointly by the author, David Williams, and Wilhelm Rutishauser on November 20, 2003.

138 **Fiedotin visits Gruentzig:** Author interview in Fiedotin's suburban Atlanta home, February 4, 2004.

139 **Michaela Gruentzig reminiscences:** Author interview, November 15, 2003.

140 **Cash payments required for first balloon catheters, Stertzer and Sones visit Zürich:** Niederhauser interview by the author, September, 1994, and Heliana Canepa interview by David Williams and the author, November 2003.

141 **Sharon Myler quotations about Stertzer:** Author and David Williams interview with Sharon and her husband Richard in Carmel, California, March 22, 2004.

141 **Stertzer's strip clubs:** "Nude club revenue to fund research," Mathis Winkler, *Las Vegas Sun*, September 6, 2001; "Bad publicity for Stertzer's clubs is unfortunate," editorial, *Stanford Daily*, November 28, 2001.

142 **Organization of first Gruentzig teaching course:** Maria Schlumpf interviews.

142 **Gruentzig anger at Felix Mahler's first kidney angioplasty:** Bollinger interview September 9, 2003. Maria Schlumpf maintains Gruentzig actually did first such procedure but did not report it until later. Mahler himself maintained differently in November 22, 2003 interview with David Williams.

142 **Entrepreneurs eyeing Schneider company:** Interview by the author with John Cvinar, former president of Bard's USCI Division at Hotel@MIT, Cambridge, Massachusetts, May 22, 2004.

143 **Catheter industry sales in late 1970s, and first Prigmore quotations:** Author and David Williams joint interview with Dave Prigmore, former USCI President and Bard Group Vice President, on July 21, 2003. This interview was conducted at Prigmore's Misquamicut, Rhode Island, beach house, with panoramic views of the Atlantic Ocean just beyond. The now-retired medical-device industry executive at first observed, "I think doctors have no interest in history, at least in my experience . . . I've always been amazed at how short an interest span the medical community has for their predecessors."

143 **Bard/USCI later runs afoul of the U.S. Food and Drug Administration:** The author, having produced medical news publications sponsored by USCI for twelve years (1986–1994), retains numerous files on the company's troubles beginning in 1989. A sampling includes: "Manufacturer Admits Selling Untested Devices for Heart," Philip J. Hilts, *New York Times*, October 16, 1993, p. 1; "Heart catheter became killer: Mass-made unit blamed in fraud," Mitchell Zuckoff and John H. Kennedy, *Boston Globe*, Sunday, October 31, 1993, pp. 1, 28; and "Ex-Bard executives sentenced," *Boston Globe*, August 9, 1996; and "Manufacturer Admits Selling Untested Devices for Heart," by Philip J. Hilts, *New York Times*, October 16, 1993, p. 1.

Although Dave Prigmore, John Cvinar, and Lee Leichter, director of regulatory affairs at Bard's USCI division, were indeed sentenced to eighteen months of imprisonment, followed by two years of supervised release, for "conspiracy to defraud and impair the functioning of the United States Food and Drug Administration," they managed to avoid imprisonment upon appeal. Their parent C.R. Bard corporation settled the case with payment of a $61 million fine.

Some observers in th early 1990s thought Prigmore, Cvinar, and three other co-defendants (initially including an executive named Kenneth Thurston and George T. Maloney, chief executive officer of Bard—against whom charges were later dropped) were being made public whipping boys by an attention-hungry chief of the United States Food and Drug Administration, David Kessler. Ironically, Kessler's own regulatory body had barely scrutinized the new balloon procedure when advocates like Richard Myler were clamoring to introduce it in the United States. Furthermore, other major device companies would in time fall into similar loggerheads with the FDA, without the federal government attempting to put their chief executives behind bars. But Prigmore and co-defendants did eventually file a guilty plea for withholding significant, and in at least two cases lethal, device-failure data from the FDA.

Of further interest is a *Boston Globe* article on July 12, 1996, by Judy Rakowsky, which cites statements by Prigmore and Cvinar's high-pressure boss, George T. Maloney, chief executive of C. R. Bard, Inc., in a videotaped 1990 speech to employees in Glens Falls, New York. As introduced into court testimony by federal prosecutors, Maloney is heard saying:

> I mean cutting a corner, not bad . . . We lost a little bit in focusing on integrity. That's where we've lost out. Because we feel, well, the government is lax. They're not looking; they're not going to check, what does it matter; it's not a big deal . . . next the corner is a little bigger and the next thing you know you're hooked. And it became a way of life . . . The pressure from competition didn't get easier . . . and so we had to get it out quicker and we'll catch up with the paperwork later, or we'll do the testing on human beings . . . You see the government is labeled, you know, they're the bad guys and the bureaucracy, and that's the way the doctors feel about them, right? Because they've got to do a lot of paperwork and it drives them crazy and we can understand what you're going through—they don't really know how guilty we are, right?

144 **First USCI meeting with Hugo Schneider:** The emissary was John Shiphard, vice president of marketing; and the account of USCI's various entreaties to Schneider was provided alternately by Dave Prigmore and John Cvinar.

144 **Schneider allegedly doing business behind the Iron Curtain:** Heliana Canepa generally supported the Prigmore and Cvinar comments, saying that her boss had previously done telecommunications business in Yugoslavia nearly two weeks a month.

145 **Assertions about problematic Gruentzig procedure at Harvard demonstration:** John Abele interview, September 30, 2003; also corroborated by Richard Myler, March 22, 2004.

146 **Stertzer on New York TV:** Transcript from Boston Scientific archives of interview with Chuck Scarborough, WNBC-TV, June 15, 1978.

146 **Stertzer holding a June press conference:** Abele letter to Gruentzig June 28, 1978.

146 **Stertzer on Walter Cronkite:** Documentation from Boston scientific archives.

146 **Stertzer in *Time* Magazin:** "Blowup in the Arteries: Tiny balloon unclogs heart blood vessels," *Time*, July 3, 1978, pp. 54–55. Also see *Newsweek*, "Substitute Scalpel," October 30, 1978, p. 111.

146 **Gruentzig's reaction to Time magazine article:** Richard and Sharon Myler interview March 22, 2004; John Abele interview, September 30, 2003; David Williams interview.

146 **Stertzer's contrasting account.** Stertzer interview by Burt Cohen, August 26, 1986. Also, the author interviewed Stertzer (and heard him speak) on a number of occasions over a twenty-year span. Stertzer waxed humble in a telephone interview by the author in July 1986, saying:

> Medicine is a very fickle and not historically oriented profession. There are a lot of things that go by the wayside. I think in fifty years nobody will remember us. If they remember Andreas, I think it will be remarkable in and of itself. If you go and look at the medical students and the residents and interns and even our fellows, they don't

have any idea of any of this. It's just, they think it was here forever and they don't really attribute any of this to anyone. And people who are starting medical school [now], I'd say 8 out of 10 of them when they graduate from medical school won't even know who Andreas Gruentzig is. I mean, they'll be lucky to know that angioplasty is a form of therapy, but they don't really care or register in their minds who discovered it. I don't find that a condemnation of the system. I couldn't tell you who developed the hematic grid. I don't know who that was. It wasn't Mr. Hematic Grid, that I can tell you.

147 **David Williams background and quotations:** Numerous interviews with the author beginning in 1986.

148 **Dave Prigmore visits Schneider:** David Williams and author interview, July 21, 2003.

149 **Schneider contrasting characterization of his USCI relationship:** Author interview, November 15, 2003.

149 **Description of the first teaching course:** Maria Schlumpf, Richard Myler, David Williams, Lamberto Bentivoglio, Marko Turina interviews.

150 **Account of USCI executives' reception at the course:** Personal interviews with Dave Prigmore and John Cvinar, and various conversations by the author with both parties between 1986–1994.

152 **Peter Block reference:** Interview by the author of David Williams, July, 1987, printed in *PTCA: News for Practitioners*, September-October 1987.

152 **Garden party after teaching course:** Bentivoglio, Myler, Schlumpf, and various other interviews; visit to the site with Michaela Gruentzig; and review of photographs from the event.

152 **General background on procedure's early results:** "Long-term Follow-up After Percutaneous Transluminal Coronary Angioplasty: The Early Zurich Experience," Andreas R. Gruentzig, M.D.,

Spencer B. King III, M.D., Maria Schlumpf, and Walter Siegenthaler, M.D, *New England Journal of Medicine* 1987, Vol. 316, pp. 1127–32. "Is the Honeymoon Over for Angioplasty? Gruntzig Answers Critics," *Medical Tribune*, August 22, 1979, pp. 6–7.

CHAPTER TEN:

154 **Terms of USCI's original agreement with Gruentzig:** The primary source was the Dave Prigmore interview on July 21, 2003. John Cvinar's recollection was consonant in most respects, although he remembered the terms of the deal to be $35,000 up front, as opposed to $27,000.

155 **Medtronic $1.35 billion collaboration deal:** Also relevant is a detailed letter, dated February 24, 1997, from Dave Prigmore to Spencer King, which was made available to the author. "Medtronic to Pay $1.35 Billion to Inventor," Andrew Pollack, *New York Times* article about payment to Los Angeles spine surgeon Gary K. Michelson, April 23, 2005.

155 **Schneider company's agreement to pay Gruentzig 20 percent of sales:** Interview by the author and David Williams with Heliana Canepa (who rose from "gal Friday" to leadership of the exponentially growing company) in the Bauer au Lac hotel, Zürich, November 21, 2003. Canepa, then fifty-five, has a crackling, seen-it-all personality, and eventually succeeded in a late 1990s attempt to raise $2.1 billion to buy Schneider Medintag, but lost out to a rival bid from Pfizer.

155 **John Simpson visits USCI:** Dave Prigmore interview, July 21, 2003.

155 **Background on Simpson's rapid progress:** *Saving the heart*, Chapter 10, pp. 154–168, and Chapter 12, pp. 184–198.

156 **Gruentzig rejects Simpson:** Richard Myler interview, March 22, 2004.

156 **Hans Gleichner characterization:** Interviews with Heliana Canepa and John Cvinar. His suicide was corroborated by Maria Schlumpf and Hugo Schneider, who described Gleichner as having turned so paranoid that he was seeing visions. None of these sources can recollect how Gleichner killed himself.

156 **USCI studies technology in Zürich:** The technician involved was Denis Kokernack.

157 **Conflict between USCI and Schneider over Gruentzig patent:** A core legal issue, according to both the Schneider company's Heliana Canepa on November 21, 2003, and USCI's Dave Prigmore, July 21, 2003, was Gruentzig's early description of intricate details of his catheter system in *Rofo: Fortschritte auf dem Gebiete der Rontgenstrahlen und der Nuklearmedizin* (or "Progress in the Practice of Radiology and Nuclear medicine"), January 1976, Vol. 124, No.1, pp. 80–86. Adding fuel to the patent conflict was the fact that six months later he published a more up-to-date description of his coronary catheter system in which he released detailed pre-patent drawings, in a June 1, 1976 article in *Klinische Wochen-Shrift*, entitled "Perkutane Dilatation von Coronarstenosen—Beschreibung eines neuen Kathetersystems" (percutaneous dilation of coronary stenoses—description of a new catheter system), Vol. 54, No. 11, pp. 543–545.

An archival document reveals that Gruentzig, together with Hans Gleichner, in fact filed for a Swiss patent on October 21, 1977, and this ultimately contested patent (no. 616337) was granted on March 31, 1980. A Dave Prigmore letter to Spencer B. King, III, dated February 24, 1997, and made available to the author and David Williams, reviews this situation, as well as the historic and somewhat comic NHLBI meeting in which USCI and the powers of cardiology attempted to map out a program for exploration of the Gruentzig balloon procedure. The allegedly compromised Grüntzig-Schneider Medintag patent was published on April 1,

1980, with the United States Patent Office and numbered 4,195,637.

157 **Pricing of first USCI balloon catheters:** John Cvinar interview, May 22, 2004.

157 **Interest in balloon angioplasty by the U.S. National Heart, Lung, and Blood Institute (NHLBI):** Interview by the author with Suzanne Cowley, nee Mullin, formerly a research assistant with the NHLBI, and her husband Michael Cowley, M.D., an early practitioner of balloon angioplasty, at an Italian restaurant outside Richmond, Virginia, February 7, 2003. "Historical Background of the National Heart, Lung, and Blood Institute Registry for Percutaneous Transluminal Coronary Angioplasty," Suzanne M. Mullin, R.N., Eugene R. Passamani, M.D., and Michael Mock, M.D., *Journal of American College of Cardiology,* June 15, 1984, Vol. 53, pp. 3c-6c.

157 **Gruentzig's attendance at exploratory meetings:** As above.

158 **Restraint:** Myler interview, March 22, 2004.

158 **Controversy at NHLBI registry meeting:** Dave Prigmore interview with the author and David Williams, July 21, 2003, and with the author, August 1987. "Percutaneous Transluminal Angioplasty (PTA) in Patients with Relative Contraindication: Results of the National Heart, Lung, and Blood Institute PTCA Registry," Lamberto G. Bentivoglio, M.D., Mark J. Van Raden, M.D., Sheryl F. Kelsey, Ph.D., and Katherine M. Detre, M.D., *Journal of American College of Cardiology* 1984; 53:82c–88c.

160 **Gruentzig's continuing difficulties in Zürich, and the nature of European academic hospitals:** Meier, Rutishauser, Schlumpf interviews.

160 **Katrin Bauben background and quotation:** Interview by the author and by her boss at that time, Marko Turina, in Zürich, September 2003.

160 **Gruentzig's Nobel Prize nomination:** An October 20, 1978, letter to the Nobel Prize Committee from William Foley, M.D., emeritus professor at Cornell University in New York City, recommended that Andreas Gruentzig and Charles Dotter jointly receive the Nobel Prize for medicine. Foley's October 9, 1978 draft had also recommended the inclusion of Eberhard Zeitler. (Boston Scientific archives.)

161 **Gruentzig's confrontation with Robicsek:** Author and David Williams's interview with Spencer B. King III, in his Druid Hills, Georgia, home, December 2003, and author interview with Joe Craver, at his Emory Clinic office on February 4, 2004.

162 **Second Gruentzig teaching course:** Ralph Lach interview by the author in Arlington Heights, Ohio, on October 22, 2003. Lach, then seventy-one, was still playing jazz piano as well as participating in a "summer fantasy" baseball league, although he had recently given up the stress of performing balloon angioplasty.

164 **Bruno Lorenzetti quotation:** From the Boston Scientific-sponsored videotape: *PTCA: A History: The 20th Anniversary of PTCA Project,* by Burt Cohen Productions, 1997.

166 **USCI's difficulties in meeting balloon catheter demand:** John Cvinar interview, May 22, 2004; Dave Prigmore interview July 21, 2003; and author and David Williams interview with Marcia Schallehn, long-term clinical education director at USCI, on April 16, 2004.

167 **Lach's first procedure:** Author interview with the Ohio cardiologist, October 22, 2003.

CHAPTER ELEVEN:

169 **Gruentzig difficulties in finding European appointment:** Kaltenbach, Meier, Schlumpf interviews.

170 **Growing American interest in recruiting Gruentzig:** Interviews with Richard Myler, Dave Prigmore, John Abele, Michael Cowley, Willis Hurst, Spencer King, others. Also, Cleveland Clinic archives.

Letter of February 12, 2004, faxed from Harry L. Page, M.D., Vanderbilt University Medical Center, to David Williams; and letter of June 3, 1980, to Gruentzig from Lazar J. Greenfield, Medical College of Virginia (Boston Scientific archives).

170 **Michaela Gruentzig's interest in Rudolf Steiner–like Zürich group:** Bernhard Meier interview by David Williams, November 2003. The notion of increasing marital friction was corroborated by many sources, and tacitly acknowledged by Michaela Gruentzig on November 15, 2003.

171 **Gruentzig's white shoes during teaching courses:** A videotape in the possession of Spencer B. King; various photographs.

171 **Emmental Valley celebration after the course:** Interviews with Spencer King, December 15, 2003, and February 4, 2004, corroborated by David Williams.

173 **Gruentzig at Snowmass:** The February 12, 2004 letter from Harry Page to David Williams describes Gruentzig traveling, along with his wife and daughter, at Page's expense. According to Page, a nervous Gruentzig asked for a double shot of Jack Daniel's on the morning before his presentation to the conclave.

173 **Emory University's new wealth from Coca-Cola and Woodruff story:** "Associates praise 'Coca-Cola man,' Close friends mourn death of Robert Woodruff," by Mike Christensen and Debbie Newby, *Atlanta Journal-Constitution*, March 8, 1985, page A1; "Master of anonymity speaks out, Boisfeuillet Jones saw to it that Woodruff's gifts went to best use," by Helen C. Smith, *Atlanta Journal-Constitution*, March 11, 1985, page B/1; "Woodruff's love for plantation lives on," by David Beasley, *Atlanta Journal-Constitution*, March 18, 1985, page A1.

It is important to note that Spencer King, Charles Hatcher, and some other Emory cardiovascular specialists dispute that the Woodruff brothers' 1979 gift of $105 million in Coca-Cola stock—estimated to be still generating $16 million a year a decade ago—

had any direct bearing on the Emory Clinic's ability to recruit Andreas Gruentzig, since the bequest was made to the university per se, and not the semi-autonomous Emory Clinic. It is the author's position that this bequest, the largest in the world to a university to that time, is indeed relevant in that it had a buoyant affect on the entire institution and was widely trumpeted by many staff members. Indeed, Gruentzig was well aware that his teaching courses at Emory would be held in the state-of-the-art new auditorium in the Woodruff Health Sciences Center and was evidently seduced in his own mind by the Coca-Cola connection. For example, various Emory Clinic staff heard Gruentzig, shortly after his arrival there, demand that intercessions somehow be made with Robert Woofruff when he longed for expanded support.

173 **Larry Flynt in Woodruff's domain:** Charles Hatcher interview, February 4, 2004.

174 **Gruentzig arrives at the Kings':** Interview by the author and David Williams with Spencer and Gail King in the couple's home in Druid Hills, Georgia, December 15, 2003.

174 **Craver quotations:** The author interviewed Joe Craver in his large end-of-corridor office at the Emory Clinic on February 4, 2004. Still a singular physical presence at sixty-five, the former all-American linebacker from the University of North Carolina had a desk surrounded by stuffed game birds and trophy fish he had landed. Craver was later quoted as saying that "several thousand" patients underwent treatment by Andreas Gruentzig before one died, but records indicate that Gruentzig performed an aggregate of about 2,500 procedures at Emory.

175 **Hatcher background and quotations:** Charles Hatcher was interviewed by the author on February 4, 2004, in his emeritus office at the Emory Clinic.

176 **Gruentzig's lunch at the Piedmont Driving Club and meeting with Griffin Bell:** The chief source for this anecdote was Spencer King; however, Willis Hurst, Charles Hatcher, and Joe Craver provided elaborations.

176 **Tom Wolfe quotations:** *A Man in Full*, Farrar, Straus and Giroux, New York, 1998, pp. 373 and 22.

177 **Gruentzig's initial fee structure at Emory:** A clarifying letter from Spencer King to David Williams dated June 13, 2006, explains that Gruentzig's initial revenue structure was to share in a strict one-third each return with King and John Douglas on whatever pooled net income the group generated from their various procedures within the Emory Clinic, less fixed percentages of those revenues that were returned to the Emory University School of Medicine to further the workings of the larger institution and to cover the group's debt for staff support, physical plant, and other overhead.

178 **Hurst background and quotations:** The author and David Williams jointly interviewed Willis Hurst (who was later joined by cardiologist Steve Clements) at Emory Hospital on December 16, 2003. Hurst, who was then eighty-two, and was busy writing novels in that period, after a long tenure as the editor of one of the foremost cardiology textbooks in the world.

180 **Gruentzig's starting salary:** Spencer King interview December 15, 2003; further discussion with Charles Hatcher, February 4, 2004. Also King e-mail discussions with the author, June 2006.

180 **Gruentzig final Zürich teaching course and the night of the torches:** Archival material regarding the session from Maria Schlumpf. Also, interviews with Dave Prigmore, Spencer King, Richard Myler, Lamberto Bentivoglio, Bernhard Meier, Michaela Gruentzig, and Maria Schlumpf. An initial interview on this subject was conducted by the author with King on July 7, 1987. Also, Alfred Bollinger on November 19, 2003, escorted the author to the scene of the night's festivities. Further background was obtained from transcript of a Burt Cohen interview with Spencer King on August 21, 1986.

184 **Gail King recollections:** Interviewed with her husband Spencer in their home, December 15, 2003.

CHAPTER TWELVE:

185 **Medical Camelot:** The former research nurse Claire Rice was interviewed in Atlanta by the author on December 19, 2003.

185 **"He came into our lives like a comet":** from "Memorial Tribute to Andreas R. Grüntzig: Given at the American Heart Association Annual Scientific Sessions in Washington, D.C." November 12, 1985, by Spencer B. King.

185 **Gruentzig's first Atlanta house shopping:** Gail King interview, December 15, 2003. Tours to Gruentzig's first two houses were led by Spencer King for the author and David Williams that same morning. Elaboration on the profiles of these houses was provided by Gruentzig's second mother-in-law, Mary Jane Thornton of Macon, Georgia.

186 **First impressions of Gruentzig's arrival at Emory:** Telephone interview by the author with Michael Kutcher, September 10, 2003.

186 **Gruentzig's magnificent hands:** Linda Green tribute to Gruentzig upon his death.

187 **Early balloon angioplasty with Gruentzig at Emory:** Interviews by the author and David Williams with Spencer King and John Douglas (in Gruentzig's former sixth-floor Emory Hospital office) on December 16, 2003. Also see interviews with Spencer King, posted on Boston Scientific-funded Web site, www.ptca.org, by Burt Cohen.

187 **William Casarella statement:** This pioneer of catheter procedures throughout the reaches of the body shifted from New York to

Emory, where he became chief of radiology and a close associate of Gruentzig. He was interviewed in his office there as his personal retirement party was beginning on December 19, 2003.

187 **Hurst on Gruentzig's élan after arriving at Emory:** Elaboration was provided in "In Memory of Andreas Roland Gruentzig and Margaret Anne Thornton Grüntzig," J. Willis Hurst, MD, Spencer B. King, III, MD, and Linda Green, *Journal of American College of Cardiology,* 1986, Feb. 1, Vol. 57, No. 4, pp. 333–336. Hurst's effusive section, entitled "On the Tilt of His Cap," reads in part:

> As I walked into the laboratory, I noted the large, blousy cap . . . . He looked like a master chef or 18th century artist. His brown eyes, marked with long eyelashes, pierced over the mask that covered the face and hid the small, neatly trimmed moustache. You could actually see his extraordinary intellect. He spoke softly to the patient. The words made the patient and everyone in the entire room comfortable. He then proceeded with extraordinary skill and confidence to manipulate the catheter that he had invented. A moment later the obstruction in the coronary artery was eliminated. He was the master of the moment. The scene was like a great drama unfolding in front of your eyes and the room was filled with beautiful music.
>
> Let me list the characteristics that, when added together, made what I saw as Gruentzig. He was intelligent, creative, persistent, charming, charismatic, kind, thoughtful, confident but never arrogant, hard-working, tireless, honest, filled with integrity, happy, secure as a person and loving. He extracted more joy out of this world per unit of time than anyone I know. His cap was tilted to the side of his head as a signal to the world that he was accustomed to defying life's obstacles—he was ready for whatever might come. . . .

188 **Gruentzig's chafing around the Kings' house:** Interviews with Spencer and Gail King, December 15, 2003.

189 **Gruentzig's dialogues with Rees:** The German-born cardiologist was interviewed by the author on December 18, 2003, over lunch in a Buckhead restaurant.

190 **Anecdotes of Gruentzig's womanizing:** Heliana Canepa interview by the author and David Williams on November 21, 2003; also, interviews with Spencer and Gail King, John Douglas, Michael Kutcher, Claire Rice, Dave Prigmore, Marcia Schallehn, and others.

192 **"Elephant in a porcelain shop":** Bernhard Meier interview, Bern, Switzerland, November 2003.

192 **Michaela Gruentzig's laments about the move to America:** Interview by the author, November 15, 2003.

195 **Background on Stone Mountain:** This monument amounts to the American South's rival showing to the massive South Dakota bas relief on Mount Rushmore of the great American presidents Washington, Jefferson, Lincoln, and Roosevelt. It may have provided a more unusual setting than Gruentzig himself realized, in that Stone Mountain was long identified with staunch, enduring Southern bitterness over the Union victory in the American Civil War.

A cross was burned at Stone Mountain's peak in 1915 by hooded and robed zealots in a ceremony that is widely regarded as inaugurating the rebirth of the Ku Klux Klan. For the next forty-five years, Stone Mountain served as an iconic gathering point for further Klan cross-burnings.

Work on this memorial to the Confederate heroes Jefferson Davis, Robert E. Lee, and Stonewall Jackson began in 1923, with some of the private funding coming directly from the Klan. The 40,000-square-foot bas-relief sculpture is said to be the largest in the world, and the work in-long-progress was eventually taken over by the United States Federal Government, and then the State of Georgia. Its completion required decades of labor by three master carvers, and the monument was finally dedicated in 1970, with Richard Nixon's notorious Vice President, Spiro Agnew, serving as a federal representative at the opening ceremony.

195 **Gruentzig's flamboyant behavior at Stone Mountain:** Interviews with Spencer and Gail King, Michael Kutcher, and others. Marcia Schallehn elaboration on attitudes Gruentzig expressed at this time.

196 **Increasing unhappiness of Gruentzig marriage:** Zeitler, Meier, King, Myler, and Craver interviews.

197 **Gruentzig animosity to Kaltenbach innovation:** Kaltenbach interview, September 4, 2003.

198 **Simpson's entrepreneurial advances:** *Saving the Heart*, "The Interventionalist as Entrepreneur," Chapter 10, pp. 154–168.

199 **Background profile of Geoffrey Hartzler:** Author interview with Hartzler, March 25, 2004, as described in text; also author telephone interviews with Hartzler in July of 1986, and on numerous occasions between that date and 1994. Hartzler, who graciously reviewed an advance draft of the manuscript, in an e-mail on January 16 offered this comment about having been perceived as a kind of brash anti-Gruentzig:

> Clearly, the Gruentzig/USCI/Myler/Stertzer/etc. axis viewed me as a loose cannon. . . . Still, I believe much of this resulted from competitive business pressures between USCI and ACS. My personal motives never included "business development." Rather, I wanted to deliver the best clinical and interventional care to each patient I cared for, and to do so with skill, with cumulative insight derived from growing experience, and with improved devices relative to the crude and limited instruments initially available from Schneider and USCI. My experiences and activities were constantly scrutinized by hundreds of visitors to our laboratories, thousands of physicians attending our teaching courses, and countless peers who analyzed the hundreds of scientific abstracts and manuscripts submitted by myself and colleagues. Was I seeking "approval" by doing all of this—absolutely! But there were additional motives, most notably a feeling of obligation to share with the medical community what my colleagues and I were learning through the practice of in-

terventional cardiology. Controversy frequently resulted when my personal observations led to a practice that differed from, or challenged, the current "party line."

199 **Background on his Mayo Clinic rivalries:** Hartzler interview and detailed comments by him in an e-mail to the author on January 16, 2006. In this communication, Hartzler does say that his actual superiors never personally admonished him, but that they did create an untenable working environment as the controversy with other staff simmered.

201 **Sending roses to USCI:** Hartzler interview and Marcia Schallehn interview, April 16, 2004.

202 **Controversy about whether Hartzler had harmed early patients:** David Williams's interviews.

202 **Hartzler's early aggressiveness with the procedure:** Interviews by the author, July 1986, and March 25, 2004. "The Zeal to Heal," by Philip Stephens, *Ingram's*, May 1984, pp. 31–36

203 **"Flash from Kansas City":** Hartzler interview in Kansas City, March 25, 2004.

CHAPTER THIRTEEN:

204 **The exponentially growing market for heart treatments:** Dan Lematrie interview in Boston by the author, April 14, 2004. Also Halperin & Levine, in *Bypass*, noted that by 1984, heart disease and related vascular disorders accounted for 989,610 deaths in the U.S. alone, as opposed to 422,720 from cancer and 102,130 from road accidents. These authors asserted that by 1984, cardiovascular disease was costing the American economy $64.4 billion that year, and that therapeutic devices were then generating $196 million in sales. They further observed that heart drugs were already generating $4 billion in annual sales, as against $400 million in annual sales in the early 1970s. The total procedural charges for coronary bypass surgery, the authors stated, was then reaching $4 billion annually, with aggregate hospital costs for the procedure hitting $28.7 billion in the U.S. alone.

See also, *Time*, "Heart Attacks: New Insights, New Treatments," cover story, pp. 52–58, June 1, 1981, with Gruentzig the lead doctor photographed. A decade later, in April of 1993, *MD* magazine ran a cover story by Charles Bankhead whose title said. "After 25 years and 3 million operations, doubts still fuel the CABG Controversy." The next year *Forbes* picked up the thread with an April 25, 1994 article entitled "The male hysterectomy" and beginning, "What is one of the most overused operations in the U.S.? Coronary bypass surgery. Before you sign up, ask a lot of questions."

206 **Sally Dineen on proper wearing of surgical caps:** From Linda Green tribute at Gruentzig memorial service.

206 **Background on "Sarge":** Interviews with Spencer and Gail King, December 15, 2003, and Gary Roubin, October 24–25, 2003.

206 **Margaret Anne Thornton meets Gruentzig:** The primary source for information about the relationship with Gruentzig's new beau was a day-long interview by the author with her parents, Earl and Mary Jane Thornton, in their Macon, Georgia, condominium on December 17, 2003. This interview included lunch at the nearby Idle Hour Country Club, where Andreas and Margaret Anne were eventually married, and a visit to the cemetery where they were buried. Various sources indicated that the couple's first dancing together occurred at the first of several Gruentzig parties at Stone Mountain, according to their memories long afterward.

206 **"Sunshine came into my life" and other quotations on early romance:** The Thorntons dispute much of the scuttlebutt that was raised about Gruentzig's relationship with their daughter. Mary Jane Thornton maintains that the couple remained so close that she actually wanted to write a book herself about their love and their fall from the sky. "They were the sweetest, most thoughtful, dearest people I ever met, and I am a realtor and have seen lots of couples," Mrs. Thornton said during the interview. "I have been doing this since 1976 and am a life member of the Million Dollar Club of Macon, Georgia [real estate agents who generate above $1 million in sales per year], and see all kinds of people, and have never known two people as close as they were."

210 **Hawaii escapade:** Interviews with Spencer King, John Douglas, Joe Craver, and Willis Hurst.

212 **Dan Lematrie background and quotations:** Dan Lematrie interviewed at the Merrill Lynch office tower, Boston, April 14, 2004.

212 **Dissolving relationship between Gruentzig and the Cook Group and inter-corporate acrimony regarding patents:** A September 18, 1980 letter from Ross Jennings, vice president of Cook, to the firm's attorneys—Woodward, Weikart, Emhardt & Naughton of Indianapolis—explains that Schneider Medintag and Andreas Gruentzig jointly filed a patent on April 1, 1980, and that USCI is licensed under this patent for manufacturing Gruentzig balloon catheters in the United States. Jennings then asked for a reading whether this filing abrogated the 1975 letter of agreement Cook signed with Gruentzig. A responding letter [from Cook archives] of October 3, 1980, to Jennings from C. David Emhardt states:

> It seems to me that you should decide what it is that you want to achieve here. Do you want to put USCI out of business on this product? If so, you may want to file a lawsuit to have the patent assigned to you and to terminate the USCI licence. We may have problems with this because Dr. Grüntzig would probably not cooperate. If you decide to adopt this course of action, you should not delay or else you may be blocked by laches [renderings that a claim is moot] or the expiration of a statute of limitations . . .
>
> Do you want to allow Dr. Grüntzig to save face and extricate himself from what appears to be conflicting agreements he has made? It appears that we do let Dr. Grüntzig off the hook if we con-

> sider the May 12, 1975 agreement as conveying only a license or shop right to us.

A handwritten memo from WAC [William A. Cook] to Jennings on October 13, 1980, said:

> Ross, it would help if you and Chris [Simonsgaard] could meet with Grüntzig to determine his wishes. He appears to be in a box and we should let him decide if we should take over the litigation to protect our right to produce (licence others?) etc. We've been had by USCI but I want to do what is best for Grüntzig. Know [sic] USCI's track record, I would say he will get the shaft from them. As important as this product is to us, a trip to Zurich with Chris might be in order. He also could be contacted perhaps at AHA; I'll go with you if you want. Bill . . .

A November 1980 letter from Chris Simonsgaard to Jennings refers to Gruentzig publishing photos and description of his device "before anyone could stop him." Simonsgaard then remonstrated that Gruentzig never bothered to consult his old correspondence before filing his U.S. patent, and defends his own actions in minute detail.

Following receipt of the Gruentzig telex on November 6, 1981, Bill Cook telexed an employee named Geoff Reeves:

> GEOFF, HERE GOES A NIFTY MAKE WORK PROJECT. GRUNTZIG TELEXED US TODAY TO QUIT USING HIS NAME. IT LOOKS LIKE GOOD OLD USCI GOT US ON THIS ONE. QUESTION—DID YOU HEAR ANYTHING FROM THE VENICE MEETING ON PROBLEMS OR DO WE HAVE SOME LARGE MOUTH COMPLAINERS IN EUROPE ON DILATING CATHETERS.

212 **Bard's Maloney applying pressure on Schneider, 1982:** Hugo Schneider interview, Zürich, November 15, 2003. Canepa interview, November 21, 2003.

212 **Schneider Medintag visit to potential news suppliers under fresh USCI pressure:** Heliana Canepa interview.

CHAPTER FOURTEEN:

215 **Gruentzig's new house purchase:** Mary Jane Thornton interview; and author visit with Spencer King. Warranty deeds in Book 4658, p. 371; Dekalb County Courthouse; Book 5042, pp. 775–76; Book 4573, pp. 520–527; and Book 4619, pp. 811–813.

216 **Gruentzig share in Schneider Medintag:** Heliana Canepa interview, November 21, 2003.

216 **Descriptions of Margaret Anne Thornton:** Sources from numerous medical circles, most quoted directly in the text, disparaged Gruentzig's relationship with the younger woman. Yet few cardiologists came to know her well, whereas radiologists actually working with Margaret Anne seem to have appreciated her more fully. The quotation about Margaret Anne's glowing attributes are excerpted from a handwritten letter by Claudia R. Adkinson from the Emory Department of Anatomy to Earl and Mary Jane Thornton, February 20, 1986, and were enthusiastically seconded during a personal interview with her department chairman, William Casarella, while a first-year medical resident in radiology at Emory in 1982–1983. In her brief medical career, Margaret A. Thornton was listed as the first author of an article that appeared in the prestigious medical journal *Circulation*, Volume 69, No. 4, pp. 721–727, 1984, entitled, "Coumadin and aspirin in prevention of recurrence after transluminal coronary angioplasty: a randomized study." She was also admitted to a total of four medical schools, a rare achievement in itself, and received high academic honors at every stage of her young life.

217 **Gruentzig burgeoning travel:** Thorntons interviews and reviews of hundreds of photographs from thirteen volumes of Gruentzig's personal photographic albums.

218 **Clements quotations:** Interviewed alongside Willis Hurst, December 16, 2003.

218 **Gruentzig's impact at The Cloister:** Author interviews with staff on Sea Island, including concierge Huston Visage, February 1, 2003.

218 **Gruentzig brainstorming at USCI:** John Cvinar interview, May 22, 2004.

219 **Hartzler getting more aggressive:** A December 1982 cover story entitled "Percutaneous Balloon Angioplasty," by Richard S. Gubner, M.D., in the trade journal *Therapaeia* describes Geoffrey Hartzler as having achieved a 90 percent success rate in 720 patients, and notes that Hartzler had already dilated as many as five arterial obstructions in a single patient. Fourteen months later in "PTCA Trials in Progress," by Sheila Stavish, *Cardio* Magazine, February, 1984, pp. 50–55, Gruentzig was quoted as saying, "But Geoffrey Hartzler has done as many as 10 inflations in 10 different places in one session. He's got a good deal of experience. We may have to learn from him to do it differently."

221 **Hartzler quotations:** From the Kansas City interview, March 25, 2004.

223 **USCI wanting to collaborate with Hartzler:** Dave Prigmore interview, July 21, 2003.

225 **Gruentzig's extravagant parties:** Interviews with Heliana Canepa, Bernhard Meier, Spencer King and others.

225 **Margaret Anne upset over Sonja's disciplining:** Interviews with Bernhard Meier, as cited.

226 **History of Sea Island:** *Georgia's Land of the Golden Isles*, Burnette Vanstory, University of Georgia Press, Athens, Georgia, 1956. *This Happy Isle: The Story of Sea Island and the Cloister,* Harold H. Martin, Sea Island Company, 1978. *Sea Island*, Sea Island Company, Sea Island, Georgia, 2003. Eugene O'Neill's description of this island paradise: *This Happy Isle,* p. 91.

227 **Addison Mizner background:** Mizner was a self-trained architect whose Spanish revival buildings became prized by the rich and famous of the American South. With backing from the likes of Irving Berlin, he designed and developed the Boca Raton community and moved on to create many of the grandest villas in Palm Beach. When the market for his exclusive properties evaporated during the Great Depression, he became despondent and died of a heart attack in 1933 at the age of sixty-three.

230 **Background on Sidney Lanier:** An acclaimed nineteenth-century poet, Sidney Lanier was born in Macon in 1841, and served in the Confederate army in Virginia before devoting his life to teaching, orchestral performances on thc flute, and writing verse, novels, literary theory, and historical works. One of his most famous and haunting poems, "The Marshes of the Glynn," is about the same brooding, Spanish moss–dripping South Georgia estuary scenery Gruentzig drove past before his fateful last plane flight.

230 **Gruentzig prenuptial agreement:** From Bibb County Courthouse files.

231 **Hay House history**: The building is a designated historic site and maintained by the Georgia Trust with numerous explications of its history available on the Internet.

231 **Wedding entertainment:** The bandleader Jack Hurd's cousin Milton also happened to be an undertaker in Macon, Georgia. He would bury the Gruentzigs little more than two years after their golden day.

CHAPTER FIFTEEN:

233 **Gruentzig running cash to Switzerland:** Mary Jane Thornton interview December 17, 2003; rumors of same mentioned in interview by the author with Gary Roubin in Jackson Hole, Wyoming, October 24–25, 2003. On November 21, 2003, Heliana Canepa said that she never paid Gruentzig in cash, but did funnel money discreetly to his brother Johannes, as cited in text. It should be noted that Johannes authored many learned articles about eye disease in Africa and Asia and therefore requests for any assistance from the younger brother were likely inspired by a desire to help with important research.

234 **René Favaloro's demise:** Roubin interview above; Internet sources.

235 **Dotter's languishing and his drug abuse:** Asked by the author how long Dotter had been taking amphetamines, Bill Cook responded in the October 23, 2003 interview:

> As long as I had known him.... He took uppers and he took downers. I can't answer the question why because I never talked to him about it. He was very discreet. I never saw him take a pill . . . I only saw him take a glass of wine and that's it. He was not a boozer. He was not a smoker . . . Amphetamines probably would keep him up, and he took downers to put himself to sleep, I guess. We've never told anybody about the amphetamines.

In this period of suffering, Charles Dotter nonetheless managed to continue to generate a flood of new ideas. A November 20, 1981 Dotter letter to E.T. Feldsted, M.D., (from Cook Group archives) mentions a projected trip in a camping van with Dotter's loyal wife to the mountains of southern California in search of the nests of the then-vanishing rare condor bird species. Dotter paints the scene as follows: "I'm sure Pamela would enjoy a trip such as this, condors or no condors. When we travel in the Blazer, we are equipped to sleep in it, for example in the backwoods when there are no motels handy. What do you think?"

Dotter then went on to worry that rare condor eggs might be rolling off cliffs and smashing themselves into nothing, and therefore he offered a drawing to suggest a trap device that could be placed below their nests to save the incubating birds.

On August 22, 1980, Dotter wrote Andreas Gruentzig a letter (from Cook files), saying: "In view of your accomplishments and competency it seems inconceivable that Zürich has not given you a full professorship with tenure. Our gain will be their loss. I will admit that in the past I have been at times jealous, especially when I heard people talking about the Grüntzig procedure. I find no jealousy now, only admiration at the thorough and innovative job you've done."

In an editorial in the journal *Applied Radiology*, May/June 1982, entitled, "Why Not?" Dotter observed:

> Living patients are dynamic and three-dimensional. Would it not be fine if we could visualize them that way? Holography, laser imaging, and microcomputer data-processing should enable us to view the patient in a moving, three-dimensional fluoroscopic display, a virtual image in space at and through which we could look and point (as all radiologists must) and around which we could talk. . . . Sounds like something out of *Star Trek*? It is not, for in the present electronic data processing era, all the aforementioned can happen. . . .

Scarcely finished, Dotter rambled on to talk about his visions for devices that might pare away potentially stroke-inducing accretions of plaque from neck arteries [a vision realized in the mid 1990s], coiled stents that could scaffold about-to-burst aneurysms of the aorta [another hot medical development in the 1990s], and the possibility of repairing ruined heart valves via catheter and without surgery [a concept whose chief later advocate, Alain Cribier, introduced his first results in a Paris cardiology meeting attended by the author in May of 2002].

236 **Gruentzig's continuing exotic travels and sometimes foolish accidents:** Mary Jane and Earl Thornton interview, December 17, 2003.

236 **Gruentzig in his sheik demeanor:** Thorntons interview, and Spencer and Gail King interviews, viewing of photos.

239 **Rick and Kristina McLees quotations:** Rick McClees and Kristina Gedgaudas were interviewed by the author in their suburban Atlanta townhouse on February 5, 2004.

240 **The "doctor killer":** The term was used loosely in the early 1980s about the Beech Bonanza primarily, but also the faster Beech Baron. In an eerie foreshadowing of the subject of this book, Thomas Thompson's *Hearts* annotates the love of flying of Michael DeBakey's protégé Ted Dietrich on page 148 as follows:

> Among surgeons who fly their own airplanes—and there are many—the rate of crash and death is *four times* that of the businessman pilot. One reason is the surgeon's rush to return to the hospital on Monday morning. Another, I suspect, is the surgeon's desperate bear hug on life, paired with the feeling—as some Las Vegas gamblers have—that God sits on their shoulder and will not allow His blessed ones to lose.

241 **Dan Emin background and quotations:** McClees interview, and phone interview by the author with Dan Emin on February 16, 2004. The author also visited Peachtree-Dekalb Airport on February 7, 2004, and in the airport bar met with various "plane spotters" who knew of the Gruentzig story.

242 **Pfizer purchase of Schneider Medintag:** Heliana Canepa interview, November 2004.

242 **Gruentzig buys West Andrews house:** Court records as cited below.

242 **History of the house:** Mary Jane Thornton interview, December 17, 2003.

243 **Description of the neighborhood:** Author visits, and *A Man In Full*, Tom Wolfe, p. 501.

243 **Contents of the house:** Interviews with the Thorntons, Gary Roubin, Bernhard Meier, Eberhard Zeitler, Rick McClees, and numerous records viewed and copied from the Fulton County Courthouse, Bibb County Courthouse, and the United States District Court for Northern District of Georgia office tower. Many of these documents were filed attendant to lawsuits by Michaela Gruentzig on behalf of Sonja Merit Gruentzig against Beech Aircraft et al. for inheritance purposes.

These sources include February 10, 1988, "Petition of Leave to Sell Property," Fulton County Courthouse records, Book 951, p. 38; also, Book 951, p. 43, and also pp. 22–36; and further, Book 587, p. 043; Book 568, p. 335. Also see "Petition for Approval of Escrow Agreement," Book 827, pp. 215–225; plus Book 928, pp. 268–272. Additional documentation was discovered in the files of the U.S. District Court for the Northern District of Georgia, Civil Docket for Case 86-CV-2221, filed October 16, 1986, on behalf of Sonja Gruentzig and Katrina Hoffman, plaintiffs, versus Beech Aircraft Corporation and King Radio Corporation. Further pertinent documents, again from Fulton County Courthouse, are referenced in Book 1423, pp. 188–193; Book 587, pp. 039–053; and Book 476, p. 57. Also viewed were Book 950, pp. 269–270; Book 586, p. 116; Book 557, pp. 346–347; Book 1579, pp. 6–11; Book 921, pp. 331–335, and pp. 314–318.

244 **Gruentzig's visit to Denton Cooley:** Interview with the Thorntons, December 17, 2003.

244 **Art buying in New York and Hilde Gerst:** Thorntons interview and various Internet sources regarding Hilde Gerst.

245 **Gala party at West Andrews:** Interviews with the Thorntons, and the Kings.

CHAPTER SIXTEEN:

246 **Gruentzig's pursuit of randomized trial pitting angioplasty against bypass surgery:** Gary Roubin interview, Jackson Hole, Wyoming, October 24–25, 2003. "Balloon angioplasty challenges bypass surgery," Lawrence M. Boxt, M.D., and David C. Levin, M.D., *Diagnostic Imaging*, July 1985, pp. 76–80.

246 **Gary Roubin background and quotations:** Two-day Roubin interview above, and numerous author interviews with Roubin beginning in 1986. Also, Roubin interview with Burt Cohen, August 21, 1986.

247 **Gruentzig's occasional racist comments:** David Williams's personal recollections; King, Myler, and Thorntons interviews. In a November 15, 2003 interview, Michaela Gruentzig disclosed that her late husband had remarked on his father having been an anti-Semite although she said it was never made clear to her whether Wilmar Gruentzig was an actual Nazi Party member.

248 **Gruentzig's increasing bitterness toward his colleagues:** Interviews with Gary Roubin, Richard and Sharon Myler, Bernhard Meier, Eberhard Zeitler, Rick McClees, and others.

249 **Gruentzig in Macon:** Thorntons interview, December 17, 2003.

249 **Gruentzig's drinking:** His penchant for heavy drinking was cited by many sources, including Klaus Rees, Spencer and Gail King, Gary Roubin, and Sharon Myler.

256 **"Gruentzig's rules":** John Douglas interview, December 16, 2003.

256 **The rising arc and burgeoning avarice of interventional cardiology:** *Saving the Heart*, and hundreds of author interviews regarding the business aspects of interventional cardiology conducted between 1986–2003. See also, "Percutaneous Transluminal Coronary Angioplasty: Current Status and Indications," Thomas Ischinger, M.D., Umit T. Aker, M.D., Harold Kennedy, M.D., *Cardiovascular Reviews and Reports*, Vol. No. 5, 8, August 1984, pp. 782–788; "Angioplasty's Role in MI Uncertain," *Medical World News,* December 9, 1985, p. 18; "The Race Against Myocardial Infarction," *Therapaeia*, March 1982, pp. 8–21; and "A. Gruentzig Reveals PTCA Progress, Future Aspirations," *Cardiology Times*, Vol. 2, No. 7, July 1983.

259 **Eli Lilly purchase of Simpson's ACS company:** *Saving the Heart,* page 167.

260 **Genetech and TPA:** *Heart Failure,* pp. 170–179; also numerous author interviews, 1986–1991.

261 **Cook Group exploration of stents:** Archival material from the corporation.

262 **The feverish pursuit of stents in general:** *Saving the Heart.* By 1993, the Cleveland Clinic cardiologist Eric Topol would be quoted in *MD* magazine (April 1983, p. 32) as ruing the 800,000 angioplasty and/or stent procedures projected to be performed in America that year as being grossly excessive. "There are too many revascularizations throughout the country. We have a ten-fold higher ratio of revascularization for the population than any other country. It's not as though we have to stamp out all angina in the world; some of these patients don't even have angina."

261 **Bard's McCafferty visiting Gruentzig:** Dave Prigmore interview, July 21, 2003.

262 **Bard/USCI's downfall:** Author association with the company from 1985–1994, extensive personal files.

CHAPTER SEVENTEEN:

263 **Gruentzig undergoing heart catheterization:** Spencer King, Willis Hurst, and Thornton interviews.

264 **Gruentzig's continuing zeal.** Gary Roubin interview by the author and transcript of interview with Burt Cohen, August 21, 1986.

266 **Tante Alf collapse:** Thornton, December 17, 2003.

266 **Myler and Stertzer recruiting Gruentzig:** King interviews.

266 **Stertzer's entrepreneurial activities:** The April 2, 1996 prospectus by Cowen and Company; Bear, Stearns & Co. Inc.; and J.P. Morgan & Co. for a public stock offering of shares in stent-developer A.V.E. (Advanced Vascular Systems, Inc.) at an initial price of $21 each. Page 58 of that prospectus lists Dr. Simon Stertzer of Woodside, California, as "beneficially owning" 3,048,424 shares prior to the offering, which represented a potential windfall of $64 million. Dr.

Gerald Dorros of Milwaukee, Wisconsin, another early disciple of Gruentzig and the early steward of Ralph Lach, was listed as "beneficially owning" 2,896,058 shares, therefore amounting to a potential return of $60.8 million. Among other enterprises, Stertzer was a director in the mid-1990s of a California start-up stent manufacturer called Quantam Medical Corporation, which was subsequently purchased by Boston Scientific.

269 **Gruentzig visits Beechcraft:** National Transportation and Safety Board report; Thorntons interview.

269 **$3 million from Bard:** Dave Prigmore interview, July 21, 2003.

269 **Tornado hits West Andrews:** Thornton interview, December 17, 2003.

270 **Gruentzig visits East Germany:** Thornton interview.

271 **Death of Charles Dotter:** In the October 23, 2003 interview with the author, Bill Cook said of Dotter:

> When he couldn't function anymore, he killed himself—it was that simple. He just pulled the plugs. He stopped the breathing apparatus and died. By the time they got into the room he was gone . . . We saw him five days before. I took Caesar Gianturco, Kurt Amplatz, myself, and George Tallage? [uncertain reference, type garbled] The four of us flew out there five days before he killed himself. He pulled out the breathing and monitoring equipment. I guess he wasn't monitored or he was able to short-circuit the monitoring equipment, I don't know. I just know he did it himself.
>
> He was on amphetamines and when he went to Milwaukee to have the surgery, he didn't tell anybody, so he went into this shock state after the surgery. They didn't know what was wrong with him until Josef Rösch came out and told them that he was on amphetamines. They fed him on amphetamines and a week later he was recovering, but he had lost about two thirds of his breathing capacity . . . His lungs just disintegrated from a lack of oxygen.

At a memorial ceremony in Portland on February 15, 1985, Leonard Laster, M.D., president of Oregon Health Sciences University, paid this tribute:

> Rarely in the course of a medical career is an individual granted the opportunity to alter forever the course of medicine for the human good. Charles Dotter was one of those happy few. Building on his personal attributes of creative imagination, individualistic courage and commitment to the betterment of the human condition, Dr. Dotter forged into a new field of medical care that was given the name "interventional radiography." Thanks to Dr. Dotter, new procedures are now available for treating human disease in such a way that the need for complex surgical intervention is obviated. He was honored internationally for his achievements. He was respected by his colleagues and valued by his students. Above all, for generations to come a host of patients will benefit from his contributions even though they will never know Dr. Dotter or what he did for them. His true reward will be the fact that because he passed this way, life for many of us will be far better than it would have been otherwise. A pause, and a moment of quiet appreciation are in order for this man who, among his other activities, climbed every mountain in Oregon to its peak, alone.

271 **Staff idyll on Sea Island:** Interviews with Gary Roubin, October 24, 2003, and Claire Rice, December 19, 2003.

272 **Andreas and Margaret Anne wanting a child:** Thorntons interview, December 17, 2003.

272 **Gruentzig purchases Beech Baron, eschewing flight instruction:** NTSB report.

274 **Bill Walker quotations:** The owner of Golden Isles Aviation, and a former barnstorming pilot with 16,000 hours of flying experience, was interviewed by the author at his office at St. Simons's Malcolm McKinnon Airport on February 2, 2004. Joel West, the assistant airport manager, was interviewed the day before, and various devotees of that airport's subculture were spoken to on the phone.

275 **Gruentzig's final teaching course and High Museum description:** Claire Rice interview, and *A Man in Full*, Tom Wolfe, p. 420.

276 **The "Island Princess" background:** Internet sources.

276 **Inspection of the Beech Baron after Gruentzig's complaints:** National Transportation and Safety Board records.

276 **Final procedure gone wrong:** Interviews with Gary Roubin, Spencer King, and Claire Rice.

CHAPTER EIGHTEEN:

277 **Background on Hurricane Juan:** NTSB records, and newspaper accounts such as "Late-season hurricane hits Gulf Coast," *Atlanta Journal-Constitution*, October 28, 1985, p. 1; also "Juan sinks offshore oil rig, forces rescue of 146 people," *Macon Telegraph and News*, October 29, 1985, pp. 1, 4.

279 **Account of beginning of final flight:** A preliminary description, based on an interview with National Transportation and Safety Board investigator Preston Hicks, appeared in an *Atlanta Journal-Constitution* cover story, "2 Emory doctors are killed: Heart specialist, wife in air crash," October 29, 1985.

280 **Quotations from air traffic controllers and Gruentzig in the air:** NTSB line-by-line account from tapes.

282 **Witnesses perceptions of the crash:** NTSB report and *Macon Telegraph and News* cover story, "Crash killed famed heart doctor, wife," October 29, 1985, by Don Schanche, Jr.

284 **The scene at the crash site:** Photos accompanying *Atlanta Journal-Constitution* article, p. 17, "2 Emory doctors are killed: Heart specialist, wife in air crash," October 29, 1985, and "Cardiologist from Emory feared killed in air crash," Charles Monte Plott and Charles Seabrook, *Atlanta Journal-Constitution*, October 28, 1985, p. 1. Factual Report: Aviation Accident/Incident, National Transportation

Safety Board, Washington, D.C., 20594, cites the Monroe County, Georgia, sheriff department as having found, at the scene the following items: a wallet, a pair pearl earrings, piece of gold jewelery, bank cards; credit cards from Saks Fifth Avenue, Bergdorf Goodman, Neiman Marcus, American Express, Fidelity, and Delta airlines; a medical license and pilot's license; and $540 in cash.

The *Macon Telegraph and News* account on October 29, 1985, also incorporated an interview with NTSB investigator Preston Hicks, and the article read in part:

> Hicks said debris from the plane's structure and the Gruentzigs' personal belonging were strewn across a 300-foot area. Among the debris were two dogs, which apparently were aboard the plane and died in the crash. There was no fire during the crash. "The plane is basically disintegrated," he said.

The crash was also described in the United Press International (UPI) wire service with versions of this report running in newspapers across the United States, including the *New York Times*—"Andreas Gruentzig Dies in Air Crash: Developed Technique to Clear Arteries by Using Balloon—Wife Is also Killed," October 29, 1985.

Transcripts from NTSB report, pp. 128–138, air controller interviews, many pages, flight logs, schematics.

EPILOGUE:

286 **Additional tributes to the couple**: In "Tribute: Andreas Roland Gruentzig (1939–1985): A private perspective," *Circulation*, Vol. 73, No. 3, March 1986, pp. 606–610, Willis Hurst wrote a moving description of the scene at the cemetery during Gruentzig's burial, with particular reference to his grieving mother Charlotta:

> She stood up, with a son on her left and a friend on her right, and walked toward the two caskets. The smile on her face was sweet and her dark eyes were concentrating on the task she was about to perform. She opened the plastic bag she clutched in her hand. She reached inside of it and began to sprinkle soil on the top of *his* cas-

> ket and when she was through there, she sprinkled soil on *her* [Margaret Anne's] casket. The others, including *her* parents, did likewise. She [Charlotta Gruentzig] and the soil were from Germany. She had come to bury her son Andreas and his wife Margaret Ann in the middle Georgia town of Macon. The date was November 1, 1985. The smile on her lovely face and her soft whispering of *auf Wiedersehen* as she slowly performed her task was simultaneously sad and beautiful, and the crowd wept. *Auf Wiedersehen* means more than goodbye. It means, until we see each other again.

Hurst also added:

> My birthday was in the early part of the week before he died. He ordered flowers for my office and bounced into the room to wish me well. He kissed my wife on the cheek and bounced out on his way to the laboratory. That was the last time I saw him. My last note from him was about a conference on lasers that he hoped one of our trainees could attend . . . There is a message on the door of the Alamo in San Antonio, Texas, which fits his unique free spirit:
>
> *"So silent friends, here heroes died,*
> *to blaze a trail for others."*

288 **Irony in Gruentzig's death:** Eberhard Zeitler remarked on September 5, 2003, that the very tragedy of Gruentzig's death may have contributed to his fame:

> One of the best known men in all of history was Achilles. He died when he was eighteen. If Kennedy hadn't died young, maybe he would have been regarded as mediocre. If you die at an early time in life you have also left less bad things behind . . . If you have done some great and good things everybody remembers you overwhelmingly as having succeeded. You have won. One click of the spark, and you have won. I think Andreas when he died was still a great inventor and a great doctor, but if he reached the age of eighty, think about how he would have dwindled. Think about Max Planck, the first man to harness atomic energy—now there are so many atomic energy specialists. He was the first, okay? And he changed the world

globally and totally. Or Martin Luther King, if he had not been killed so young, would he be remembered so well?

289 **Feuds after Gruentzig's death:** Unfortunately, the mourning soon became tainted by acrimony. Background references regarding this situation include: "Daughters sue over Emory doctor's death in plane crash," Gail Epstein, *Atlanta Journal-Constitution,*" October 17, 1986, p. 28, describing a $30 million lawsuit against Beech Aircraft Corporation, King Radio Corporation, and Hanger, Inc; and author interviews with various lawyers, including Arthur H. Rosenberg of Soberman & Rosenberg, New York, ultimate lead attorney on the case, on October 9, 2003. Other relevant items include "Petition of leave to sell property," Fulton County Courthouse, Book 951, February 10, 1988, pp. 22–36, 38, and 43; and various files on the www.lawskills.com Georgia Caselaw Web site.

In late August of 2003 phone call to the author, Johannes Gruentzig displayed what became lasting reluctance to entertain a personal interview, despite a series of letters of supplication. Johannes said in this phone discussion that he had suspicions that his brother did not die from an accident per se, but rather perhaps from outright subterfuge.

In his initial August 6, 2003 letter to the author, Johannes asked, "*Aber wie wäre es, wenn Ihre Recherchen mit den Flugzeug-Absturz meines Bruders beginnen? Wurde er abgestürtz? Wenn ja, warum? Falls Sie diesbezüglich auf Neuland stossen, können Sie auf meine Kooperation zählen.*" ("How would it be if you began your research with the plane accident of my brother? If he *was* an accident victim? If he was, why? If you should want to explore new ground regarding this development, you can count on my cooperation.") In a March 3, 2004 letter, Johannes kindly offered a number of specific leads regarding the circumstances of Andreas's death, although several of these had already been pursued.

At least five lawsuits followed the plane crash, with at first separate ones and later a conjoint proceeding filed on behalf of Sonja

Merit Gruentzig and Katrin Hoffman, with Beech Aircraft and King Air, manufacturer of the plane's instrumentation, being the primary targets, along with Hanger One, an aircraft maintenance firm whose representatives had inspected Gruentzig's Beech Baron in mid-October, following his complaints about malfunctions in the plane's navigational equipment. The Thorntons also became caught up in several proceedings regarding their rights to inheritance.

The acrimony that developed at the same time is an object lesson in the foolhardiness of not filling a will and entering a legal limbo termed "intestate." Eventually, after years of bitter legal jousting, millions were awarded to the three contesting parties—Sonja Gruentzig via Michaela, Katrin Hoffman, and the Thorntons.

In any case, the remains of Andreas and Margaret Anne Thornton lay forever mute, even if the impact of their lives resonates still.

Acknowledgments

Journey into the Heart is about an epic quest of twentieth-century exploration. As happens with works of this nature, the telling became a miniature epic in its own right, and ultimately needed many helping hands.

I became interested in the story of this great quest in the course of medical reporting in the early 1980s, and in the middle of that decade I began producing a series of newsletters, principally dealing with breaking developments in cardiovascular medicine. Over the course of thousands of interviews over the next years, certain names began to resonate—especially those of Werner Forssmann, Mason Sones, Charles Dotter, and Andreas Gruentzig, along with their storied surgical counterparts.

In time, I started to sketch out the bones of their collective quest, and in 1994 I ventured to Switzerland to first interview several of the closest sources involved with the life of Andreas Gruentzig, who is the centerpiece of *Journey into the Heart.* Draft chapters were written, but, alas, other demands intervened, and I moved on in both my work and locale—the latter now being Cork, Ireland.

As it happened, a thread lingered through my connection with Dr. David O. Williams of Brown University/Rhode Island Hospital, whom I had interviewed many times over the course of fifteen years. And by the by, it turned out that he shared my dream of relating the saga of heart exploration as it had never been told before.

Moreover, in 2002, Dr. Williams began making arrangements for a no-strings-attached research grant from the Cordis division of the Johnson & Johnson corporation, in order to launch this long overdue project, and inquired whether I was interested in completing the story. YES—was the answer. Thus, this book would never have come to fruition without the support of Dr. Williams and the generous backing of J & J, who never asked to read a single paragraph as the work evolved.

Therefore, my first acknowledgment must be to Dave Williams, who has been tireless in his support, while offering many leads, participating in countless discussions, joining in on a number of the most far-flung interviews, and reviewing the evolving text through many stages. His knowledge of the subject is second to no one's. His lovely wife, Carole, was also an avid and perceptive reader.

J & J deserves credit, too, for both their largesse and their large-mindedness in respecting the project's independence and historical importance. Marcia Schallehn there, who has devoted her working life to disseminating vital information on the breakthroughs described in this book, provided helpful but unobtrusive encouragement, as did Jesse Penn and Rick Anderson.

Looming just as large in the book's creation was the serendipity of finding not only the perfect agent for it, Fred Hill, but also a most devoted publisher at the Gotham Books imprint of the Penguin Group (USA), Bill Shinker, and my sensitive and highly skilled editor there, Erin Moore. Behind the scenes all the while labored an accomplished writer friend with a fine editing brush indeed, Barnaby Conrad. Here stood four musketeers from the old school—and what a gift.

The research conducted for this book involved roughly a hundred personal interviews in journeys of many thousands of miles across the breadth of the United States, and across Switzerland and Germany. Many subjects became emotional in relating their perspectives on this inspiring, yet ultimately tragic human history. Hopefully, none of these sources will be offended if only a few figures can be singled out, due to space limitations.

ACKNOWLEDGMENTS

Those closest to the tragic life of Andreas Gruentzig shared not only observations and facts, but their own heartfelt memories. Michaela Seebrunner Gruentzig, the first wife of the book's chief protagonist, was profoundly generous in relating her personal recollections on an inspiring but troubled legacy. Special gratitude is also owed to the willingness of Earl and Mary Jane Thornton, parents-in-law of Andreas Gruentzig and parents of their beloved Margaret Anne, to participate in a long and often emotional interview, and also for taking me to the scenes of their only daughter and son-in-law's wedding and eventual final resting ground. Maria Schlumpf, a soul mate to Andreas Gruentzig, provided prodigious help at numerous stages over the course of several years. None of these contributions could have been easy, because they were all fraught with anguish.

Many physicians went to great lengths to help this book forward. At the Cleveland Clinic, several retired colleagues of Mason Sones joined in lengthy, yet vivid-as-yesterday interviews, notably William Proudfit, Earl Shirey, and Elaine Clayton. In Portland, Oregon, the delightful Josef Rösch took every measure to ensure that his old colleague Charles Dotter was revealed as a man in full.

A number of former colleagues of Andreas Gruentzig generously contributed to the research. One was Eberhard Zeitler, in Nürnberg, Germany, who gave two charming days to interviews. In Switzerland, Bernhard Meier consented to be interviewed three times, and Alfred Bollinger not only joined in interviews but drove the author around the Swiss country side to important sites in Gruentzig's life. Ulrich Sigwart avidly read early chapters.

Within the U.S., Spencer B. King III, of Atlanta, was unfailingly generous with his help, which included a thoughtful review of the manuscript and considerable later correspondence. In Wyoming, Gary Roubin did the same, and provided particularly soulful support for the entire project. In Kansas City, Geoffrey Hartzler was tremendously helpful and also reviewed the text with a knowing but gentle hand. Richard and Sharon Myler were similarly welcoming in Carmel, California, as was Dave Prigmore in Rhode Island and Ralph Lach in Ohio. The list could go on.

Of particular value was archival material from the Dotter Vascular Institute of Portland, Oregon; the Cleveland Clinic Foundation; the Fulton County, Georgia courthouse; the Boston Scientific Corporation in Natick, Massachusetts; and the Cook Group in Indianapolis, Indiana. Also of value were extensive personal files and photographs provided by Maria Schlumpf, Martin Kaltenbach, Wilhelm Rutishauser, Gary Roubin, and Earl and Mary Jane Thornton.

Within Ireland, thanks are due to Brian Hartnett and Eamonn Coughlan for their keen interest and assistance. The most particular gratitude is due to my wife and in-house editor, Jamie, and to my children, for their support and forbearance for a project that took so much more time than any of us bargained on.

Truly, a book like this is a sum of many parts.

One final note: A close reader will notice that German spellings are observed in almost every case, save the possibly perplexing choice to preserve the Anglicized spelling, sans umlaut, of Andreas Grüntzig's German name. The rationale was that, with few exceptions, the Anglicized spelling of his name is observed globally and should be the first choice for any reader pursuing bibliographic or Internet research.

Thanks then, to the reader, for putting up with the stylistic inconsistency and participating in this journey into the human heart.

David Monagan
Cork, Ireland

INDEX

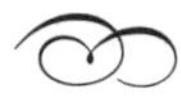